CAMBRIDGE LIBRARY COLLECTION

Books of enduring scholarly value

Mathematics

From its pre-historic roots in simple counting to the algorithms powering modern desktop computers, from the genius of Archimedes to the genius of Einstein, advances in mathematical understanding and numerical techniques have been directly responsible for creating the modern world as we know it. This series will provide a library of the most influential publications and writers on mathematics in its broadest sense. As such, it will show not only the deep roots from which modern science and technology have grown, but also the astonishing breadth of application of mathematical techniques in the humanities and social sciences, and in everyday life.

Mathematical Papers of the Late George Green

A miller's son, George Green (1793–1841) received little formal schooling yet managed to acquire significant knowledge of modern mathematics, especially French work. In 1828 he published his *Essay on the Application of Mathematical Analysis to the Theories of Electricity and Magnetism*, the work for which he is now celebrated. Admitted to Cambridge in 1833 as a mature student, Green went on to become a fellow of Gonville and Caius College. His early death, however, cut short a promising career as a mathematical physicist. While English contemporaries saw what he might have achieved, they did not understand what he had actually achieved. Only when William Thomson (later Lord Kelvin) rediscovered Green's first publication and shared it with the French mathematical elite was his greatness truly appreciated. Edited by the Cambridge mathematician Norman Macleod Ferrers (1829–1903) and published in 1871, this collection comprises Green's influential essay and nine further papers.

Cambridge University Press has long been a pioneer in the reissuing of out-of-print titles from its own backlist, producing digital reprints of books that are still sought after by scholars and students but could not be reprinted economically using traditional technology. The Cambridge Library Collection extends this activity to a wider range of books which are still of importance to researchers and professionals, either for the source material they contain, or as landmarks in the history of their academic discipline.

Drawing from the world-renowned collections in the Cambridge University Library and other partner libraries, and guided by the advice of experts in each subject area, Cambridge University Press is using state-of-the-art scanning machines in its own Printing House to capture the content of each book selected for inclusion. The files are processed to give a consistently clear, crisp image, and the books finished to the high quality standard for which the Press is recognised around the world. The latest print-on-demand technology ensures that the books will remain available indefinitely, and that orders for single or multiple copies can quickly be supplied.

The Cambridge Library Collection brings back to life books of enduring scholarly value (including out-of-copyright works originally issued by other publishers) across a wide range of disciplines in the humanities and social sciences and in science and technology.

Mathematical Papers of the Late George Green

EDITED BY N.M. FERRERS

CAMBRIDGE
UNIVERSITY PRESS

CAMBRIDGE
UNIVERSITY PRESS

University Printing House, Cambridge, CB2 8BS, United Kingdom

Published in the United States of America by Cambridge University Press, New York

Cambridge University Press is part of the University of Cambridge.
It furthers the University's mission by disseminating knowledge in the pursuit of education, learning and research at the highest international levels of excellence.

www.cambridge.org
Information on this title: www.cambridge.org/9781108065603

This edition first published 1871
This digitally printed version 2014

ISBN 978-1-108-06560-3 Paperback

MATHEMATICAL PAPERS

OF THE LATE

GEORGE GREEN.

Cambridge:
PRINTED BY C. J. CLAY, M.A.
AT THE UNIVERSITY PRESS.

MATHEMATICAL PAPERS

OF THE LATE

GEORGE GREEN,

FELLOW OF GONVILLE AND CAIUS COLLEGE, CAMBRIDGE.

EDITED BY

N. M. FERRERS, M.A.,

FELLOW AND TUTOR OF GONVILLE AND CAIUS COLLEGE.

London:

MACMILLAN AND CO.

1871.

PREFACE.

HAVING been requested by the Master and Fellows of Gonville and Caius College to superintend an edition of the mathematical writings of the late George Green, I have fulfilled the task to the best of my ability. The publication may be opportune at present, as several of the subjects with which they are directly or indirectly concerned, have recently been introduced into the course of mathematical study at Cambridge. They have also an interest as being the work of an almost entirely self-taught mathematical genius.

George Green was born at Sneinton, near Nottingham, in 1793. He commenced residence at Gonville and Caius College, in October, 1833, and in January, 1837, took his degree of Bachelor of Arts as Fourth Wrangler. It is hardly necessary to say that this position, distinguished as it was, most inadequately represented his mathematical power. He laboured under the double disadvantage of advanced age, and of inability to submit entirely to the course of systematic training needed for the highest places in the Tripos. He was elected to a fellowship of his College in 1839, but did not long enjoy this position, as he died in 1841. The contents of the following pages will sufficiently shew the heavy loss which the scientific world sustained by his premature death.

A slight sketch of the papers comprised in this volume may not be uninteresting.

The first paper, which is also the longest and perhaps the most important, was published by subscription at Nottingham in 1828. It was in this paper that the term *potential* was first introduced to denote the result obtained by adding together the masses of all the particles of a system, each divided by its distance from a given point. In this essay, which is divided into three parts, the properties of this function are first considered, and they are then applied, in the second and third parts, to the theories of magnetism and electricity respectively. The full analysis of this essay which the author has given in his Preface, renders any detailed description in this place unnecessary. In connexion with this essay, the corresponding portions of Thomson and Tait's *Natural Philosophy* should be studied, especially Appendix A. to Chap. I., and Arts. 482—550, inclusive.

The next paper, "On the Laws of the Equilibrium of Fluids analogous to the Electric Fluid," was laid before the Cambridge Philosophical Society by Sir Edward Ffrench Bromhead, in 1832. The law of repulsion of the particles of the supposed fluid here considered is taken to be inversely proportional to the nth power of the distance. This paper, though displaying great analytical power, is perhaps rather curious than practically interesting; and a similar remark applies to that which succeeds it, "On the determination of the attractions of Ellipsoids of variable Densities," which, like its predecessor, was communicated to the Cambridge Philosophical Society by Sir E. F. Bromhead. Space of n dimensions is here considered, and the surfaces of the attracting bodies are supposed to be represented by equations formed by equating to unity the sums of the squares of the n variables, each divided by an appropriate coefficient. It is of course possible to adapt the formula of this paper to the case of nature by supposing $n = 3$.

The next paper, "On the Motion of Waves in a variable canal of small depth and width," though short, is interesting. It was read before the Cambridge Philosophical Society, on May 15, 1837, and a Supplement to it on Feb. 18, 1839. On Dec. 11, 1837, were communicated two of his most valuable memoirs, "On the Reflexion and Refraction of Sound," and "On the Reflexion and Refraction of Light at the common surface of two non-crystallized media." These two papers should be studied together. The question discussed in the first is, in fact, that of the propagation of normal vibrations through a fluid. Particular attention should be paid to the mode in which, from the differential equations of motion, is deduced an explanation of a phenomenon analogous to that known in Optics as Total internal reflection when the angle of incidence exceeds the critical angle. By supposing that there are propagated, in the second medium, vibrations which rapidly diminish in intensity, and become evanescent at sensible distances, the change of phase which accompanies this phenomenon is clearly brought into view.

The immediate object of the next paper, "On the Reflexion and Refraction of Light at the common surface of two non-crystalline media," is to do for the theory of light what in the former paper has been done for that of sound. This is done in a manner which will present little difficulty to one who has mastered the former paper. But this paper has an interest extending far beyond this subject. For the purpose of explaining the propagation of transversal vibrations through the luminiferous ether, it becomes necessary to investigate the equations of motion of an elastic solid. It is here that Green for the first time enunciates the principle of the Conservation of work, which he bases on the assumption of the impossibility of a perpetual motion. This principle he enunciates in the following words: "In whatever manner the elements of any material system may act upon each other, if all the internal forces be multiplied by the elements of their respective directions, the total sum for any assigned portion of the mass will always be the exact differential of some function." This function, it will be seen, is what is now known under the name of Potential Energy, and the above principle is in fact equivalent to stating that the sum of the Kinetic and Potential Energies of the system is constant. This function, supposing the displacements so small that powers above the second may be neglected, is shewn for the most general constitution of the medium to involve twenty-one coefficients, which reduce to nine in the case of a medium symmetrical with respect to three rectangular planes, to five in the case of a medium symmetrical around an axis, and to two in the case of an isotropic or uncrystallized medium. The present paper is devoted to the consideration of the propagation of vibrations from one of two media of this nature. The two coefficients above mentioned, called respectively A and B, are shewn to be proportional to the squares of the velocities of propagation of normal and transversal vibrations respectively. It is to be regretted that the *statical* interpretation is not also given. It may however be shewn (see Thomson and Tait's *Natural Philosophy*, p. 711 (*m.*)) that $A - \frac{4}{3}B$ measures the resistance of the medium to compression or dilatation, or its *elasticity of volume*, while B measures its resistance to distortion, or its *rigidity*. The equilibrium of the medium, it may be shewn, cannot be stable, unless both of these quantities are positive*. A Supplement to this paper supplying certain omissions, immediately follows it.

In the next paper, "On the Propagation of Light in Crystalline Media," the principle of Conservation of Work is again assumed as a starting-point and applied to a medium of any description. It is first assumed that the medium is symmetrical with respect to three planes at right angles to one another, by which supposition the twenty-one coefficients previously mentioned are reduced to nine. Fresnel's supposition, that the vibrations affecting the eye are accurately in front of the wave, is then introduced, and a complete explanation of the phenomena of polarization is shewn to follow, on the hypothesis that the vibrations constituting a plane-polarized ray are *in* the plane of polarization. The hypothesis adopted in the former paper—that these vibrations are *perpendicular* to the plane of polarization—is then resumed, and an explanation arrived at, by the aid of a subsidiary assumption—unfortunately not of the same simple character as those previously introduced—that for the three principal waves the wave-velocity depends on the direction of the disturbance only, and is independent of the position of the wave's front. The paper concludes by taking the case of a perfectly general medium, and it is shewn that Fresnel's supposition of the vibrations being accurately in the wave-front, gives rise to fourteen re-

* In comparing Green's paper with the passage in Thomson and Tait's *Natural Philosophy* above referred to, it should be remarked that the A of the former is equal to the $m - \frac{1}{3}n$ of the latter, and that $B = n$.

lations among the twenty-one coefficients, which virtually reduce the medium to one symmetrical with respect to three planes at right angles to one another.

This paper, read May 20, 1839, was his last production. Another, "On the Vibrations of Pendulums in Fluid Media," read before the Royal Society of Edinburgh, on Dec. 16, 1833, will be found at the end of this collection. The problem here considered is that of the motion of an inelastic fluid agitated by the small vibrations of a solid ellipsoid, moving parallel to itself.

I have to express my thanks to the Council of the Cambridge Philosophical Society, and to that of the Royal Society of Edinburgh, for the permission to reproduce the papers published in their respective Transactions which they have kindly given.

N. M. FERRERS.

GONVILLE AND CAIUS COLLEGE,
Dec. 1870.

CONTENTS.

ERRATA.

Page 23, line 11, *for* there *read* these.
,, 23, ,, 25, *for* $dzdy$ *read* $dydz$.
,, 29, ,, 7, *for* $\frac{d\overline{V}}{dx}$ *read* $\frac{d\overline{V}}{dw}$.
,, 29, ,, 16, *for* $\frac{d\overline{V}}{dx'}$ *read* $\frac{d\overline{V}}{dw'}$.
,, 36, ,, 22, *after* co-ordinates, *insert* of.
,, 37, ,, 2 from bottom, *for* $\delta\overline{V}$, *read* $\delta'\overline{V}$.
,, 43, ,, 8, *for* axes, *read* axis.
,, 46, ,, 19, *after* radius, *insert* is.
,, 53, ,, 7, *for* ρ *read* (g).
,, 54, ,, 11, *for* $4\pi f^3$ *read* $2\pi f^3$.
,, 54, ,, 16, *for* $4\pi a f^2$ *read* $4\pi a f^3$.
,, 56, ,, 19, *for* $Q\left(\frac{1}{a}-\frac{1}{a}\right)=Q'\left(\frac{1}{a'}-\frac{1}{b}\right)$ *read* $Q\left(\frac{1}{a}-\frac{1}{b}\right)=Q'\left(\frac{1}{a'}-\frac{1}{b}\right)$.
,, 60, ,, 13, *for* $\int\frac{d\sigma}{f^3}V$ *read* $\int\frac{ds}{f^3}\overline{V}$.
,, 64, ,, 4 from bottom, *before* a potential *insert* of.
,, 71, throughout *for* dw and $d\omega$ read $d\varpi$.
,, 72, ,, 18, *for* $d\omega$ *read* $d\varpi$.
,, 74, ,, 24, *for* his *read* this.
,, 84, ,, 11, *for* $\frac{U^{(2)}}{r'^3}$ *read* $\frac{U^{(1)}}{r'^2}$.
,, 88, ,, 20, *for* r^2 *read* r^3.
,, 89, ,, 17, *for* $\sin\theta'$ *read* $\sin\theta$.
,, 89, ,, 18, *for* $U^{(0)}$ *read* $U^{(1)}$.
,, 90, ,, 2, *for* $\frac{d}{dx}$ *read* $\frac{d}{dx'}$.
,, 92 ,, 24, *for* $\frac{3}{4}\pi$ *read* $\frac{3}{4\pi}$.
,, 107 ,, 20, *for* $r^2\frac{d^2\phi}{dx^2}=$ *read* $r^2\frac{d^2\phi}{dx^2}+$
,, 123 ,, 19, *for* these *read* thus.
,, 129 ,, 24, *for* $\sin\varpi'_2$ *read* $\sin\varpi'$.

AN ESSAY

ON THE APPLICATION OF MATHEMATICAL ANALYSIS

TO THE THEORIES

OF ELECTRICITY AND MAGNETISM.*

* *Published at Nottingham, in* 1828.

PREFACE.

AFTER I had composed the following Essay, I naturally felt anxious to become acquainted with what had been effected by former writers on the same subject, and, had it been practicable, I should have been glad to have given, in this place, an historical sketch of its progress; my limited sources of information, however, will by no means permit me to do so; but probably I may here be allowed to make one or two observations on the few works which have fallen in my way, more particularly as an opportunity will thus offer itself, of noticing an excellent paper, presented to the Royal Society by one of the most illustrious members of that learned body, which appears to have attracted little attention, but which, on examination, will be found not unworthy the man who was able to lay the foundations of pneumatic chymistry, and to discover that water, far from being according to the opinions then received, an elementary substance, was a compound of two of the most important gases in nature.

It is almost needless to say the author just alluded to is the celebrated CAVENDISH, who, having confined himself to such simple methods, as may readily be understood by any one possessed of an elementary knowledge of geometry and fluxions, has rendered his paper accessible to a great number of readers; and although, from subsequent remarks, he appears dissatisfied with an hypothesis which enabled him to draw some important conclusions, it will readily be perceived, on an attentive perusal of his paper, that a trifling alteration will suffice to render the whole perfectly legitimate*.

* In order to make this quite clear, let us select one of CAVENDISH'S propositions, the twentieth for instance, and examine with some attention the method

Little appears to have been effected in the mathematical theory of electricity, except immediate deductions from known formulæ, that first presented themselves in researches on the figure of the earth, of which the principal are,—the determination of the law of the electric density on the surfaces of conducting bodies differing little from a sphere, and on those of ellipsoids, from 1771, the date of CAVENDISH'S paper, until about 1812, when M. POISSON presented to the French Institute two memoirs of singular elegance, relative to the distribution of electricity on the surfaces of conducting spheres, previously electrified and put in presence of each other. It would be quite

there employed. The object of this proposition is to show, that when two similar conducting bodies communicate by means of a long slender canal, and are charged with electricity, the respective quantities of redundant fluid contained in them, will be proportional to the $n-1$ power of their corresponding diameters: supposing the electric repulsion to vary inversely as the n power of the distance. This is proved by considering the canal as cylindrical, and filled with incompressible fluid of uniform density: then the quantities of electricity in the interior of the two bodies are determined by a very simple geometrical construction, so that the total action exerted on the whole canal by one of them, shall exactly balance that arising from the other; and from some remarks in the 27th proposition, it appears the results thus obtained, agree very well with experiments in which real canals are employed, whether they are straight or crooked, provided, as has since been shown by COULOMB, n is equal to two. The author however confesses he is by no means able to demonstrate this, although, as we shall see immediately, it may very easily be deduced from the propositions contained in this paper.

For this purpose, let us conceive an incompressible fluid of uniform density, whose particles do not act on each other, but which are subject to the same actions from all the electricity in their vicinity, as real electric fluid of like density would be; then supposing an infinitely thin canal of this hypothetical fluid, whose perpendicular sections are all equal and similar, to pass from a point a on the surface of one of the bodies, through a portion of its mass, along the interior of the real canal, and through a part of the other body, so as to reach a point A on its surface, and then proceed from A to a in a right line, forming thus a closed circuit, it is evident from the principles of hydrostatics, and may be proved from our author's 23d proposition, that the whole of the hypothetical canal will be in equilibrium, and as every particle of the portion contained within the system is necessarily so, the rectilinear portion aA must therefore be in equilibrium. This simple consideration serves to complete CAVENDISH'S demonstration, whatever may be the form or thickness of the real canal, provided the quantity of electricity in it is very small compared with that contained in the bodies. An analogous application of it will render the demonstration of the 22d proposition complete, when the two coatings of the glass plate communicate with their respective conducting bodies, by fine metallic wires of any form.

impossible to give any idea of them here: to be duly appretiated they must be read. It will therefore only be remarked, that they are in fact founded upon the consideration of what have, in this Essay, been termed potential functions, and by means of an equation in variable differences, which may immediately be obtained from the one given in our tenth article, serving to express the relation between the two potential functions arising from any spherical surface, the author deduces the values ot these functions belonging to each of the two spheres under consideration, and thence the general expression of the electric density on the surface of either, together with their actions on any exterior point.

I am not aware of any material accessions to the theory of electricity, strictly so called, except those before noticed; but since the electric and magnetic fluids are subject to one common law of action, and their theory, considered in a mathematical point of view, consists merely in developing the consequences which flow from this law, modified only by considerations arising from the peculiar constitution of natural bodies with respect to these two kinds of fluid, it is evident the mathematical theory of the latter, must be very intimately connected with that of the former; nevertheless, because it is here necessary to consider bodies as formed of an immense number of insulated particles, all acting upon each other mutually, it is easy to conceive that superior difficulties must, on this account, present themselves, and indeed, until within the last four or five years, no successful attempt to overcome them had been published. For this farther extension of the domain of analysis, we are again indebted to M. Poisson, who has already furnished us with three memoirs on magnetism: the first two contain the general equations on which the magnetic state of a body depends, whatever may be its form, together with their complete solution in case the body under consideration is a hollow spherical shell, of uniform thickness, acted upon by any exterior forces, and also when it is a solid ellipsoid subject to the influence of the earth's action. By supposing magnetic changes to require time, although an exceedingly short one, to complete them, it had been suggested that M. Arago's discovery relative to the magnetic effects developed in copper, wood, glass, *etc.*, by rotation, might be explained. On this hypothesis M. POISSON has founded his third memoir, and thence deduced formulæ applicable to magnetism in a state of motion. Whether the preceding hypothesis will serve to explain the singular phenomena observed by M. ARAGO or not, it would ill become me to decide; but it is probably quite adequate to account for those produced by the rapid rotation of iron bodies.

We have just taken a cursory view of what has hitherto been written, to the best of my knowledge, on subjects connected with the mathematical theory of electricity; and although many of the artifices employed in the works before mentioned are remarkable for their elegance, it is easy to see they are adapted only to particular objects, and that some general method, capable of being employed in every case, is still wanting. Indeed M. POISSON, in the commencement of his first memoir (*Mém. de l'Institut*, 1811), has incidentally given a method for determining the distribution of electricity on the surface of a spheroid of any form, which would naturally present itself to a person occupied in these researches, being in fact nothing more than the ordinary one noticed in our introductory observations, as requiring the resolution of the equation (a). Instead however of supposing, as we have done, that the point p must be upon the surface, in order that the equation may subsist, M. POISSON availing himself of a general fact, which was then supported by experiment only, has conceived the equation to hold good wherever this point may be situated, provided it is within the spheroid, but even with this extension the method is liable to the same objection as before.

Considering how desirable it was that a power of universal agency, like electricity, should, as far as possible, be submitted to calculation, and reflecting on the advantages that arise in the solution of many difficult problems, from dispensing altogether with a particular examination of each of the forces which actuate the various bodies in any system, by confining the attention solely to that peculiar function on whose differentials they all depend, I was induced to try whether it would be possible to discover any general relations, existing between this function and the quantities of electricity in the bodies producing it. The advantages LAPLACE had derived in the third book of the *Mécanique Céleste*, from the use of a partial differential equation of the second order, there given, were too marked to escape the notice of any one engaged with the present subject, and naturally served to suggest that this equation might be made subservient to the object I had in view. Recollecting, after some attempts to accomplish it, that previous researches on partial differential equations, had shown me the necessity of attending to what have, in this Essay, been denominated the singular values of functions, I found, by combining this consideration with the preceding, that the resulting method was capable of being applied with great advantage to the electrical theory, and was thus, in a short time, enabled to demonstrate the general formulæ contained in the preliminary part of the Essay. The remaining part ought to be regarded principally as furnishing particular examples of the use of these general formulæ; their number might with great ease have been increased, but those which are given, it is hoped, will suffice to point out to mathematicians, the mode of applying the preliminary results to any case they may wish to investigate. The hypotheses on which the received theory of magnetism is founded, are by no means so certain as the facts on which the electrical theory rests; it is however not the less necessary to have the means of submitting them to calculation, for the only way that appears open to us in the investigation of these subjects, which seem as it were desirous to conceal themselves from our view, is to form the most probable hypotheses we can, to deduce rigorously the consequences which flow from them, and to examine whether such consequences agree numerically with accurate experiments.

The applications of analysis to the physical Sciences, have the double advantage of manifesting the extraordinary powers of this wonderful instrument of thought, and at the same time of serving to increase them; numberless are the instances of the truth of this assertion. To select one we may remark, that M. FOURIER, by his investigations relative to heat, has not only discovered the general equations on which its motion depends, but has likewise been led to new analytical formulæ, by whose aid MM. Cauchy and Poisson have been enabled to give the complete theory of the motion of the waves in an indefinitely extended fluid. The same formulæ have also put us in possession of the solutions of many other interesting problems, too numerous to be detailed here. It must certainly be regarded as a pleasing prospect to analysts, that at a time when astronomy, from the state of perfection to which it has attained, leaves little room for farther applications of their art, the rest of the physical sciences should show themselves daily more and more willing to submit to it; and, amongst other things, probably the theory that supposes light to depend on the undulations of a luminiferous fluid, and to which the celebrated Dr T. Young has given such plausibility, may furnish a useful subject of research, by affording new opportunities of applying the general theory of the motion of fluids. The number of these opportunities can scarcely be too great, as it must be evident to those who have examined the subject, that, although we have long been in possession of the general equations on which this kind of motion depends, we are not yet well acquainted with the various limitations it will be necessary to introduce, in order to adapt them to the different physical circumstances which may occur.

Should the present Essay tend in any way to facilitate the application of analysis to one of the most interesting of the physical sciences, the author will deem himself amply repaid for any labour he may have bestowed upon it; and it is hoped the difficulty of the subject will incline mathematicians to read this work with indulgence, more particularly when they are informed that it was written by a young man, who has been obliged to obtain the little knowledge he possesses, at such intervals and by such means, as other indispensable avocations which offer but few opportunities of mental improvement, afforded.

INTRODUCTORY OBSERVATIONS.

THE object of this Essay is to submit to Mathematical Analysis the phenomena of the equilibrium of the Electric and Magnetic Fluids, and to lay down some general principles equally applicable to perfect and imperfect conductors; but, before entering upon the calculus, it may not be amiss to give a general idea of the method that has enabled us to arrive at results, remarkable for their simplicity and generality, which it would be very difficult if not impossible to demonstrate in the ordinary way.

It is well known, that nearly all the attractive and repulsive forces existing in nature are such, that if we consider any material point p, the effect, in a given direction, of all the forces acting upon that point, arising from any system of bodies S under consideration, will be expressed by a partial differential of a certain function of the co-ordinates which serve to define the point's position in space. The consideration of this function is of great importance in many inquiries, and probably there are none in which its utility is more marked than in those about to engage our attention. In the sequel we shall often have occasion to speak of this function, and will therefore, for abridgement, call it the potential function arising from the system S. If p be a particle of positive electricity under the influence of forces arising from any electrified body, the function in question, as is well known, will be obtained by dividing the quantity of electricity in each element of the body, by its distance from the particle p, and taking the total sum of these quotients for the whole body, the quantities of electricity in those elements which are negatively electrified, being regarded as negative.

It is by considering the relations existing between the density of the electricity in any system, and the potential functions thence arising, that we have been enabled to submit many electrical phenomena to calculation, which had hitherto resisted the attempts of analysts; and the generality of the consideration here employed, ought necessarily, and does, in fact, introduce a great generality into the results obtained from it. There is one consideration peculiar to the analysis itself, the nature and utility of which will be best illustrated by the following sketch.

Suppose it were required to determine the law of the distribution of the electricity on a closed conducting surface A without thickness, when placed under the influence of any electrical forces whatever: these forces, for greater simplicity, being reduced to three, X, Y, and Z, in the direction of the rectangular co-ordinates, and tending to increase them. Then ρ representing the density of the electricity on an element $d\sigma$ of the surface, and r the distance between $d\sigma$ and p, any other point of the surface, the equation for determining ρ which would be employed in the ordinary method, when the problem is reduced to its simplest form, is known to be

$$\text{cons} = a = \int \frac{\rho d\sigma}{r} - \int (Xdx + Ydy + Zdz) \quad \ldots\ldots\ldots \quad (a);$$

the first integral relative to $d\sigma$ extending over the whole surface A, and the second representing the function whose complete differential is $Xdx + Ydy + Zdz$, x, y and z being the co-ordinates of p.

This equation is supposed to subsist, whatever may be the position of p, provided it is situate upon A. But we have no general theory of equations of this description, and whenever we are enabled to resolve one of them, it is because some consideration peculiar to the problem renders, in that particular case, the solution comparatively simple, and must be looked upon as the effect of chance, rather than of any regular and scientific procedure.

We will now take a cursory view of the method it is proposed to substitute in the place of the one just mentioned.

Let us make $B=\int(Xdx+Ydy+Zdz)$ whatever may be the position of the point p, $V=\int\frac{\rho d\sigma}{r}$ when p is situate any where within the surface A, and $V'=\int\frac{\rho d\sigma}{r}$ when p is exterior to it: the two quantities V and V', although expressed by the same definite integral, are essentially distinct functions of x, y, and z, the rectangular co-ordinates of p; these functions, as is well known, having the property of satisfying the partial differential equations

$$0=\frac{d^2V}{dx^2}+\frac{d^2V}{dy^2}+\frac{d^2V}{dz^2},$$

$$0=\frac{d^2V'}{dx^2}+\frac{d^2V'}{dy^2}+\frac{d^2V'}{dz^2}.$$

If now we could obtain the values of V and V' from these equations, we should have immediately, by differentiation, the required value of ρ, as will be shown in the sequel.

In the first place, let us consider the function V, whose value at the surface A is given by the equation (a), since this may be written

$$a=\overline{V}-\overline{B},$$

the horizontal line over a quantity indicating that it belongs to the surface A. But, as the general integral of the partial differential equation ought to contain two arbitrary functions, some other condition is requisite for the complete determination of V. Now since $V=\int\frac{\rho d\sigma}{r}$, it is evident that none of its differential coefficients can become infinite when p is situate any where within the surface A, and it is worthy of remark, that this is precisely the condition required: for, as will be afterwards shown, when it is satisfied we shall have generally

$$V=-\int(\rho)\,d\sigma\,\overline{V};$$

the integral extending over the whole surface, and (ρ) being a quantity dependent upon the respective positions of p and $d\sigma$.

All the difficulty therefore reduces itself to finding a function V which satisfies the partial differential equation, becomes equal to the known value of V at the surface, and is moreover such that none of its differential coefficients shall be infinite when p is within A.

In like manner, in order to find V', we shall obtain $\overline{V}'$, its value at A, by means of the equation (a), since this evidently becomes

$$a = \overline{V}' - \overline{B}, \text{ i.e. } \overline{V}' = \overline{V}.$$

Moreover it is clear, that none of the differential coefficients of $V' = \int \frac{\rho d\sigma}{r}$ can be infinite when p is exterior to the surface A, and when p is at an infinite distance from A, V' is equal to *zero*. These two conditions combined with the partial differential equation in V', are sufficient in conjunction with its known value $\overline{V}'$ at the surface A for the complete determination of V', since it will be proved hereafter, that when they are satisfied we shall have

$$V' = -\int (\rho)\, d\sigma\, \overline{V}';$$

the integral, as before, extending over the whole surface A, and (ρ) being a quantity dependent upon the respective position of p and $d\sigma$.

It only remains therefore to find a function V' which satisfies the partial differential equation, becomes equal to $\overline{V}'$ when p is upon the surface A, vanishes when p is at an infinite distance from A, and is besides such, that none of its differential coefficients shall be infinite, when the point p is exterior to A.

All those to whom the practice of analysis is familiar, will readily perceive that the problem just mentioned, is far less difficult than the direct resolution of the equation (a), and therefore the solution of the question originally proposed has been rendered much easier by what has preceded. The peculiar consideration relative to the differential coefficients of V and V', by restricting the generality of the integral of the partial differential equation, so that it can in fact contain only one arbitrary function, in the place of two which it ought otherwise to have contained, and, which has thus enabled us to effect the simplification in question, seems worthy of the attention of analysts, and may be of use in other researches where equations of this nature are employed.

We will now give a brief account of what is contained in the following Essay. The first seven articles are employed in demonstrating some very general relations existing between the density of the electricity on surfaces and in solids, and the corresponding potential functions. These serve as a foundation to the more particular applications which follow them. As it would be difficult to give any idea of this part without employing analytical symbols, we shall content ourselves with remarking, that it contains a number of singular equations of great generality and simplicity, which seem capable of being applied to many departments of the electrical theory besides those considered in the following pages.

In the eighth article we have determined the general values of the densities of the electricity on the inner and outer surfaces of an insulated electrical jar, when, for greater generality, these surfaces are supposed to be connected with separate conductors charged in any way whatever; and have proved, that for the same jar, they depend solely on the difference existing between the two constant quantities, which express the values of the potential functions within the respective conductors. Afterwards, from these general values the following consequences have been deduced:—

When in an insulated electrical jar we consider only the electricity accumulated on the two surfaces of the glass itself, the total quantity on the inner surface is precisely equal to that on the outer surface, and of a contrary sign, notwithstanding the great accumulation of electricity on each of them: so that if a communication were established between the two sides of the jar, the sum of the quantities of electricity which would manifest themselves on the two metallic coatings, after the discharge, is exactly equal to that which, before it had taken place, would have been observed to have existed on the surfaces of the coatings farthest from the glass, the only portions then sensible to the electrometer.

If an electrical jar communicates by means of a long slender wire with a spherical conductor, and is charged in the ordinary way, the density of the electricity at any point of the interior surface of the jar, is to the density on the conductor itself, as the radius of the spherical conductor to the thickness of the glass in that point.

The total quantity of electricity contained in the interior of any number of equal and similar jars, when one of them communicates with the prime conductor and the others are charged *by cascade*, is precisely equal to that, which one only would receive, if placed in communication with the same conductor, its exterior surface being connected with the common reservoir. This method of charging batteries, therefore, must not be employed when any great accumulation of electricity is required.

It has been shown by M. Poisson, in his first Memoir on Magnetism (Mém. de l'Acad. de Sciences, 1821 et 1822), that when an electrified body is placed in the interior of a hollow spherical conducting shell of uniform thickness, it will not be acted upon in the slightest degree by any bodies exterior to the shell, however intensely they may be electrified. In the ninth article of the present Essay this is proved to be generally true, whatever may be the form or thickness of the conducting shell.

In the tenth article there will be found some simple equations, by means of which the density of the electricity induced on a spherical conducting surface, placed under the influence of any electrical forces whatever, is immediately given; and thence the general value of the potential function for any point either within or without this surface is determined from the arbitrary value at the surface itself, by the aid of a definite integral. The proportion in which the electricity will divide itself between two insulated conducting spheres of different diameters, connected by a very fine wire, is afterwards considered; and it is proved, that when the radius of one of them is small compared with the distance between their surfaces, the product of the mean density of the electricity on either sphere, by the radius of that sphere, and again by the shortest distance of its surface from the centre of the other sphere, will be the same for both. Hence when their distance is very great, the densities are in the inverse ratio of the radii of the spheres.

When any hollow conducting shell is charged with electricity, the whole of the fluid is carried to the exterior surface, without leaving any portion on the interior one, as may be immediately shown from the fourth and fifth articles. In the experimental verification of this, it is necessary to leave a small orifice in the shell: it became therefore a problem of some interest to determine the modification which this alteration would produce. We have, on this account, terminated the present article, by investigating the law of the distribution of electricity on a thin spherical conducting shell, having a small circular orifice, and have found that its density is very nearly constant on the exterior surface, except in the immediate vicinity of the orifice; and the density at any point p of the inner surface, is to the constant density on the outer one, as the product of the diameter of a circle into the cube of the radius of the orifice, is to the product of three times the circumference of that circle into the cube of the distance of p from the centre of the orifice; excepting as before those points in its immediate vicinity. Hence, if the diameter of the sphere were twelve inches, and that of the orifice one inch, the density at the point on the inner surface opposite the centre of the orifice, would be less than the hundred and thirty thousandth part of the constant density on the exterior surface.

In the eleventh article some of the effects due to atmospherical electricity are considered; the subject is not however insisted upon, as the great variability of the cause which produces them, and the impossibility of measuring it, gives a degree of vagueness to these determinations.

The form of a conducting body being given, it is in general a problem of great difficulty, to determine the law of the distribution of the electric fluid on its surface: but it is possible to give different forms, of almost every imaginable variety of shape, to conducting bodies; such, that the values of the density of the electricity on their surfaces may be rigorously assignable by the most simple calculations: the manner of doing this is explained in the twelfth article, and two examples of its use are given. In the last, the resulting form of the conducting body is an oblong spheroid, and the density of the electricity on its surface, here found, agrees with the one long since deduced from other methods.

Thus far perfect conductors only have been considered. In order to give an example of the application of theory to bodies which are not so, we have, in the thirteenth article, supposed the matter of which they are formed to be endowed with a constant coercive force equal to β, and analogous to friction in its operation, so that when the resultant of the electric forces acting upon any one of their elements is less than β, the electrical state of this element shall remain unchanged; but, so soon as it begins to exceed β, a change shall ensue. Then imagining a solid of revolution to turn continually about its axis, and to be subject to a constant electrical force f acting in parallel right lines, we determine the permanent electrical state at which the body will ultimately arrive. The result of the analysis is, that in consequence of the coercive force β, the solid will receive a new polarity, equal to that which would be induced in it if it were a perfect conductor and acted upon by the constant force β, directed in lines parallel to one in the body's equator, making the angle $90^0+\gamma$, with a plane passing through its axis and parallel to the direction of f: f being supposed resolved into two forces, one in the direction of the body's axis, the other b directed along the intersection of its equator with the plane just mentioned, and γ being determined by the equation

$$\sin\gamma = \frac{\beta}{b}.$$

In the latter part of the present article the same problem is considered under a more general point of view, and treated by a different analysis: the body's progress from the initial, towards that permanent state it was the object of the former part to determine is exhibited, and the great rapidity of this progress made evident by an example.

The phenomena which present themselves during the rotation of iron bodies, subject to the influence of the earth's magnetism, having lately engaged the attention of experimental philosophers, we have been induced to dwell a little on the solution of the preceding problem, since it may serve in some measure to illustrate what takes place in these cases. Indeed, if there were any substances in nature whose magnetic powers, like those of iron and nickel, admit of considerable developement, and in which moreover the coercive force was, as we have here supposed it, the same for all their elements, the results of the preceding theory ought scarcely to differ from what would be observed in bodies formed of such substances, provided no one of their dimensions was very small, compared with the others. The hypothesis of a constant coercive force was adopted in this article, in order to simplify the calculations: probably, however, this is not exactly the case of nature, for a bar of the hardest steel has been shown (I think by Mr Barlow) to have a very considerable degree of magnetism induced in it by the earth's action, which appears to indicate, that although the coercive force of some of its particles is very great, there are others in which it is so small as not to be able to resist the feeble action of the earth. Nevertheless, when iron bodies are turned slowly round their axes, it would seem that our theory ought not to differ greatly from observation; and in particular, it is very probable the angle γ might be rendered sensible to experiment, by sufficiently reducing b the component of the force f.

The remaining articles treat of the theory of magnetism. This theory is here founded on an hypothesis relative to the constitution of magnetic bodies, first proposed by COULOMB, and afterwards generally received by philosophers, in which they are considered as formed of an infinite number of conducting elements, separated by intervals absolutely impervious to the magnetic fluid, and by means of the general results contained in the former part of the Essay, we readily obtain the necessary equations for determining the magnetic state induced in a body of any form, by the action of exterior magnetic forces. These equations accord with those M. POISSON has found by a very different method. (Mém. de l'Acad. des Sciences, 1821 et 1822.)

If the body in question be a hollow spherical shell of constant thickness, the analysis used by LAPLACE (Méc. Cél. Liv. 3) is applicable, and the problem capable of a complete solution, whatever may be the situation of the centres of the magnetic forces acting upon it. After having given the general solution, we have supposed the radius of the shell to become infinite, its thickness remaining unchanged, and have thence deduced formulæ belonging to an indefinitely extended plate of uniform thickness. From these it follows, that when the point p, and the centres of the magnetic forces are situate on opposite sides of a soft iron plate of great extent, the total action on p will have the same direction as the resultant of all the forces, which would be exerted on the points p, p', p'', p''', *etc. in infinitum* if no plate were interposed, and will be equal to this resultant multiplied by a very small constant quantity: the points p, p', p'', p''', &c. being all on a right line perpendicular to the flat surfaces of the plate, and receding from it so, that the distance between any two consecutive points may be equal to twice the plate's thickness.

What has just been advanced will be sensibly correct, on the supposition of the distances between the point p and the magnetic centres not being very great, compared with the plate's thickness, for, when these distances are exceedingly great, the interposition of the plate will make no sensible alteration in the force with which p is solicited.

When an elongated body, as a steel wire for instance, has, under the influence of powerful magnets, received a greater degree of magnetism than it can retain alone, and is afterwards left to itself, it is said to be magnetized to saturation. Now if in this state we consider any one of its conducting elements, the force with which a particle p of magnetism situate within the element tends to move, will evidently be precisely equal to its coercive force f, and in equilibrium with it. Supposing therefore this force to be the same for every element, it is clear that the degree of magnetism retained by the wire in a state of saturation, is, on account of its elongated form, exactly the same as would be induced by the action of a constant force, equal to f, directed along lines parallel to its axis, if all the elements were perfect conductors; and consequently, may readily be determined by the general theory. The number and accuracy of COULOMB's experiments on cylindric wires magnetized to saturation, rendered an application of theory to this particular case very desirable, in order to compare it with experience. We have therefore effected this in the last article, and the result of the comparison is of the most satisfactory kind.

GENERAL PRELIMINARY RESULTS.

(1.) THE function which represents the sum of all the electric particles acting on a given point divided by their respective distances from this point, has the property of giving, in a very simple form, the forces by which it is solicited, arising from the whole electrified mass.—We shall, in what follows, endeavour to discover some relations between this function, and the density of the electricity in the mass or masses producing it, and apply the relations thus obtained, to the theory of electricity.

Firstly, let us consider a body of any form whatever, through which the electricity is distributed according to any given law, and fixed there, and let x', y', z', be the rectangular co-ordinates of a particle of this body, ρ' the density of the electricity in this particle, so that $dx'dy'dz'$ being the volume of the particle, $\rho' dx'dy'dz'$ shall be the quantity of electricity it contains: moreover, let r' be the distance between this particle and a point p exterior to the body, and V represent the sum of all the particles of electricity divided by their respective distances from this point, whose co-ordinates are supposed to be x, y, z, then shall we have

$$r' = \sqrt{(x'-x)^2 + (y'-y)^2 + (z'-z)^2},$$

and

$$V = \int \frac{\rho' dx'dy'dz'}{r'};$$

the integral comprehending every particle in the electrified mass under consideration.

2—2

LAPLACE has shown, in his Méc. Céleste, that the function V has the property of satisfying the equation

$$0 = \frac{d^2V}{dx^2} + \frac{d^2V}{dy^2} + \frac{d^2V}{dz^2},$$

and as this equation will be incessantly recurring in what follows, we shall write it in the abridged form $0 = \delta V$; the symbol δ being used in no other sense throughout the whole of this Essay.

In order to prove that $0 = \delta V$, we have only to remark, that by differentiation we immediately obtain $0 = \delta \frac{1}{r}$, and consequently each element of V substituted for V in the above equation satisfies it; hence the whole integral (being considered as the sum of all these elements) will also satisfy it. This reasoning ceases to hold good when the point p is within the body, for then, the coefficients of some of the elements which enter into V becoming infinite, it does not therefore necessarily follow that V satisfies the equation

$$0 = \delta V,$$

although each of its elements, considered separately, may do so.

In order to determine what δV becomes for any point within the body, conceive an exceedingly small sphere whose radius is a inclosing the point p at the distance b from its centre, a and b being exceedingly small quantities. Then, the value of V may be considered as composed of two parts, one due to the sphere itself, the other due to the whole mass exterior to it: but the last part evidently becomes equal to zero when substituted for V in δV, we have therefore only to determine the value of δV for the small sphere itself, which value is known to be

$$\delta\,(2\pi a^2\rho - \tfrac{2}{3}\pi b^2\rho);$$

ρ being equal to the density within the sphere and consequently to the value of ρ' at p. If now $x_{,}$, $y_{,}$, $z_{,}$ be the co-ordinates of the centre of the sphere, we have

$$b^2 = (x_{,} - x)^2 + (y_{,} - y)^2 + (z_{,} - z)^2,$$

and consequently

$$\delta\,(2\pi a^2\rho - \tfrac{2}{3}\pi b^2\rho) = -4\pi\rho.$$

Hence, throughout the interior of the mass

$$0 = \delta V + 4\pi\rho\,;$$

of which, the equation $0 = \delta V$ for any point exterior to the body is a particular case, seeing that, here $\rho = 0$.

Let now q be any line terminating in the point p, supposed without the body, then $-\left(\frac{dV}{dq}\right)$ = the force tending to impel a particle of positive electricity in the direction of q, and tending to increase it. This is evident, because each of the elements of V substituted for V in $-\left(\frac{dV}{dq}\right)$, will give the force arising from this element in the direction tending to increase q, and consequently, $-\left(\frac{dV}{dq}\right)$ will give the sum of all the forces due to every element of V, or the total force acting on p in the same direction. In order to show that this will still hold good, although the point p be within the body; conceive the value of V to be divided into two parts as before, and moreover let p be at the surface of the small sphere or $b = a$, then the force exerted by this small sphere will be expressed by

$$\tfrac{4}{3}\,\pi a\rho\left(\frac{da}{dq}\right);$$

da being the increment of the radius a, corresponding to the increment dq of q, which force evidently vanishes when $a = 0$: we need therefore have regard only to the part due to the mass exterior to the sphere, and this is evidently equal to

$$V - \frac{4\pi}{3}\,a^2\rho.$$

But as the first differentials of this quantity are the same as those of V when a is made to vanish, it is clear, that whether the point p be within or without the mass, the force acting upon it in the direction of q increasing, is always given by $-\left(\frac{dV}{dq}\right)$.

Although in what precedes we have spoken of one body only, the reasoning there employed is general, and will apply equally to a system of any number of bodies whatever, in those cases even, where there is a finite quantity of electricity spread over their surfaces, and it is evident that we shall have for a point p in the interior of any one of these bodies

$$0 = \delta V + 4\pi\rho \quad\text{.....................}\quad (1).$$

Moreover, the force tending to increase a line q ending in any point p within or without the bodies, will be likewise given by $-\left(\frac{dV}{dq}\right)$; the function V representing the sum of all the electric particles in the system divided by their respective distances from p. As this function, which gives in so simple a form the values of the forces by which a particle p of electricity, any how situated, is impelled, will recur very frequently in what follows, we have ventured to call it the potential function belonging to the system, and it will evidently be a function of the co-ordinates of the particle p under consideration.

(2.) It has been long known from experience, that whenever the electric fluid is in a state of equilibrium in any system whatever of perfectly conducting bodies, the whole of the electric fluid will be carried to the surface of those bodies, without the smallest portion of electricity remaining in their interior: but I do not know that this has ever been shown to be a necessary consequence of the law of electric repulsion, which is found to take place in nature. This however may be shown to be the case for every imaginable system of conducting bodies, and is an immediate consequence of what has preceded. For let x, y, z, be the rectangular co-ordinates of any particle p in the interior of one of the bodies; then will $-\left(\frac{dV}{dx}\right)$ be the force with which p is impelled in the direction of the co-ordinate x, and tending to increase it. In the same way $-\frac{dV}{dy}$ and $-\frac{dV}{dz}$ will be the forces in y and z, and since the fluid is in equilibrium all these forces are equal to *zero:* hence

$$0 = \frac{dV}{dx}\,dx + \frac{dV}{dy}\,dy + \frac{dV}{dz}\,dz = dV,$$

which equation being integrated gives

$$V = \text{const.}$$

This value of V being substituted in the equation (1) of the preceding number gives

$$\rho = 0,$$

and consequently shows, that the density of the electricity at any point in the interior of any body in the system is equal to *zero*.

The same equation (1) will give the value of ρ the density of the electricity in the interior of any of the bodies, when there are not perfect conductors, provided we can ascertain the value of the potential function V in their interior.

(3.) Before proceeding to make known some relations which exist between the density of the electric fluid at the surfaces of bodies, and the corresponding values of the potential functions within and without those surfaces, the electric fluid being confined to them alone, we shall in the first place, lay down a general theorem which will afterwards be very useful to us. This theorem may be thus enunciated:

Let U and V be two continuous functions of the rectangular co-ordinates x, y, z, whose differential co-efficients do not become infinite at any point within a solid body of any form whatever; then will

$$\int dx\,dz\,dy\,U\delta V + \int d\sigma\,U\left(\frac{dV}{dw}\right) = \int dx\,dy\,dz\,V\delta U + \int d\sigma\,V\left(\frac{dU}{dw}\right);$$

the triple integrals extending over the whole interior of the body, and those relative to $d\sigma$, over its surface, of which $d\sigma$ represents an element: dw being an infinitely small line perpendicular to the surface, and measured from this surface towards the interior of the body.

To prove this let us consider the triple integral

$$\int dx\,dy\,dz \left\{\left(\frac{dV}{dx}\right)\left(\frac{dU}{dx}\right)+\left(\frac{dV}{dy}\right)\left(\frac{dU}{dy}\right)+\left(\frac{dV}{dz}\right)\left(\frac{dU}{dz}\right)\right\}.$$

The method of integration by parts, reduces this to

$$\int dy\,dz\,V''\frac{dU''}{dx}-\int dy\,dz\,V'\frac{dU'}{dx}$$
$$+\int dx\,dz\,V''\frac{dU''}{dy}-\int dx\,dz\,V'\frac{dU'}{dy}$$
$$+\int dx\,dy\,V''\frac{dU''}{dz}-\int dx\,dy\,V'\frac{dU'}{dz}$$
$$-\int dx\,dy\,dz\,V\left\{\frac{d^2U}{dx^2}+\frac{d^2U}{dy^2}+\frac{d^2U}{dz^2}\right\};$$

the accents over the quantities indicating, as usual, the values of those quantities at the limits of the integral, which in the present case are on the surface of the body, over whose interior the triple integrals are supposed to extend.

Let us now consider the part $\int dy\,dz\,V''\frac{dU''}{dx}$ due to the greater values of x. It is easy to see since dw is every where perpendicular to the surface of the solid, that if $d\sigma''$ be the element of this surface corresponding to $dy\,dz$, we shall have

$$dy\,dz=-\frac{dx}{dw}\,d\sigma'',$$

and hence by substitution

$$\int dy\,dz\,V''\frac{dU''}{dx}=-\int d\sigma''\frac{dx}{dw}V''\frac{dU''}{dx}.$$

In like manner it is seen, that in the part

$$-\int dy\,dz\,V'\frac{dU'}{dx}$$

due to the smaller values of x, we shall have

$$dy\,dz=+\frac{dx}{dw}\,d\sigma',$$

and consequently

$$-\int dy\,dz\,V'\frac{dU'}{dx} = -\int d\sigma'\,\frac{dx}{dw}\,V'\frac{dU'}{dx}.$$

Then, since the sum of the elements represented by $d\sigma'$, together with those represented by $d\sigma''$, constitute the whole surface of the body, we have by adding these two parts

$$\int dy\,dz\left(V''\frac{dU''}{dx} - V'\frac{dU'}{dx}\right) = -\int d\sigma\,\frac{dx}{dw}\,V\frac{dU}{dx}:$$

where the integral relative to $d\sigma$ is supposed to extend over the whole surface, and dx to be the increment of x corresponding to the increment dw.

In precisely the same way we have

$$\int dx\,dz\left(V''\frac{dU''}{dy} - V'\frac{dU'}{dy}\right) = -\int d\sigma\,\frac{dy}{dw}\,V\frac{dU}{dy},$$

and $$\int dx\,dy\left(V''\frac{dU''}{dz} - V'\frac{dU'}{dz}\right) = -\int d\sigma\,\frac{dz}{dw}\,V\frac{dU}{dz};$$

therefore, the sum of all the double integrals in the expression before given will be obtained by adding together the three parts just found; we shall thus have

$$-\int d\sigma\,V\left\{\frac{dU}{dx}\frac{dx}{dw} + \frac{dU}{dy}\frac{dy}{dw} + \frac{dU}{dz}\frac{dz}{dw}\right\} = -\int d\sigma\,V\frac{dU}{dw};$$

where V and $\frac{dU}{dw}$ represent the values at the surface of the body. Hence, the integral

$$\int dx\,dy\,dz\left\{\frac{dV}{dx}\frac{dU}{dx} + \frac{dV}{dy}\frac{dU}{dy} + \frac{dV}{dz}\frac{dU}{dz}\right\},$$

by using the characteristic δ in order to abridge the expression, becomes

$$-\int d\sigma\,V\frac{dU}{dw} - \int dx\,dy\,dz\,V\delta U.$$

Since the value of the integral just given remains unchanged when we substitute V in the place of U and reciprocally, it is clear, that it will also be expressed by

$$-\int d\sigma U \frac{dV}{dw} - \int dx dy dz \, U \delta V.$$

Hence, if we equate these two expressions of the same quantity, after having changed their signs, we shall have

$$\int d\sigma V \frac{dU}{dw} + \int dx \, dy \, dz \, V \delta U = \int d\sigma U \frac{dV}{dw} + \int dx dy dz \, U \delta V \ldots \text{ (2)}.$$

Thus the theorem appears to be completely established, whatever may be the form of the functions U and V.

In our enunciation of the theorem, we have supposed the differentials of U and V to be finite within the body under consideration, a condition, the necessity of which does not appear explicitly in the demonstration, but, which is understood in the method of integration by parts there employed.

In order to show more clearly the necessity of this condition, we will now determine the modification which the formula must undergo, when one of the functions, U for example, becomes infinite within the body; and let us suppose it to do so in one point p' only: moreover, infinitely near this point let U be sensibly equal to $\frac{1}{r}$; r being the distance between the point p' and the element $dx \, dy \, dz$. Then if we suppose an infinitely small sphere whose radius is a to be described round p', it is clear that our theorem is applicable to the whole of the body exterior to this sphere, and since, $\delta U = \delta \frac{1}{r} = 0$ within the sphere, it is evident, the triple integrals may still be supposed to extend over the whole body, as the greatest error that this supposition can induce, is a quantity of the order a^2. Moreover, the part of $\int d\sigma U \frac{dV}{dw}$, due to the surface of the small sphere is only an infinitely small quantity of the order a; there only remains therefore to consider the part of $\int d\sigma V \frac{dU}{dw}$ due to this same surface, which, since we have here

$$\frac{dU}{dw} = \frac{dU}{dr} = \frac{d\frac{1}{r}}{dr} = \frac{-1}{dr^2} = \frac{-1}{a^2},$$

becomes $-4\pi V'$

when the radius a is supposed to vanish. Thus, the equation (2) becomes

$$\int dxdydz\, U\delta V + \int d\sigma\, U\frac{dV}{dw} = \int dxdydz\, V\delta U + \int d\sigma\, V\frac{dU}{dw} - 4\pi V' \ldots (3);$$

where, as in the former equation, the triple integrals extend over the whole volume of the body, and those relative to $d\sigma$, over its exterior surface: V' being the value of V at the point p'.

In like manner, if the function V be such, that it becomes infinite for any point p'' within the body, and is moreover, sensibly equal to $\frac{1}{r'}$, infinitely near this point, as U is infinitely near to the point p', it is evident from what has preceded that we shall have

$$\int dxdydz\, U\delta V + \int d\sigma\, U\frac{dV}{dw} - 4\pi U'' = \int dxdydz\, V\delta U + \int d\sigma\, V\frac{dU}{dw} - 4\pi V' \ldots (3');$$

the integrals being taken as before, and U'' representing the value of U, at the point p'' where V becomes infinite. The same process will evidently apply, however great may be the number of similar points belonging to the functions U and V.

For abridgment, we shall in what follows, call those singular values of a given function, where its differential coefficients become infinite, and the condition originally imposed upon U and V will be expressed by saying, that neither of them has any singular values within the solid body under consideration.

(4.) We will now proceed to determine some relations existing between the density of the electric fluid at the surface of a body, and the potential functions thence arising, within and without this surface. For this, let $\rho d\sigma$ be the quantity of electricity on an element $d\sigma$ of the surface, and V, the value of the potential function for any point p within it, of which the co-ordinates are

x, y, z. Then, if V' be the value of this function for any other point p' exterior to this surface, we shall have

$$V = \int \frac{\rho d\sigma}{\sqrt{(\xi - x)^2 + (\eta - y)^2 + (\zeta - z)^2}};$$

ξ, η, ζ being the co-ordinates of $d\sigma$, and

$$V' = \int \frac{\rho d\sigma}{\sqrt{(\xi - x')^2 + (\eta - y')^2 + (\zeta - z')^2}}:$$

the integrals relative to $d\sigma$ extending over the whole surface of the body.

It might appear at first view, that to obtain the value of V' from that of V, we should merely have to change x, y, z into x', y', z': but, this is by no means the case; for, the form of the potential function changes suddenly, in passing from the space within to that without the surface. Of this, we may give a very simple example, by supposing the surface to be a sphere whose radius is a and centre at the origin of the co-ordinates; then, if the density ρ be constant, we shall have

$$V = 4\pi\rho a \text{ and } V' = \frac{4\pi a^2}{\sqrt{x'^2 + y'^2 + z'^2}};$$

which are essentially distinct functions.

With respect to the functions V and V' in the general case, it is clear that each of them will satisfy LAPLACE'S equation, and consequently

$$0 = \delta V \text{ and } 0 = \delta' V':$$

moreover, neither of them will have singular values; for any point of the spaces to which they respectively belong, and at the surface itself, we shall have

$$\overline{V} = \overline{V'}$$

the horizontal lines over the quantities indicating that they belong to the surface. At an infinite distance from this surface, we shall likewise have

$$V' = 0.$$

We will now show, that if any two functions whatever are taken, satisfying these conditions, it will always be in our power to assign one, and only one value of ρ, which will produce them for corresponding potential functions. For this we may remark, that the Equation (3) art. 3 being applied to the space within the body, becomes, by making $U = \frac{1}{r}$,

$$\int \frac{d\sigma}{r} \left(\frac{d\overline{V}}{dx}\right) = \int d\sigma \overline{V} \left(\frac{d\frac{1}{r}}{dw}\right) - 4\pi V;$$

since $U = \frac{1}{r}$, has but one singular point, viz. p; and, we have also $\delta V = 0$ and $\delta \frac{1}{r} = 0$: r being the distance between the point p to which V belongs, and the element $d\sigma$.

If now, we conceive a surface inclosing the body at an infinite distance from it, we shall have, by applying the formula (2) of the same article to the space between the surface of the body and this imaginary exterior surface (seeing that here $\frac{1}{r} = U$ has no singular value)

$$\int \frac{d\sigma}{r} \left(\frac{d\overline{V'}}{dx}\right) = \int d\sigma \overline{V'} \left(\frac{d\frac{1}{r}}{dw'}\right):$$

since the part due to the infinite surface may be neglected, because V' is there equal to *zero*. In this last equation, it is evident that dw' is measured from the surface, into the exterior space, and hence

$$\left(\frac{d\frac{1}{r}}{dw}\right) = -\left(\frac{d\frac{1}{r}}{dw'}\right) \quad \text{i.e.} \quad 0 = \left(\frac{d\frac{1}{r}}{dw}\right) + \left(\frac{d\frac{1}{r}}{dw'}\right);$$

which equation reduces the sum of the two just given to

$$\int \frac{d\sigma}{r} \left\{\left(\frac{d\overline{V}}{dw}\right) + \left(\frac{d\overline{V'}}{dw'}\right)\right\} = -4\pi V.$$

In exactly the same way, for the point p' exterior to the surface, we shall obtain

$$\int \frac{d\sigma}{r'}\left\{\left(\overline{\frac{dV}{dw}}\right)+\left(\overline{\frac{dV'}{dw'}}\right)\right\}=-4\pi V'.$$

Hence it appears, that there exists a value of ρ, viz.

$$\rho=\frac{-1}{4\pi}\left\{\left(\overline{\frac{dV}{dw}}\right)+\left(\overline{\frac{dV'}{dw'}}\right)\right\},$$

which will give V and V', for the two potential functions, within and without the surface.

Again, $-\left(\overline{\frac{dV}{dw}}\right)=$ force with which a particle of positive electricity p, placed within the surface and infinitely near it, is impelled in the direction dw perpendicular to this surface, and directed inwards; and $-\left(\overline{\frac{dV'}{dw'}}\right)$ expresses the force with which a similar particle p' placed without this surface, on the same normal with p, and also infinitely near it, is impelled outwards in the direction of this normal: but the sum of these two forces is equal to double the force that an infinite plane would exert upon p, supposing it uniformly covered with electricity of the same density as at the foot of the normal on which p is; and this last force is easily shown to be expressed by $2\pi\rho$, hence by equating

$$4\pi\rho=-\left(\overline{\frac{dV}{dw}+\frac{dV'}{dw'}}\right) \quad\text{.....................} \quad (4),$$

and consequently there is only one value of ρ, which can produce V and V' as corresponding potential functions.

Although in what precedes, we have considered the surface of one body only, the same arguments apply, how great soever may be their number; for the potential functions V and V' would still be given by the formulæ

$$V=\int\frac{\rho d\sigma}{r} \text{ and } V'=\int\frac{\rho d\sigma}{r'};$$

the only difference would be, that the integrations must now extend over the surface of all the bodies, and, that the number of functions represented by V, would be equal to the number of the bodies, one for each. In this case, if there were given a value of V for each body, together with V' belonging to the exterior space; and moreover, if these functions satisfied to the above mentioned conditions, it would always be possible to determine the density on the surface of each body, so as to produce these values as potential functions, and there would be but one density, viz. that given by

$$0 = 4\pi\rho + \frac{\overline{dV}}{dw} + \frac{\overline{dV'}}{dw'} \text{ .... } (4'),$$

which could do so: ρ, $\frac{\overline{dV}}{dw}$ and $\frac{\overline{dV'}}{dw'}$ belonging to a point on the surface of any of these bodies.

(5.) From what has been before established (art. 3), it is easy to prove, that when the value of the potential function $\overline{V}$ is given on any closed surface, there is but one function which can satisfy at the same time the equation

$$0 = \delta V,$$

and the condition, that V shall have no singular values within this surface. For the equation (3) art. 3, becomes by supposing $\delta U = 0$,

$$\int d\sigma \overline{U} \frac{\overline{dV}}{dw} = \int d\sigma \overline{V} \frac{\overline{dU}}{dw} - 4\pi V'.$$

In this equation, U is supposed to have only one singular value within the surface, viz. at the point p', and, infinitely near to this point, to be sensibly equal to $\frac{1}{r}$; r being the distance from p'. If now we had a value of U, which, besides satisfying the above written conditions, was equal to zero at the surface itself, we should have $\overline{U} = 0$, and this equation would become

$$0 = \int d\sigma \overline{V} \frac{\overline{dU}}{dw} - 4\pi V' \text{ .... } (5),$$

which shows, that V' the value of V at the point p' is given, when $\overline{V}$ its value at the surface is known.

To convince ourselves that there does exist such a function as we have supposed U to be; conceive the surface to be a perfect conductor put in communication with the earth, and a unit of positive electricity to be concentrated in the point p', then the total potential function arising from p' and from the electricity it will induce upon the surface, will be the required value of U. For, in consequence of the communication established between the conducting surface and the earth, the total potential function at this surface must be constant, and equal to that of the earth itself, i. e. to *zero* (seeing that in this state they form but one conducting body). Taking, therefore, this total potential function for U, we have evidently $0 = \overline{U}$, $0 = \delta U$, and $U = \frac{1}{r}$ for those parts infinitely near to p'. As moreover, this function has no other singular points within the surface, it evidently possesses all the properties assigned to U in the preceding proof.

Again, since we have evidently $U' = 0$, for all the space exterior to the surface, the equation (4) art. 4 gives

$$0 = 4\pi\,(\rho) + \overline{\frac{dU}{dw}};$$

where (ρ) is the density of the electricity induced on the surface, by the action of a unit of electricity concentrated in the point p'. Thus, the equation (5) of this article becomes

$$V' = -\int d\sigma\,(\rho)\,\overline{V} \;\ldots\ldots\ldots\ldots\ldots\ldots\ldots\ldots\ldots\; (6).$$

This equation is remarkable on account of its simplicity and singularity, seeing that it gives the value of the potential for any point p', within the surface, when $\overline{V}$, its value at the surface itself is known, together with (ρ), the density that a unit of electricity concentrated in p' would induce on this surface, if it conducted electricity perfectly, and were put in communication with the earth.

Having thus proved, that V' the value of the potential function V, at any point p' within the surface is given, provided its value $\overline{V}$ is known at this surface, we will now show, that whatever the value of $\overline{V}$ may be, the general value of V deduced from it by the formula just given shall satisfy the equation

$$0 = \delta V.$$

For, the value of V at any point p whose co-ordinates are x, y, z, deduced from the assumed value of $\overline{V}$, by the above written formula, is

$$4\pi V = \int d\sigma \overline{V} \left(\frac{dU}{dw}\right),$$

U being the total potential function within the surface, arising from a unit of electricity concentrated in the point p, and the electricity induced on the surface itself by its action. Then, since $\overline{V}$ is evidently independent of x, y, z, we immediately deduce

$$4\pi\delta V = \int d\sigma \overline{V} \delta \left(\frac{dU}{dw}\right).$$

Now the general value of U will depend upon the position of the point p producing it, and upon that of any other point p' whose co-ordinates are x', y', z', to which it is referred, and will consequently be a function of the six quantities, x, y, z, x', y', z'. But we may conceive U to be divided into two parts, one $= \frac{1}{r}$ (r being the distance pp') arising from the electricity in p, the other, due to the electricity induced on the surface by the action of p, and which we shall call $U_{,}$. Then since U has no singular values within the surface, we may deduce its general value from that at the surface, by a formula similar to the one just given. Thus

$$4\pi U_{,} = \int d\sigma \overline{U}_{,} \left(\frac{\overline{dU'}}{dw}\right);$$

where U' is the total potential function, which would be produced by a unit of electricity in p', and therefore, $\left(\frac{\overline{dU'}}{dw}\right)$ is independent of the co-ordinates x, y, z, of p, to which δ refers.

Hence

$$4\pi\delta U_{,} = \int d\sigma \left(\frac{\overline{dU'}}{dw}\right) \delta \overline{U}_{,}.$$

We have before supposed

$$U = \frac{1}{r} + U_{,},$$

and as $\delta \frac{1}{r} = 0$, we immediately obtain

$$\delta U = \delta U_{,}.$$

Again, since we have at the surface itself

$$0 = \overline{U} = \frac{1}{\overline{r}} + \overline{U}_{,};$$

$\overline{r}$ being the distance between p and the element $d\sigma$, we hence deduce

$$0 = \delta \overline{U}_{,};$$

this substituted in the general value of $\delta U_{,}$ before given, there arises $\delta U_{,} = 0$, and consequently $0 = \delta U$. The result just obtained being general, and applicable to any point p'' within the surface, gives immediately

$$0 = \delta \left(\frac{\overline{dU}}{dw}\right),$$

and we have by substituting in the equation determining δV,

$$0 = \delta V.$$

In a preceding part of this article, we have obtained the equation

$$0 = 4\pi (\rho) + \left(\frac{\overline{dU}}{dw}\right),$$

which combined with $0 = \delta \left(\frac{\overline{dU}}{dw}\right)$, gives

$$0 = \delta (\rho),$$

and therefore the density (ρ) induced on any element $d\sigma$, which is evidently a function of the co-ordinates x, y, z, of p, is also such a function as will satisfy the equation $0 = \delta(\rho)$: it is moreover evident, that (ρ) can never become infinite when p is within the surface.

It now remains to prove that the formula

$$V = \frac{1}{4\pi}\int d\sigma \overline{V}\left(\frac{\overline{dU}}{dw}\right) = -\int d\sigma\,(\rho)\overline{V},$$

shall always give $V = \overline{V}$, for any point within the surface and infinitely near it, whatever may be the assumed value of $\overline{V}$.

For this, suppose the point p to approach infinitely near the surface; then it is clear that the value of (ρ), the density of the electricity induced by p, will be insensible, except for those parts infinitely near to p, and in these parts it is easy to see, that the value of (ρ) will be independent of the form of the surface, and depend only on the distance $p, d\sigma$. But, we shall afterwards show (art. 10), that when this surface is a sphere of any radius whatever, the value of (ρ) is

$$(\rho) = \frac{-a}{2\pi\,.\,f^3};$$

a being the shortest distance between p and the surface, and f representing the distance $p, d\sigma$. This expression will give an idea of the rapidity with which (ρ) decreases, in passing from the infinitely small portion of the surface in the immediate vicinity of p, to any other part situate at a finite distance from it, and when substituted in the above written value of V, gives, by supposing a to vanish,

$$V = \overline{V}.$$

It is also evident, that the function V, determined by the above written formula, will have no singular values within the surface under consideration.

What was before proved, for the space within any closed surface, may likewise be shown to hold good, for that exterior to a number of closed surfaces, of any forms whatever, provided we introduce the condition, that V' shall be equal to zero at an infinite distance from these surfaces. For, conceive a surface at an infinite distance from those under consideration; then, what we have before said, may be applied to the whole space within the infinite surface and exterior to the others; consequently

$$4\pi V' = \int d\sigma \overline{V'} \left(\frac{d\overline{U}}{dw}\right) \dots\dots\dots\dots\dots\dots (5'),$$

where the sign of integration must extend over all the surfaces, (seeing that the part due to the infinite surface is destroyed by the condition, that V' is there equal to *zero*), and dw must evidently be measured from the surfaces, into the exterior space to which V' now belongs.

The form of the equation (6) remains also unaltered, and

$$V' = -\int (\rho)\, d\sigma \overline{V'} \dots\dots\dots\dots\dots\dots (6');$$

the sign of integration extending over all the surfaces, and (ρ) being the density of the electricity which would be induced on each of the bodies, in presence of each other, supposing they all communicated with the earth by means of infinitely thin conducting wires.*

(6.) Let now A be any closed surface, conducting electricity perfectly, and p a point within it, in which a given quantity of electricity Q is concentrated, and suppose this to induce an electrical state in A; then will $\overline{V}$, the value of the potential function arising from the surface only, at any other point p', also within it, be such a function of the co-ordinates of p and p'; that we may change the co-ordinates of p, into those of p', and reciprocally, without altering its value. Or, in other words, the value of the potential function at p', due to the surface alone, when the inducing electricity Q is concentrated in p, is equal to that which would have place at p, if the same electricity Q were concentrated in p'.

For, in consequence of the equilibrium at the surface, we have evidently, in the first case, when the inducing electricity is concentrated in p,

$$\frac{Q}{r} + \overline{V} = \beta;$$

* In connexion with the subject of this article, see a paper by Professor Thomson, *Cambridge and Dublin Mathematical Journal*, Vol. VI. p. 109.

r being the distance between p and $d\sigma'$ an element of the surface A, and β a constant quantity dependent upon the quantity of electricity originally placed on A. Now the value of V at p' is

$$V = -\int(\rho')\, d\sigma'\, \overline{V},$$

by what has been shown (art. 5); (ρ') being, as in that article, the density of the electricity which would be induced on the element $d\sigma'$ by a unit of electricity in p', if the surface A were put in communication with the earth. This equation gives

$$\delta V = -\int(\rho')\, d\sigma'\, \delta\overline{V} = 0;$$

since $\delta\overline{V} = -\delta\frac{Q}{r} = 0$; the symbol δ referring to the co-ordinates x, y, z, of p. But we know that $0 = \delta' V$; where δ' refers in a similar way to the co-ordinates x', y', z', of p' only. Hence we have simultaneously

$$0 = \delta V \text{ and } 0 = \delta' V;$$

where it must be remarked, that the function V has no singular values, provided the points p and p' are both situate within the surface A. This being the case the first equation evidently gives (art. 5)

$$V = -\int(\rho)\, d\sigma\, \overline{V};$$

$\overline{V}$ being what V would become, if the inducing point p were carried to $d\sigma$, p' remaining fixed. Where $\overline{V}$ is a function of x', y', z', and ξ, η, ζ, the co-ordinates of $d\sigma$, whereas (ρ) is a function of $x, y, z, \xi, \eta, \zeta$, independent of x', y', z'; hence by the second equation

$$0 = \delta' V = -\int(\rho)\, d\sigma\, \delta'\overline{V},$$

which could not hold generally whatever might be the situation of p, unless we had

$$0 = \delta'\overline{V};$$

where we must be cautious, not to confound the present value of

$\overline{V}$, with that employed at the beginning of this article in proving the equation $0 = \delta V$, which last, having performed its office, will be no longer employed.

The equation $0 = \delta' V'$ gives in the same way

$$V = -\int (\rho')\, d\sigma' \overline{\overline{V}}';$$

$\overline{\overline{V}}$ being what $\overline{\overline{V}}$ becomes by bringing the point p' to any other element $d\sigma'$ of the surface A. This substituted for $\overline{V}$, in the expression before given, there arises

$$V = + \iint (\rho)\, (\rho')\, d\sigma\, d\sigma' \overline{\overline{V}}' :$$

in which double integral, the signs of integration, relative to each of the independent elements $d\sigma$ and $d\sigma'$, must extend over the whole surface.

If now, we represent by V', the value of the potential function at p arising from the surface A, when the electricity Q is concentrated in p', we shall evidently have

$$V_{\prime} = + \int (\rho')\, (\rho)\, d\sigma'\, d\sigma \overline{\overline{V}}_{\prime},$$

where the order of integrations alone is changed, the limits remaining unaltered: $\overline{\overline{V}}_{\prime}$ being what $V_{\prime}$ would become, by first bringing the electrical point p' to the surface, and afterward the point p to which $V_{\prime}$ belongs. This being done, it is clear that $\overline{\overline{V}}'$ and $\overline{\overline{V}}_{\prime}$ represent but one and the same quantity, seeing that each of them serves to express the value of the potential function, at any point of the surface A, arising from the surface itself, when the electricity is induced upon it by the action of an electrified point, situate in any other point of the same surface, and hence we have evidently

$$V = V_{\prime},$$

as was asserted at the commencement of this article.

It is evident from art. 5, that our preceding arguments will be equally applicable to the space exterior to the surfaces of any number of conducting bodies, provided we introduce the condition, that the potential function V, belonging to this space, shall be equal to *zero*, when either p or p' shall remove to an infinite distance from these bodies, which condition will evidently be satisfied, provided all the bodies are originally in a natural state. Supposing this therefore to be the case, we see that the potential function belonging to any point p' of the exterior space, arising from the electricity induced on the surfaces of any number of conducting bodies, by an electrified point in p, is equal to that which would have place at p, if the electrified point were removed to p'.

What has been just advanced, being perfectly independent of the number and magnitude of the conducting bodies, may be applied to the case of an infinite number of particles, in each of which the fluid may move freely, but which are so constituted that it cannot pass from one to another. This is what is always supposed to take place in the theory of magnetism, and the present article will be found of great use to us when in the sequel we come to treat of that theory.

(7.) These things being established with respect to electrified surfaces; the general theory of the relations between the density of the electric fluid and the corresponding potential functions, when the electricity is disseminated through the interior of solid bodies as well as over their surfaces, will very readily flow from what has been proved (art. 1).

For this let V' represent the value of the potential function at a point p', within a solid body of any form, arising from the whole of the electric fluid contained in it, and ρ' be the density of the electricity in its interior; ρ' being a function of the three rectangular co-ordinates x, y, z: then if ρ be the density at the surface of the body, we shall have

$$V' = \int \frac{dx\,dy\,dz\,\rho'}{r'} + \int \frac{d\sigma\rho}{r};$$

r' being the distance between the point p' whose co-ordinates are

x', y', z', and that whose co-ordinates are x, y, z, to which ρ' belongs, also r the distance between p' and $d\sigma$, an element of the surface of the body: V' being evidently a function of x', y', z'. If now V be what V' becomes by changing x', y', z' into x, y, z, it is clear from art. 1, that ρ' will be given by

$$0 = 4\pi\rho' + \delta V.$$

Substituting for ρ', the value which results from this equation, in that immediately preceding we obtain

$$V' = -\int \frac{dx\, dy\, dz\, \delta V}{4\pi r'} + \int \frac{\rho d\sigma}{r},$$

which, by means of the equation (3) art. 3, becomes

$$\int \frac{\rho d\sigma}{r} = \frac{1}{4\pi} \left\{ \int d\sigma \overline{V} \left(\frac{d\frac{1}{r}}{dw} \right) - \int \frac{d\sigma}{r} \left(\overline{\frac{dV}{dw}} \right) \right\};$$

the horizontal lines over the quantities indicating that they belong to the surface itself.

Suppose $V_{,}$ to be the value of the potential function in the space exterior to the body, which, by art. 5, will depend on the value of V at the surface only; and the equation (2) art. 3, applied to this exterior space, will give, since $\delta V_{,} = 0$ and $\delta \frac{1}{r} = 0$,

$$\int d\sigma \overline{V} \left(\frac{d\frac{1}{r}}{dw'} \right) = \int d\sigma \overline{V}_{,} \left(\frac{d\frac{1}{r}}{dw'} \right) = \int \frac{d\sigma}{r} \left(\overline{\frac{dV_{,}}{dw'}} \right);$$

where dw' is measured from the surface into the exterior space to which $V_{,}$ belongs, as dw is, into the interior space. Consequently $dw = -dw'$, and therefore

$$\int d\sigma \overline{V} \left(\frac{d\frac{1}{r}}{dw} \right) = -\int d\sigma \overline{V} \left(\frac{d\frac{1}{r}}{dw'} \right) = -\int \frac{d\sigma}{r} \left(\overline{\frac{dV_{,}}{dw'}} \right).$$

Hence the equation determining ρ becomes, by substituting for

$$\int d\sigma \overline{V} \left(\frac{d\frac{1}{r}}{dw} \right)$$

its value just given,

$$\int \frac{\rho d\sigma}{r} = \frac{-1}{4\pi} \int \frac{d\sigma}{r} \left\{ \left(\frac{\overline{dV}}{dw} \right) + \left(\frac{\overline{dV_{\prime}}}{dw'} \right) \right\},$$

an equation which could not subsist generally, unless

$$\rho = \frac{-1}{4\pi} \left\{ \frac{\overline{dV}}{dw} + \frac{\overline{dV_{\prime}}}{dw'} \right\} \text{ .... (7).}$$

Thus the whole difficulty is reduced to finding the value $V_{\prime}$ of the potential function exterior to the body.

Although we have considered only one body, it is clear that the same theory is applicable to any number of bodies, and that the values of ρ and ρ' will be given by precisely the same formulæ, however great that number may be: $V_{\prime}$ being the exterior potential function common to all the bodies.

In case the bodies under consideration are all perfect conductors, we have seen (art. 1), that the whole of the electricity will be carried to their surfaces, and therefore there is here no place for the application of the theory contained in this article; but as there are probably no perfectly conducting bodies in nature, this theory becomes indispensably necessary, if we would investigate the electrical phenomena in all their generality.

Having in this, and the preceding articles, laid down the most general principles of the electrical theory, we shall in what follows apply these principles to more special cases; and the necessity of confining this Essay within a moderate extent, will compel us to limit ourselves to a brief examination of the more interesting phenomena.

APPLICATION OF THE PRECEDING RESULTS TO THE THEORY OF ELECTRICITY.

(8.) THE first application we shall make of the foregoing principles, will be to the theory of the Leyden phial. For this, we will call the inner surface of the phial A, and suppose it to be of any form whatever, plane or curved, then, B being its outer surface, and θ the thickness of the glass measured along a normal to A; θ will be a very small quantity, which, for greater generality, we will suppose to vary in any way, in passing from one point of the surface A to another. If now the inner coating of the phial be put in communication with a conductor C, charged with any quantity of electricity, and the outer one be also made to communicate with another conducting body C', containing any other quantity of electricity, it is evident, in consequence of the communications here established, that the total potential function, arising from the whole system, will be constant throughout the interior of the inner metallic coating, and of the body C. We shall here represent this constant quantity by

$$\beta.$$

Moreover, the same potential function within the substance of the outer coating, and in the interior of the conductor C', will be equal to another constant quantity

$$\beta'.$$

Then designating by V, the value of this function, for the whole of the space exterior to the conducting bodies of the system, and consequently for that within the substance of the glass itself; we shall have (art. 4)

$$\overline{V} = \beta \text{ and } \overline{\overline{V}} = \beta'.$$

One horizontal line over any quantity indicating that it belongs to the inner surface A, and two showing that it belongs to the outer one B.

At any point of the surface A, suppose a normal to it to be drawn, and let this be the axes of $\overline{w}$: then $\overline{w}'$, $\overline{w}''$, being two other rectangular axes, which are necessarily in the plane tangent to A at this point; V may be considered as a function of $\overline{w}$, $\overline{w}'$ and $\overline{w}''$, and we shall have by TAYLOR's theorem, since $\overline{w}' = 0$ and $\overline{w}'' = 0$ at the axis of $\overline{w}$ along which θ is measured,

$$\overline{\overline{V}} = \overline{V} + \frac{d\overline{V}}{d\overline{w}} \cdot \frac{\theta}{1} + \frac{d^2\overline{V}}{d\overline{w}^2} \cdot \frac{\theta^2}{1 \cdot 2} + \text{\&c.};$$

where, on account of the smallness of θ, the series converges very rapidly. By writing in the above, for $\overline{V}$ and $\overline{\overline{V}}$ their values just given, we obtain

$$\beta' - \beta = \frac{d\overline{V}}{d\overline{w}} \cdot \frac{\theta}{1} + \frac{d^2\overline{V}}{d\overline{w}^2} \cdot \frac{\theta^2}{1 \cdot 2} + \text{\&c.}$$

In the same way, if $\overline{\overline{w}}$ be a normal to B, directed towards A, and $\theta_{,}$ be the thickness of the glass measured along this normal, we shall have

$$\beta - \beta' = \frac{d\overline{\overline{V}}}{d\overline{\overline{w}}} \cdot \frac{\theta_{,}}{1} + \frac{d^2\overline{\overline{V}}}{d\overline{\overline{w}}^2} \cdot \frac{\theta_{,}^2}{1 \cdot 2} + \text{\&c.}$$

But, if we neglect quantities of the order θ, compared with those retained, the following equation will evidently hold good,

$$\frac{d^n\overline{\overline{V}}}{d\overline{\overline{w}}^n} = (-1)^n \frac{d^n\overline{V}}{d\overline{w}^n};$$

n being any whole positive number, the factor $(-1)^n$ being introduced because $\overline{w}$ and $\overline{\overline{w}}$ are measured in opposite directions. Now by article 4

$$-4\pi\overline{\rho} = \frac{d\overline{V}}{d\overline{w}} \text{ and } -4\pi\overline{\overline{\rho}} = \frac{d\overline{\overline{V}}}{d\overline{\overline{w}}};$$

$\bar{\rho}$ and $\bar{\bar{\rho}}$ being the densities of the electric fluid at the surfaces A and B respectively. Permitting ourselves, in what follows, to neglect quantities of the order θ^2 compared with those retained, it is clear that we may write θ for $\theta_{,,}$ and hence by substitution

$$\beta' - \beta = -4\pi\bar{\rho}\theta + \left(\frac{d^2\bar{V}}{d\bar{w}}\right)\frac{\theta^2}{1.2},$$

$$\beta - \beta' = -4\pi\bar{\bar{\rho}}\theta + \left(\frac{d^2\bar{\bar{V}}}{d\bar{\bar{w}}^2}\right)\frac{\theta^2}{1.2};$$

where V and ρ are quantities of the order $\frac{1}{\theta}$; β' and β being the order θ^0 or unity. The only thing which now remains to be determined, is the value of $\frac{d^2\bar{V}}{d\bar{w}^2}$ for any point on the surface A.

Throughout the substance of the glass, the potential function V will satisfy the equation $0 = \delta V$, and therefore at a point on the surface of A, where of necessity w, w' and w'' are each equal to zero, we have

$$0 = \frac{d^2\bar{V}}{dw^2} + \frac{d^2\bar{V}}{dw'^2} + \frac{d^2\bar{V}}{dw''^2} = \delta\bar{V};$$

the horizontal mark over w, w' and w'' being, for simplicity, omitted. Then since $w' = 0$,

$$\frac{d^2\bar{V}}{dw'^2} = (V_0 - 2V_{dw'} + V_{2dw'}) : dw'^2,$$

and as V is constant and equal to β at the surface A, there hence arises

$$V_0 = \beta; \quad V_{dw'} = \beta + \frac{d\bar{V}}{dw}\frac{dw'^2}{2R}, \quad V_{2dw'} = \beta + \frac{d\bar{V}}{dw}\frac{4dw'^2}{2R};$$

R being the radius of curvature at the surface A, in the plane (w, w'). Substituting these values in the expression immediately preceding, we get

$$\frac{d^2\bar{V}}{dw'^2} = \frac{1}{R}\frac{d\bar{V}}{dw} = \frac{-4\pi\bar{\rho}}{R}.$$

In precisely the same way we obtain, by writing R' for the radius of curvature in the plane (w, w''),

$$\frac{d^2\overline{V}}{dw''^2} = \frac{-4\pi\overline{\rho}}{R'}:$$

both rays being accounted positive on the side where w, i.e. $\overline{w}$, is negative. These values substituted in $0 = \delta\overline{V}$, there results

$$\frac{d^2\overline{V}}{dw^2} = 4\pi\overline{\rho}\left(\frac{1}{R} + \frac{1}{R'}\right)$$

for the required value of $\dfrac{d^2V}{dw^2}$, and thus the sum of the two equations into which it enters, yields

$$\overline{\rho}\left\{1 + \left(\frac{1}{R} + \frac{1}{R'}\right)\theta\right\} = -\overline{\overline{\rho}},$$

and the difference of the same equations gives

$$\beta - \beta' = 2\pi\,(\overline{\rho} - \overline{\overline{\rho}})\,\theta;$$

therefore the required values of the densities $\overline{\rho}$ and $\overline{\overline{\rho}}$ are

$$\left.\begin{aligned} \overline{\rho} &= \frac{\beta - \beta'}{4\pi\theta}\left\{1 + \frac{\theta}{2}\left(\frac{1}{R} + \frac{1}{R'}\right)\right\} \\ \overline{\overline{\rho}} &= \frac{\beta' - \beta}{4\pi\theta}\left\{1 - \frac{\theta}{2}\left(\frac{1}{R} + \frac{1}{R'}\right)\right\} \end{aligned}\right\}\text{ .... (8)}$$

which values are correct to quantities of the order $\theta^2\overline{\rho}$, or, which is the same thing, to quantities of the order θ; these having been neglected in the latter part of the preceding analysis, as unworthy of notice.

Suppose $d\sigma$ is an element of the surface A, the corresponding element of B, cut off by normals to A, will be $d\sigma\left\{1 + \theta\left(\frac{1}{R} + \frac{1}{R'}\right)\right\}$, and therefore the quantity of fluid on this last element will be $\overline{\overline{\rho}}d\sigma\left\{1 + \theta\left(\frac{1}{R} + \frac{1}{R'}\right)\right\}$; substituting for $\overline{\overline{\rho}}$ its value before found, $\overline{\overline{\rho}} = -\overline{\rho}\left\{1 - \theta\left(\frac{1}{R} + \frac{1}{R'}\right)\right\}$, and neglecting $\theta^2\overline{\rho}$, we obtain

$$-\overline{\rho}d\sigma,$$

the same quantity as on the element $d\sigma$ of the first surface. If, therefore, we conceive any portion of the surface A, bounded by a closed curve, and a corresponding portion of the surface B, which would be cut off by a normal to A, passing completely round this curve; the sum of the two quantities of electric fluid, on these corresponding portions, will be equal to zero; and consequently, in an electrical jar any how charged, the total quantity of electricity in the jar may be found, by calculating the quantity, on the two exterior surfaces of the metallic coatings farthest from the glass, as the portions of electricity, on the two surfaces adjacent to the glass, exactly neutralise each other. This result will appear singular, when we consider the immense quantity of fluid collected on these last surfaces, and moreover, it would not be difficult to verify it by experiment.

As a particular example of the use of this general theory: suppose a spherical conductor whose radius a, to communicate with the inside of an electrical jar, by means of a long slender wire, the outside being in communication with the common reservoir; and let the whole be charged: then P representing the density of the electricity on the surface of the conductor, which will be very nearly constant, the value of the potential function within the sphere, and, in consequence of the communication established, at the inner coating A also, will be $4\pi aP$ very nearly, since we may, without sensible error, neglect the action of the wire and jar itself in calculating it. Hence

$$\beta = 4\pi aP \text{ and } \beta' = 0,$$

and the equations (8), by neglecting quantities of the order θ, give

$$\overline{\rho} = \frac{\beta}{4\pi\theta} = \frac{a}{\theta}P \text{ and } \overline{\overline{\rho}} = \frac{-\beta}{4\pi\theta} = -\frac{a}{\theta}P.$$

We thus obtain, by the most simple calculation, the values of the densities, at any point on either of the surfaces *A* and *B*, next the glass, when that on the spherical conductor is known.

The theory of the condenser, electrophorous, &c. depends upon what has been proved in this article; but these are details into which the limits of this Essay will not permit me to enter; there is, however, one result, relative to charging a number of jars *by cascade*, that appears worthy of notice, and which flows so readily from the equations (8), that I cannot refrain from introducing it here.

Conceive any number of equal and similar insulated Leyden phials, of uniform thickness, so disposed, that the exterior coating of the first may communicate with the interior one of the second; the exterior one of the second, with the interior one of the third; and so on throughout the whole series, to the exterior surface of the last, which we will suppose in communication with the earth. Then, if the interior of the first phial be made to communicate with the prime conductor of an electrical machine, in a state of action, all the phials will receive a certain charge, and this mode of operating is called charging *by cascade*. Permitting ourselves to neglect the small quantities of free fluid on the exterior surfaces of the metallic coatings, and other quantities of the same order, we may readily determine the electrical state of each phial in the series: for thus, the equations (8) become

$$\bar{\rho} = \frac{\beta - \beta'}{4\pi\theta}, \quad \bar{\bar{\rho}} = \frac{\beta' - \beta}{4\pi\theta}.$$

Designating now, by an index at the foot of any letter, the number of the phial to which it belongs, so that, $\bar{\rho}_1$ may belong to the first, $\bar{\rho}_2$ to the second phial, and so on; we shall have, by supposing their whole number to be n, since θ is the same for every one,

$$\bar{\rho}_1 = \frac{\beta_1 - \beta'_1}{4\pi\theta}, \qquad \bar{\bar{\rho}}_1 = \frac{\beta'_1 - \beta_1}{4\pi\theta},$$

$$\bar{\rho}_2 = \frac{\beta_2 - \beta'_2}{4\pi\theta}, \qquad \bar{\bar{\rho}}_2 = \frac{\beta'_2 - \beta_2}{4\pi\theta},$$

&c. &c.

$$\bar{\rho}_n = \frac{\beta_n - \beta'_n}{4\pi\theta}, \qquad \bar{\bar{\rho}}_n = \frac{\beta'_n - \beta_n}{4\pi\theta}.$$

Now β represents the value of the total potential function, within the prime conductor and interior coating of the first phial, and in consequence of the communications established in this system, we have in regular succession, beginning with the prime conductor, and ending with the exterior surface of the last phial, which communicates with the earth,

$$\beta = \beta_1;\ \beta'_1 = \beta_2;\ \beta'_2 = \beta'_3;\ \&c. \ldots \beta'_{n-1} = \beta_n;\ \beta'_n = 0$$

$$0 = \bar{\bar{\rho}}_1 + \bar{\rho}_2;\ 0 = \bar{\bar{\rho}}_2 + \bar{\rho}_3;\ \&c. \ldots 0 = \bar{\bar{\rho}}_{n-1} + \bar{\rho}_n.$$

But the first system of equations gives $0 = \bar{\rho}_s + \bar{\bar{\rho}}_s$, whatever whole number s may be, and the second line of that just exhibited is expressed by $0 = \bar{\rho}_{s-1} + \bar{\bar{\rho}}_s$; hence by comparing these two last equations,

$$\bar{\rho}_s = \bar{\rho}_{s-1},$$

which shows that every phial of the system is equally charged. Moreover, if we sum up vertically, each of the columns of the first system, there will arise in virtue of the second

$$\bar{\rho}_1 + \bar{\rho}_2 + \bar{\rho}_3 \ldots\ldots\ldots + \bar{\rho}_n = \frac{\beta}{4\pi\theta},$$

$$\bar{\bar{\rho}}_1 + \bar{\bar{\rho}}_2 + \bar{\bar{\rho}}_3 \ldots\ldots\ldots + \bar{\bar{\rho}}_n = \frac{-\beta}{4\pi\theta}.$$

We therefore see, that the total charge of all the phials is precisely the same, as that which one only would receive, if placed in communication with the same conductor, provided its exterior coating were connected with the earth. Hence this mode of charging, although it may save time, will never produce a greater accumulation of fluid than would take place if one phial only were employed.

(9.) Conceive now a hollow shell of perfectly conducting matter, of any form and thickness whatever, to be acted upon by any electrified bodies, situate without it; and suppose them to induce an electrical state in the shell; then will this induced state be such, that the total action on an electrified particle, placed any where within it, will be absolutely null.

For let V represent the value of the total potential function, at any point p within the shell, then we shall have at its inner surface, which is a closed one,

$$\overline{V} = \beta;$$

β being the constant quantity, which expresses the value of the potential function, within the substance of the shell, where the electricity is, by the supposition, in equilibrium, in virtue of the actions of the exterior bodies, combined with that arising from the electricity induced in the shell itself. Moreover, V evidently satisfies the equation $0 = \delta V$, and has no singular value within the closed surface to which it belongs: it follows therefore, from Art. 5, that its general value is

$$V = \beta,$$

and as the forces acting upon p, are given by the differentials of V, these forces are evidently all equal to *zero*.

If, on the contrary, the electrified bodies are all within the shell, and its exterior surface is put in communication with the earth, it is equally easy to prove, that there will not be the slightest action on any electrified point exterior to it; but, the action of the electricity induced on its inner surface, by the electrified bodies within it, will exactly balance the direct action of the bodies themselves. Or more generally:

Suppose we have a hollow, and perfectly conducting shell, bounded by any two closed surfaces, and a number of electrical bodies are placed, some within and some without it, at will; then, if the inner surface and interior bodies be called the interior system; also, the outer surface and exterior bodies the exterior system; all the electrical phenomena of the interior system, relative to attractions, repulsions, and densities, will be the same as would take place if there were no exterior system, and the inner surface were a perfect conductor, put in communication with the earth; and all those of the exterior system will be the same, as if the interior one did not exist, and the outer surface were a perfect conductor, containing a quantity of electricity, equal to

4

the whole of that originally contained in the shell itself, and in all the interior bodies.

This is so direct a consequence of what has been shown in articles 4 and 5, that a formal demonstration would be quite superfluous, as it is easy to see, the only difference which could exist, relative to the interior system, between the case where there is an exterior system, and where there is not one, would be in the addition of a constant quantity, to the total potential function within the exterior surface, which constant quantity must necessarily disappear in the differentials of this function, and consequently, in the values of the attractions, repulsions, and densities, which all depend on these differentials alone. In the exterior system there is not even this difference, but the total potential function exterior to the inner surface is precisely the same, whether we suppose the interior system to exist or not.

(10.) The consideration of the electrical phenomena, which arise from spheres variously arranged, is rather interesting, on account of the ease with which all the results obtained from theory may be put to the test of experiment; but, the complete solution of the simple case of two spheres only, previously electrified, and put in presence of each other, requires the aid of a profound analysis, and has been most ably treated by M. POISSON (Mém. de l'Institut. 1811). Our object, in the present article, is merely to give one or two examples of determinations, relative to the distribution of electricity on spheres, which may be expressed by very simple formulæ.

Suppose a spherical surface whose radius is a, to be covered with electric matter, and let its variable density be represented by ρ; then if, as in the Méc. Céleste, we expand the potential function V, belonging to a point p within the sphere, in the form

$$V = U^{(0)} + U^{(1)}\frac{r}{a} + U^{(2)}\frac{r^2}{a^2} + U^{(3)}\frac{r^3}{a^3} + \text{etc.};$$

r being the distance between p and the centre of the sphere, and $U^{(0)}$, $U^{(1)}$, etc. functions of the two other polar co-ordinates of p, it is clear, by what has been shown in the admirable work just mentioned, that the potential function V', arising from the same spherical surface, and belonging to a point p', exterior to this surface, at the distance r' from its centre, and on the radius r produced, will be

$$V' = U^{(0)}\frac{a}{r'} + U^{(1)}\frac{a^2}{r'^2} + U^{(2)}\frac{a^3}{r'^3} + \text{etc.}$$

If, therefore, we make $V = \phi(r)$, and $V' = \psi(r')$, the two functions ϕ and ψ will satisfy the equation

$$\psi(r) = \frac{a}{r}\,\phi\left(\frac{a^2}{r}\right) \quad \text{or} \quad \phi(r) = \frac{a}{r}\,\psi\left(\frac{a^2}{r}\right).$$

But (art. 4)

$$4\pi\rho = -\frac{d\overline{V}}{dw} - \frac{d\overline{V'}}{dw'} = +\frac{d\overline{V}}{dr} - \frac{d\overline{V'}}{dr'} = \phi'(a) - \psi'(a),$$

and the equation between ϕ and ψ, in its first form, gives, by differentiation,

$$\psi'(r) = -\frac{a}{r^2}\,\phi\left(\frac{a^2}{r}\right) - \frac{a^3}{r^3}\,\phi'\left(\frac{a^2}{r}\right).$$

Making now $r = a$ there arises

$$\psi'(a) = -\frac{\phi(a)}{a} - \phi'(a);$$

ϕ' and ψ' being the characteristics of the differential co-efficients of ϕ and ψ, according to Lagrange's notation.

In the same way the equation in its second form yields

$$\phi'(a) = -\frac{\psi(a)}{a} - \psi'(a).$$

These substituted successively, in the equation by which ρ is determined, we have the following,

$$\left.\begin{aligned} 4\pi\rho &= 2\phi'(a) + \frac{\phi(a)}{a} = 2\frac{d\overline{V}}{dr} + \frac{\overline{V}}{a} \\ 4\pi\rho &= -2\psi'(a) - \frac{\psi(a)}{a} = -2\frac{d\overline{V'}}{dr'} - \frac{\overline{V'}}{a} \end{aligned}\right\} \quad \ldots\ldots\ldots (9).$$

If, therefore, the value of the potential function be known, either for the space within the surface, or for that without it, the value of the density ρ will be immediately given, by one or other of these equations.

From what has preceded, we may readily determine how the electric fluid will distribute itself, in a conducting sphere whose radius is a, when acted upon by any bodies situate without it; the electrical state of these bodies being given. In this case, we have immediately the value of the potential function arising from them. Let this value, for any point p within the sphere, be represented by A; A being a function of the radius r, and two other polar co-ordinates. Then the whole of the electricity will be carried to the surface (art. 1), and if V be the potential function arising from this electrified surface, for the same point p, we shall have, in virtue of the equilibrium within the sphere,

$$V + A = \beta \text{ or } V = \beta - A;$$

β being a constant quantity. This value of V being substituted in the first of the equations (9), there results

$$4\pi\rho = -2\,\frac{\overline{dA}}{dr} - \frac{\overline{A}}{a} + \frac{\beta}{a}:$$

the horizontal lines indicating, as before, that the quantities under them belong to the surface itself.

In case the sphere communicates with the earth, β is evidently equal to *zero*, and ρ is completely determined by the above: but if the sphere is insulated, and contains any quantity Q of electricity, the value of β may be ascertained as follows: Let V' be the value of the potential function without the surface, corresponding to the value $V = \beta - A$ within it; then, by what precedes

$$V' = \frac{\beta}{r'} - A';$$

A' being determined from A by the following equations:

$$A = \phi_{,}(r), \quad \psi_{,}(r) = \frac{a}{r}\,\phi_{,}\left(\frac{a^2}{r}\right), \quad A' = \psi_{,}(r'),$$

and r', being the radius corresponding to the point p', exterior to the sphere, to which A' belongs. When r' is finite, we have evidently $V'=\frac{Q}{r'}$. Therefore by equating

$$\frac{Q}{r'}=\frac{\beta}{r'}-A' \text{ or } \beta=Q+r'A';$$

r' being made infinite. Having thus the value of β, the value of ρ becomes known.

To give an example of the application of the second equation in ρ; let us suppose a spherical conducting surface, whose radius is a, in communication with the earth, to be acted upon by any bodies situate within it, and B' to be the value of the potential function arising from them, for a point p' exterior to it. The total potential function, arising from the interior bodies and surface itself, will evidently be equal to *zero* at this surface, and consequently (art. 5), at any point exterior to it. Hence $V'+B'=0$; V' being due to the surface. Thus the second of the equations (9) becomes

$$4\pi\rho=2\,\overline{\frac{dB'}{dr'}}+\frac{\overline{B'}}{a}.$$

We are therefore able, by means of this very simple equation, to determine the density of the electricity induced on the surface in question.

Suppose now all the interior bodies to reduce themselves to a single point P, in which a unit of electricity is concentrated, and f to be the distance Pp': the potential function arising from P will be $\frac{1}{f}$, and hence

$$B'=\frac{1}{f};$$

r' being, as before, the distance between p' and the centre O of the shell. Let now b represent the distance OP, and θ the angle POp', then will $f^2=b^2-2br'.\cos\theta+r'^2$. From which equation we deduce successively,

$$\left(\frac{df}{dr'}\right)=\frac{r'-b\cos\theta}{f},$$

and $$2\frac{d\overline{B'}}{dr'} = -\frac{2}{f^2}\left(\frac{df}{dr'}\right) = \frac{-2r' + 2b.\cos\theta}{f^3}.$$

Making $r' = a$ in this, and in the value of B' before given, in order to obtain those which belong to the surface, there results

$$2\frac{d\overline{B'}}{dr'} + \frac{\overline{B'}}{a} = \frac{-2a^2 + 2ab.\cos\theta + f^2}{af^3} = \frac{b^2 - a^2}{af^3}.$$

This substituted in the general equation written above, there arises

$$\rho = \frac{b^2 - a^2}{4\pi af^3}.$$

If P is supposed to approach infinitely near to the surface, so that $b = a - \alpha$; α being an infinitely small quantity, this would become

$$\rho = \frac{-\alpha}{4\pi f^3}.$$

In the same way, by the aid of the equation between A and ρ, the density of the electric fluid, induced on the surface of a sphere whose radius is a, when the electrified point P is exterior to it, is found to be

$$\rho = \frac{a^2 - b^2}{4\pi af^2};$$

supposing the sphere to communicate, by means of an infinitely fine wire, with the earth, at so great a distance, that we might neglect the influence of the electricity induced upon it by the action of P. If the distance of P from the surface be equal to an infinitely small quantity α, we shall have in this case, as in the foregoing,

$$\rho = \frac{-\alpha}{2\pi.f^3}.$$

From what has preceded, we may readily deduce the general value of V, belonging to any point P, within the sphere, when $\overline{V}$ its value at the surface is known. For (ρ), the density induced upon an element $d\sigma$ of the surface, by a unit of electricity concentrated in P, has just been shown to be

$$\frac{b^2 - a^2}{4\pi af^3};$$

f being the distance P, $d\sigma$. This substituted in the general equation (6), art. 5, gives

$$V = -\int d\sigma\,(\rho)\,\overline{V} = \frac{a^2 - b^2}{4\pi a}\int \frac{d\sigma}{f^3}\,\overline{V} \;\dots\dots\dots\; (10).$$

In the same way we shall have, when the point P is exterior to the sphere,

$$V = \frac{b^2 - a^2}{4\pi a}\int \frac{d\sigma}{f^3}\,\overline{V} \;\dots\dots\dots\dots\; (11).$$

The use of these two equations will appear almost immediately, when we come to determine the distribution of the electric fluid, on a thin spherical shell, perforated with a small circular orifice.

The results just given may be readily obtained by means of LAPLACE'S much admired analysis (Méc. Cél. Liv. 3, Ch. II.), and indeed, our general equations (9), flow very easily from the equation (2) art. 10 of that Chapter. Want of room compels me to omit these confirmations of our analysis, and this I do the more freely, as the manner of deducing them must immediately occur to any one who has read this part of the Mécanique Céleste.

Conceive now, two spheres S and S', whose radii are a and a', to communicate with each other by means of an infinitely fine wire: it is required to determine the ratio of the quantities of electric fluid on these spheres, when in a state of equilibrium; supposing the distance of their centres to be represented by b.

The value of the potential function, arising from the electricity on the surface of S, at a point p, placed in its centre, is

$$\int \frac{\rho d\sigma}{a} = \frac{1}{a}\int \rho d\sigma = \frac{Q}{a};$$

$d\sigma$ being an element of the surface of the sphere, ρ the density of the fluid on this element, and Q the total quantity on the sphere. If now we represent by F', the value of the potential function for the same point p, arising from S', we shall have, by adding together both parts,

$$F' + \frac{Q}{a};$$

the value of the total potential function belonging to p, the centre of S. In like manner, the value of this function at p', the centre of S', will be

$$F + \frac{Q'}{a'}:$$

F being the part arising from S, and Q' the total quantity of electricity on S'. But in consequence of the equilibrium of the system, the total potential function throughout its whole interior is a constant quantity. Hence

$$F' + \frac{Q'}{a} = F + \frac{Q'}{a'}.$$

Although it is difficult to assign the rigorous values of F and F'; yet when the distance between the surfaces of the two spheres is considerable, compared with the radius of one of them, it is easy to see that F and F' will be very nearly the same, as if the electricity on each of the spheres producing them was concentrated in their respective centres, and therefore we have very nearly

$$F = \frac{Q}{b} \text{ and } F' = \frac{Q'}{b}.$$

These substituted in the above, there arises

$$\frac{Q}{b} + \frac{Q'}{a'} = \frac{Q'}{b} + \frac{Q}{a} \quad i.e.\ Q\left(\frac{1}{a} - \frac{1}{a}\right) = Q'\left(\frac{1}{a'} - \frac{1}{b}\right).$$

Thus the ratio of Q to Q' is given by a very simple equation, whatever may be the form of the connecting wire, provided it be a very fine one.

If we wished to put this result of calculation to the test of experiment, it would be more simple to write P and P' for the mean densities of the fluid on the spheres, or those which would be observed when, after being connected as above, they were separated to such a distance, as not to influence each other sensibly. Then since

$$Q = 4\pi a^2 P \text{ and } Q' = 4\pi a'^2 P',$$

we have by substitution, etc.

$$\frac{P}{P'} = \frac{a\,(b-a)}{a'\,(b-a')}.$$

We therefore see, that when the distance b between the centres of the spheres is very great, the mean densities will be inversely as the radii; and these last remaining unchanged, the density on the smaller sphere will decrease, and that on the larger increase in a very simple way, by making them approach each other.

Lastly, let us endeavour to determine the law of the distribution of the electric fluid, when in equilibrium on a very thin spherical shell, in which there is a small circular orifice. Then, if we neglect quantities of the order of the thickness of the shell, compared with its radius, we may consider it as an infinitely thin spherical surface, of which the greater segment S is a perfect conductor, and the smaller one s constitutes the circular orifice. In virtue of the equilibrium, the value of the potential function, on the conducting segment, will be equal to a constant quantity, as F, and if there were no orifice, the corresponding value of the density would be

$$\frac{F}{4\pi a};$$

a being the radius of the spherical surface. Moreover on this supposition, the value of the potential function for any point P, within the surface, would be F. Let therefore, $\frac{F}{4\pi a} + \rho$ represent the general value of the density, at any point on the surface of either segment of the sphere, and $F+V$, that of the corresponding potential function for the point P. The value of the potential function for any point on the surface of the sphere will be $F+\overline{V}$, which equated to F, its value on S, gives for the whole of this segment

$$0=\overline{V}.$$

Thus the equation (10) of this article becomes

$$V=\frac{a^2-b^2}{4\pi a}\int\frac{d\sigma}{f^3}\overline{V};$$

the integral extending over the surface of the smaller segment s only, which, without sensible error, may be considered as a plane.

But, since it is evident that ρ is the density corresponding to the potential function V, we shall have for any point on the segment s, treated as a plane,

$$\rho = \frac{-1}{2\pi}\frac{d\overline{V}}{dw},$$

as it is easy to see, from what has been before shown (art. 4); dw being perpendicular to the surface, and directed towards the centre of the sphere; the horizontal line always serving to indicate quantities belonging to the surface. When the point P is very near the plane s, and z is a perpendicular from P upon s, z will be a very small quantity, of which the square and higher powers may be neglected. Thus $b = a - z$, and by substitution

$$V = \frac{z}{2\pi}\int \frac{d\sigma}{f^3}\overline{V};$$

the integral extending over the surface of the small plane s, and f being, as before, the distance P, $d\sigma$. Now

$$\frac{d\overline{V}}{dw} = \frac{d\overline{V}}{dz}$$

at the surface of s, and $\frac{z}{f^3} = -\frac{d}{dz}\frac{1}{f}$; hence

$$\rho = \frac{-1}{2\pi}\frac{d\overline{V}}{dw} = \frac{-1}{2\pi}\frac{d\overline{V}}{dz} = \frac{-1}{4\pi^2}\frac{d}{dz}\int\frac{zd\sigma}{f^3}\overline{V} = \frac{1}{4\pi^2}\frac{d^2}{dz^2}\int\frac{d\sigma}{f}\overline{V};$$

provided we suppose $z = 0$ at the end of the calculus. Now the density $\frac{F}{4\pi a} + \rho$, upon the surface of the orifice s, is equal to *zero*, and therefore we have for the whole of this surface

$$\rho = -\frac{F}{4\pi a}.$$

Hence by substitution

$$\frac{-F\pi}{a} = \frac{d^2}{dz^2}\int\frac{d\sigma}{f}\overline{V} \quad \ldots\ldots\ldots\ldots\ldots\ldots (12);$$

the integral extending over the whole of the plane s, of which $d\sigma$ is an element, and z being supposed equal to *zero*, after all the operations have been effected.

It now only remains to determine the value of $\overline{V}$ from this equation. For this, let β now represent the linear radius of s, and y, the distance between its centre C and the foot of the perpendicular z: then if we conceive an infinitely thin oblate spheroid, of uniform density, of which the circular plane s constitutes the equator, the value of the potential function at the point P, arising from this spheroid, will be

$$\phi = k\int\frac{d\sigma}{f}\sqrt{(\beta^2-\eta^2)};$$

η being the distance $d\sigma$, C, and k a constant quantity. The attraction exerted by this spheroid, in the direction of the perpendicular z, will be $-\frac{d\phi}{dz}$, and by the known formulæ relative to the attractions of homogeneous spheroids, we have

$$-\frac{d\phi}{dz}=\frac{3Mz}{\beta^3}(\tan\theta-\theta);$$

M representing the mass of the spheroid, and θ being determined by the equations

$$\alpha^2=\frac{1}{2}(z^2+y^2-\beta^2)+\frac{1}{2}\sqrt{\{(z^2+y^2-\beta^2)^2+4\beta^2z^2\}}$$

$$\tan\theta=\frac{\beta}{\alpha}.$$

Supposing now z very small, since it is to vanish at the end of the calculus, and $y<\beta$, in order that the point P may fall within the limits of s, we shall have by neglecting quantities of the order z^2 compared with those retained

$$\theta=\tfrac{1}{2}\pi-\frac{z}{\sqrt{(\beta^2-y^2)}};$$

and consequently

$$-\frac{d\phi}{dz}=\frac{-d}{dz}k\int\frac{d\sigma}{f}\sqrt{(\beta^2-\eta^2)}=\frac{3M\sqrt{(\beta^2-y^2)}}{\beta^3}-\frac{3M\pi}{2\beta^3}z.$$

This expression, being differentiated again relative to z, gives

$$\frac{d^2}{dz^2}k\int\frac{d\sigma}{f}\sqrt{(\beta^2-\eta^2)}=\frac{3M\pi}{2\beta^3}.$$

But the mass M is given by

$$M = k\int d\sigma\,\sqrt{(\beta^2 - \eta^2)} = 2\pi k\int \eta d\eta\,\sqrt{(\beta^2 - \eta^2)} = \frac{2\pi k\beta^3}{3}.$$

Hence by substitution

$$\frac{d^2}{dz^2}k\int \frac{d\sigma}{f}\sqrt{(\beta^2 - \eta^2)} = \pi^2 k:$$

which expression is rigorously exact when $z = 0$. Comparing this result with the equation (12) of the present article, we see that if $\overline{V} = k\sqrt{(\beta^2 - \eta^2)}$, the constant quantity k may be always determined, so as to satisfy (12). In fact, we have only to make

$$\pi^2 k = \frac{-F\pi}{a} \quad i.e.\ \ k = \frac{-F}{a\pi}.$$

Having thus the value of $\overline{V}$, the general value of V is known, since

$$V = \frac{a^2 - b^2}{4\pi a}\int \frac{d\sigma}{f^3}V = -\frac{a^2 - b^2}{4\pi a z}\frac{d}{dz}\int \frac{d\sigma}{f}\{\overline{V} = k\sqrt{(\beta^2 - \eta^2)}\}$$

$$= \frac{a^2 - b^2}{4\pi a z} \times -\frac{d\phi}{dz} = \frac{a^2 - b^2}{4\pi a z} \times \frac{3Mz}{\beta^3}(\tan\theta - \theta)$$

$$= -\frac{a^2 - b^2}{2\pi a^2}F(\tan\theta - \theta).$$

The value of the potential function, for any point P within the shell, being $F + V$, and that in the interior of the conducting matter of the shell being constant, in virtue of the equilibrium, the value ρ' of the density, at any point on the inner surface of the shell, will be given immediately by the general formula (4) art. 4. Thus

$$\rho' = \frac{-1}{4\pi}\frac{d\overline{V}}{dw} = \frac{1}{4\pi}\frac{d\overline{V}}{db} = \frac{+F}{4\pi^2 a}(\tan\theta - \theta):$$

in which equation, the point P is supposed to be upon the element $d\sigma'$ of the interior surface, to which ρ' belongs. If now R be the distance between C, the centre of the orifice, and $d\sigma'$, we shall have $R^2 = y^2 + z^2$, and by neglecting quantities of the order $\frac{\beta^2}{R^2}$ compared with those retained, we have successively

$$\alpha = R, \quad \theta = \frac{\beta}{R}, \quad \text{and } \tan\theta - \theta = \tfrac{1}{3}\theta^3 = \frac{\beta^3}{3R^3}.$$

Thus the value of ρ' becomes

$$\rho' = \frac{F}{12\pi^2 a} \frac{\beta^3}{R^3}.$$

In the same way, it is easy to show from the equation (11) of this article, that ρ'', the value of the density on an element $d\sigma''$ of the exterior surface of the shell, corresponding to the element $d\sigma'$ of the interior surface, will be

$$\rho'' = \frac{F}{4\pi a} + \rho',$$

which, on account of the smallness of ρ' for every part of the surface, except very near the orifice s, is sensibly constant and equal to $\frac{F}{4\pi a}$, therefore

$$\frac{\rho'}{\rho''} = \frac{\beta^3}{3\pi . R^3}:$$

which equation shows how very small the density within the shell is, even when the orifice is considerable.

(11.) The determination of the electrical phenomena, which result from long metallic wires, insulated and suspended in the atmosphere, depends upon the most simple calculations. As an example, let us conceive two spheres A and B, connected by a long slender conducting wire; then $\rho dxdydz$ representing the quantity of electricity in an element $dxdydz$ of the exterior space, (whether it results from the ground in the vicinity of the wire having become slightly electrical, or from a mist, or even a passing cloud,) and r being the distance of this element from A's centre; also r' its distance from B's, the value of the potential function at A's centre, arising from the whole exterior space, will be

$$\int \frac{\rho dx dy dz}{r},$$

and the value of the same function at B's centre will be

$$\int \frac{\rho dx dy dz}{r'},$$

the integrals extending over all the space exterior to the conducting system under consideration.

If now, Q be the total quantity of electricity on A's surface, and Q' that on B's, their radii being a and a'; it is clear, the value of the potential function at A's centre, arising from the system itself, will be

$$\frac{Q}{a};$$

seeing that, we may neglect the part due to the wire, on account of its fineness, and that due to the other sphere, on account of its distance. In a similar way, the value of the same function at B's centre will be found to be

$$\frac{Q'}{a'}.$$

But (art. 1) the value of the total potential function must be constant throughout the whole interior of the conducting system, and therefore its value at the two centres must be equal; hence

$$\frac{Q}{a} + \int \frac{\rho dx dy dz}{r} = \frac{Q'}{a'} + \int \frac{\rho dx dy dz}{r'}.$$

Although ρ, in the present case, is exceedingly small, the integrals contained in this equation may not only be considerable, but very great, since they are of the second dimension relative to space. The spheres, when at a great distance from each other, may therefore become highly electrical, according to the observations of experimental philosophers, and the charge they will receive in any proposed case may readily be calculated; the value of ρ being supposed given. When one of the spheres,

B for instance, is connected with the ground, Q' will be equal to zero, and consequently Q immediately given. If, on the contrary, the whole system were insulated and retained its natural quantity of electricity, we should have, neglecting that on the wire,

$$0 = Q + Q',$$

and hence Q and Q' would be known.

If it were required to determine the electrical state of the sphere A, when in communication with a wire, of which one extremity is elevated into the atmosphere, and terminates in a fine point p, we should only have to make the radius of B, and consequently, Q', vanish in the expression before given. Hence in this case

$$\frac{Q}{a} = \int \frac{\rho dxdydz}{r'} - \int \frac{\rho dxdydz}{r};$$

r' being the distance between p and the element $dxdydz$. Since the object of the present article is merely to indicate the cause of some phenomena of atmospherical electricity, it is useless to extend it to a greater length, more particularly as the extreme difficulty of determining correctly the electrical state of the atmosphere at any given time, precludes the possibility of putting this part of the theory to the test of accurate experiment.

(12.) Supposing the form of a conducting body to be given, it is in general impossible to assign, rigorously, the law of the density of the electric fluid on its surface in a state of equilibrium, when not acted upon by any exterior bodies, and, at present, there has not even been found any convenient mode of approximation applicable to this problem. It is, however, extremely easy to give such forms to conducting bodies, that this law shall be rigorously assignable by the most simple means. The following method, depending upon art. 4 and 5, seems to give to these forms the greatest degree of generality of which they are susceptible, as, by a tentative process, any form whatever might be approximated indefinitely.

Take any continuous function V', of the rectangular co-ordinates x', y', z', of a point p', which satisfies the partial differential equation $0 = \delta V'$, and vanishes when p' is removed to an infinite distance from the origin of the co-ordinates.

Choose a constant quantity b, such that $V' = b$ may be the equation of a closed surface A, and that V' may have no singular values, so long as p' is exterior to this surface: then if we form a conducting body, whose outer surface is A, the density of the electric fluid in equilibrium upon it, will be represented by

$$\rho = \frac{-h}{4\pi} \frac{d\overline{V}'}{dw'},$$

and the potential function due to this fluid, for any point p', exterior to the body, will be

$$hV';$$

h being a constant quantity dependent upon the total quantity of electricity Q, communicated to the body. This is evident from what has been proved in the articles cited.

Let R represent the distance between p', and any point within A; then the potential function arising from the electricity upon it will be expressed by $\frac{Q}{R}$, when R is infinite. Hence the condition

$$\frac{Q}{R} = hV' \text{ (R being infinite)},$$

which will serve to determine h, when Q is given.

In the application of this general method, we may assume for V', either some analytical expression containing the co-ordinates of p', which is known to satisfy the equation $0 = \delta V'$, and to vanish when p' is removed to an infinite distance from the origin of the co-ordinates; as, for instance, some of those given by LAPLACE (Méc. Céleste, Liv. 3, Ch. 2), or, the value a potential function, which would arise from a quantity of electricity anyhow distributed within a finite space, at a point p' without that space; since this last will always satisfy the conditions to which V' is subject.

It may be proper to give an example of each of these cases. In the first place, let us take the general expression given by LAPLACE,

$$V = \frac{U^{(0)}}{r} + \frac{U^{(1)}}{r^2} + \frac{U^{(2)}}{r^3} + \&c.,$$

then, by confining ourselves to the two first terms, the assumed value of V' will be

$$V' = \frac{U^{(0)}}{r} + \frac{U^{(1)}}{r^2};$$

r being the distance of p' from the origin of the co-ordinates, and $U^{(0)}$, $U^{(1)}$, &c. functions of the two other polar co-ordinates θ and ϖ. This expression by changing the direction of the axes, may always be reduced to the form

$$V' = \frac{2a}{r} + \frac{k^2 \cos\theta}{r^2};$$

a and k being two constant quantities, which we will suppose positive. Then if b be a very small positive quantity, the form of the surface given by the equation $V' = b$, will differ but little from a sphere, whose radius is $\frac{2a}{b}$: by gradually increasing b, the difference becomes greater, until $b = \frac{a^2}{k^2}$; and afterwards, the form assigned by $V = b$, becomes improper for our purpose. Making therefore $b = \frac{a^2}{k^2}$, in order to have a surface differing as much from a sphere, as the assumed value of V' admits, the equation of the surface A becomes

$$V' = \frac{2a}{r} + \frac{k^2 \cos\theta}{r^2} = \frac{a^2}{k^2}.$$

From which we obtain

$$r = \frac{k^2}{a}\left(1 + \sqrt{2}\cos\frac{\theta}{2}\right).$$

If now ϕ represents the angle formed by dr and dw', we have

$$\frac{-dr}{rd\theta} = \frac{\sqrt{2}\sin\frac{\theta}{2}}{2 + 2\sqrt{2}\cos\frac{\theta}{2}} = \tan\phi,$$

and as the electricity is in equilibrium upon A, the force with which a particle p, infinitely near to it, would be repelled, must be directed along dw': but the value of this force is $-\frac{d\overline{V}'}{dw'}$, and consequently its effect in the direction of the radius r, and tending to increase it, will be $-\frac{d\overline{V}'}{dw'}\cos\phi$. This last quantity is equally represented by $-\frac{d\overline{V}'}{dr}$, and therefore

$$-\frac{d\overline{V}'}{dr} = -\frac{d\overline{V}'}{dw'}\cos\phi;$$

the horizontal lines over quantities, indicating, as before, that they belong to the surface itself. The value of $-\left(\frac{d\overline{V}'}{dw'}\right)$, deduced from this equation, is

$$-\frac{d\overline{V}'}{dw'} = \frac{-1}{\cos\phi}\frac{d\overline{V}'}{dr} = \frac{1}{\cos\phi}\left\{\frac{2a}{r^2} + \frac{2k^2\cos\theta}{r^3}\right\} = \frac{2a\sqrt{2}\cos\frac{\theta}{2}}{r^2\cos\phi},$$

this substituted in the general value of ρ, before given, there arises

$$\rho = \frac{-h}{4\pi}\frac{d\overline{V}'}{dw'} = \frac{ha\sqrt{2}\cos\frac{\theta}{2}}{2\pi r^2\cos\phi}.$$

Supposing Q is the quantity of electricity communicated to the surface, the condition

$$\frac{Q}{R} = hV \text{ (where } R \text{ is infinite)},$$

before given, becomes, since r may here be substituted for R, seeing that it is measured from a point within the surface,

$$\frac{Q}{r} = \frac{2ah}{r} \text{ i. e. } h = \frac{Q}{2a}.$$

We have thus the rigorous value of ρ for the surface A whose equation is $r = \frac{k^2}{a}\left(1 + \sqrt{2}\cos\frac{\theta}{2}\right)$ when the quantity Q of electricity upon it is known, and by substituting for r and h their values just given, there results

$$\rho = \frac{Qa^2 \sqrt{2} \cos \frac{\theta}{2}}{4\pi k^4 \cos \phi \left(1 + \sqrt{2} \cos \frac{\theta}{2}\right)^2}.$$

Moreover the value of the potential function for the point p' whose polar co-ordinates are r, θ, and ϖ, is

$$hV' = \frac{Q}{r} + \frac{Qk^2 \cos \theta}{2ar^2}.$$

From which we may immediately deduce the forces acting on any point p' exterior to A.

In tracing the surface A, θ is supposed to extend from $\theta = 0$ to $\theta = \pi$, and ϖ, from $\varpi = 0$ to $\varpi = 2\pi$: it is therefore evident, by constructing the curve whose equation is

$$r = \frac{k^2}{a}\left(1 + \sqrt{2} \cos \frac{\theta}{2}\right),$$

that the parts about P, where $\theta = \pi$, approximate continually in form towards a cone whose apex is P, and as the density of the electricity at P is null, in the example before us, we may make this general inference: when any body whatever has a part of its surface in the form of a cone, directed inwards; the density of the electricity in equilibrium upon it, will be null at its apex, precisely the reverse of what would take place, if it were directed outwards, for then, the density at the apex would become infinite*.

* Since this was written, I have obtained formulæ serving to express, generally, the law of the distribution of the electric fluid near the apex O of a cone, which forms part of a conducting surface of revolution having the same axis. From these formulæ it results that, when the apex of the cone is directed inwards, the density of the electric fluid at any point p, near to it, is proportional to r^{n-1}; r being the distance Op, and the exponent n very nearly such as would satisfy the simple equation $(4n+2)\,\beta = 3\pi$: where 2β is the angle at the summit of the cone. If 2β exceeds π, this summit is directed outwards, and when the excess is not very considerable, n will be given as above: but 2β still increasing, until it becomes $2\pi - 2\gamma$; the angle 2γ at the summit of the cone, which is now directed outwards, being very small, n will be given by $2n \log \frac{2}{\gamma} = 1$, and in case the conducting body is a sphere whose radius is b, on which P represents the mean density

As a second example, we will assume for V', the value of the potential function arising from the action of a line uniformly covered with electricity. Let $2a$ be the length of the line, y the perpendicular falling from any point p' upon it, x the distance of the foot of this perpendicular from the middle of the line, and x' that of the element dx' from the same point: then taking the element dx', as the measure of the quantity of electricity it contains, the assumed value of V' will be

$$V' = \int \frac{dx'}{\sqrt{\{y^2 + (x - x')^2\}}} = \log \frac{a - x + \sqrt{\{y^2 + (a - x)^2\}}}{-a - x + \sqrt{\{y^2 + (a + x)^2\}}};$$

the integral being taken from $x' = -a$ to $x' = +a$. Making this equal to a constant quantity $\log b$, we shall have, for the equation of the surface A,

$$\frac{a - x + \sqrt{\{y^2 + (a - x)^2\}}}{-a - x + \sqrt{\{y^2 + (a + x)^2\}}} = b,$$

which by reduction becomes

$$0 = y^2 (1 - b^2)^2 + x^2 . 4b (1 - b)^2 - a^2 . 4b (1 + b)^2.$$

We thus see that this surface is a spheroid produced by the revolution of an ellipsis about its greatest diameter; the semi-transverse axis being

$$a \frac{1 + b}{1 - b} = \beta,$$

and semi-conjugate

$$a \frac{2 \sqrt{b}}{1 - b} = \gamma.$$

By differentiating the general value of V', just given, and substituting for y its value at the surface A, we obtain

$$\frac{d\overline{V'}}{dx} = \frac{-2x \dfrac{1 - b}{1 + b}}{\left(\dfrac{1 + b}{1 - b}\right)^2 a^2 - \left(\dfrac{1 - b}{1 + b}\right)^2 x^2} = \frac{-2a\beta x}{\beta^4 - a^2 x^2}.$$

of the electric fluid, ρ, the value of the density near the apex O, will be determined by the formula

$$\rho = \frac{2 Pbn}{(a + b)\gamma} \left(\frac{r}{a}\right)^{n-1};$$

a being the length of the cone.

Now writing ϕ for the angle formed by dx and dw', we have

$$\frac{1}{\cos\phi} = \frac{ds}{-dy} = \frac{1-b}{2x\sqrt{b}}\sqrt{\left\{\left(\frac{1+b}{1-b}\right)^4 a^2 - x^2\right\}} = \frac{\sqrt{(\beta^4 - \alpha^2 x^2)}}{\gamma x};$$

ds being an element of the generating ellipsis. Hence, as in the preceding example, we shall have

$$\frac{d\overline{V}'}{dw'} = \frac{1}{\cos\phi}\cdot\frac{d\overline{V}'}{dx} = \frac{-2\alpha\beta}{\gamma\sqrt{(\beta^4 - \alpha^2 x^2)}}.$$

On the surface A therefore, in this example, the general value of ρ is

$$\rho = \frac{-h}{4\pi}\,\frac{d\overline{V}'}{dw'} = \frac{\alpha h\beta}{2\pi\gamma\sqrt{(\beta^4 - \alpha^2 x^2)}},$$

and the potential function for any point p', exterior to A, is

$$hV' = h\log\frac{a - x + \sqrt{\{y^2 + (a-x)^2\}}}{-a - x + \sqrt{\{y^2 + (a+x)^2\}}}.$$

Making now x and y both infinite, in order that p' may be at an infinite distance, there results

$$hV' = \frac{2ah}{\sqrt{(x^2 + y^2)}},$$

and thus the condition determining h, in Q, the quantity of electricity upon the surface, is, since R may be supposed equal to $\sqrt{(x^2 + y^2)}$,

$$\frac{Q}{R} = hV' = \frac{2ah}{\sqrt{(x^2 + y^2)}} \quad \text{i. e.} \quad h = \frac{Q}{2a}.$$

These results of our analysis agree with what has been long known concerning the law of the distribution of electric fluid on the surface of a spheroid, when in a state of equilibrium.

(13.) In what has preceded, we have confined ourselves to the consideration of perfect conductors. We will now give an example of the application of our general method, to a body that is supposed to conduct electricity imperfectly, and which will, moreover, be interesting, as it serves to illustrate the magnetic phenomena, produced by the rotation of bodies under the influence of the earth's magnetism.

If any solid body whatever of revolution, turn about its axis, it is required to determine what will take place, when the matter of this solid is not perfectly conducting, supposing it under the influence of a constant electrical force, acting parallel to any given right line fixed in space, the body being originally in a natural state.

Let β designate the coercive force of the body, which we will suppose analogous to friction in its operation, so that as long as the total force acting upon any particle within the body is less than β, its electrical state shall remain unchanged, but when it begins to exceed β, a change shall ensue.

In the first place, suppose the constant electrical force, which we will designate by b, to act in a direction parallel to a line passing through the centre of the body, and perpendicular to its axis of revolution; and let us consider this line as the axis of x, that of revolution being the axis of z, and y the other rectangular co-ordinate of a point p, within the body and fixed in space. Thus, if V be the value of the total potential function for the same point p, at any instant of time, arising from the electricity of the body and the exterior force,

$$bx + V$$

will be the part due to the body itself at the same instant: since $-bx$ is that due to the constant force b, acting in the direction of x, and tending to increase it. If now we make

$$z = r\cos\theta, \quad x = r\sin\theta\cos\varpi, \quad y = r\sin\theta\sin\varpi;$$

the angle ϖ being supposed to increase in the direction of the body's revolution, the part due to the body itself becomes

$$br\sin\theta\cos\varpi + V.$$

Were we to suppose the value of the potential function V given at any instant, we might find its value at the next instant, by conceiving, that whilst the body moves forward through the infinitely small angle $d\omega$, the electricity within it shall remain fixed, and then be permitted to move, until it is in equilibrium with the coercive force.

Now the value of the potential function at p, arising from the body itself, after having moved through the angle $d\omega$ (the electricity being fixed), will evidently be obtained by changing ϖ into $\varpi - d\omega$ in the expression just given, and is therefore

$$br \sin\theta \cos\varpi + V + br \sin\theta \sin\varpi\, dw - \frac{dV}{d\varpi} d\omega,$$

adding now the part $-bx = -br \sin\theta \cos\varpi$, due to the exterior bodies, and restoring x, y, &c. we have, since

$$\frac{dV}{d\varpi} = -y\frac{dV}{dx} + x\frac{dV}{dy},$$

$$V + d\omega\left(by + y\frac{dV}{dx} - x\frac{dV}{dy}\right)$$

for the value of the total potential function at the end of the next instant, the electricity being still supposed fixed. We have now only to determine what this will become, by allowing the electricity to move forward until the total forces acting on points within the body, which may now exceed the coercive force by an infinitely small quantity, are again reduced to an equilibrium with it. If this were done, we should, when the initial state of the body was given, be able to determine, successively, its state for every one of the following instants. But since it is evident from the nature of the problem, that the body, by revolving, will quickly arrive at a permanent state, in which the value of V will afterwards remain unchanged and be independent of its initial value, we will here confine ourselves to the determination of this permanent state. It is easy to see, by considering the forces arising from the new total potential function, whose value has just been given, that in this case the electricity will be in motion over the whole interior of the body, and consequently

$$\beta^2 = \left(\frac{dV}{dx}\right)^2 + \left(\frac{dV}{dy}\right)^2 + \left(\frac{dV}{dz}\right)^2,$$

which equation expresses that the total force to move any particle p, within the body, is just equal to β, the coercive force. Now if we can assume any value for V, satisfying the above, and such, that it shall reproduce itself after the electricity belonging to the new total potential function (Art. 7), is allowed to find its equilibrium with the coercive force, it is evident this will be the required value, since the rest of the electricity is exactly in equilibrium with the exterior force b, and may therefore be here neglected. To be able to do this the more easily, conceive two new axes X', Y', in advance of the old ones X, Y, and making the angle γ with them; then the value of the new potential function, before given, becomes

$$V + d\omega \,.\left(by' \cos\gamma + bx' \sin\gamma + y' \frac{dV}{dx'} - x' \frac{dV}{dy'}\right),$$

which, by assuming $V = \beta y'$, and determining γ by the equation

$$0 = b \sin\gamma - \beta,$$

reduces itself to

$$y'\,(\beta + b \cos\gamma\, d\omega).$$

Considering now the symmetrical distribution of the electricity belonging to this potential function, with regard to the plane whose equation is $0 = y'$, it will be evident that, after the electricity has found its equilibrium, the value of V at this plane must be equal to *zero:* a condition which, combined with the partial differential equation before given, will serve to determine, completely, the value of V at the next instant, and this value of V will be

$$V = \beta y'.$$

We thus see that the assumed value of V reproduces itself at the end of the following instant, and is therefore the one required belonging to the permanent state.

If the body had been a perfect conductor, the value of V would evidently have been equal to *zero*, seeing that it was supposed originally in a natural state: that just found is therefore due to the rotation combined with the coercive force, and we thus see that their effect is to polarise the body in the direction of y' positive, making the angle $\frac{1}{2}\pi + \gamma$ with the direction of the constant force b; and the degree of polarity will be the same as would be produced by a force equal to β, acting in this direction on a perfectly conducting body of the same dimensions.

We have hitherto supposed the constant force to act in a direction parallel to the equatorial plane of the body, but whatever may be its direction, we may conceive it decomposed into two; one equal to b as before, and parallel to this plane, the other perpendicular to it, which last will evidently produce no effect on the value of V, as this is due to the coercive force, and would still be equal to *zero* under the influence of the new force, if the body conducted electricity perfectly.

Knowing the value of the potential function at the surface of the body, due to the rotation, its value for all the exterior space may be considered as determined (Art. 5), and if the body be a solid sphere, may easily be expressed analytically; for it is evident (Art. 7), from the value of V just given, that even in the present case all the electricity will be confined to the surface of the solid; and it has been shown (Art. 10), that when the value of the potential function for the point p within a spherical surface, whose radius is a, is represented by

$$\phi(r),$$

the value of the same function for a point p', situate without this sphere, on the prolongation of r, and at the distance r' from its centre, will be

$$\frac{a}{r'}\phi\left(\frac{a^2}{r'}\right).$$

But we have seen that the value of V due to the rotation, for the point p, is

$$V = \beta y' = \beta r \cos\theta';$$

θ' being the angle formed by the ray r and the axis of y'; the corresponding value for the point p' will therefore be

$$V' = \frac{\beta a^3 \cos\theta'}{r'^2}.$$

And hence, by differentiation, we immediately obtain the value of the forces acting on any particle situate without the sphere, which arise from its rotation; but, if we would determine the total forces arising from the sphere, we must, to the value of the potential function just found, add that part which would be produced by the action of the constant force upon this sphere, when it is supposed to conduct electricity perfectly, which will be given in precisely the same way as the former. In fact, f designating the constant force, and θ'' the angle formed by r and a line parallel to the direction of f, the potential function arising from it, for the point p, will be

$$-rf\cos\theta'',$$

and consequently the part arising from the electricity, induced by its action, must be

$$+fr\cos\theta'',$$

seeing that their sum ought to be equal to zero. The corresponding value for the point p', exterior to the sphere, is therefore

$$\frac{fa^3\cos\theta''}{r'^2},$$

this added to the value of V', before found, will give the value of the total potential function for the point p', arising from the sphere itself.

It will be seen when we come to treat of the theory of magnetism, that the results of his theory, in general, agree very nearly with those which would arise from supposing the magnetic fluid at liberty to move from one part of a magnetized body to another; at least, for bodies whose magnetic powers admit of considerable developement, as iron and nickel for example; the errors of the latter supposition being of the order $1-g$ only; g being a constant quantity dependant on the nature of the body, which in those just mentioned, differs very little from unity. It is therefore evident that when a solid of revolution, formed of iron, is caused to revolve slowly round its axis, and placed under the influence of the earth's magnetic force f, the act of revolving, combined with the coercive force β of the body, will produce a new polarity, whose direction and quantity will be very nearly the same as those before determined. Now f having been supposed resolved into two forces, one equal to b in the plane of the body's equator, and another perpendicular to this plane; if β be very small compared with b, the angle γ will be very small, and the direction of the new polarity will be very nearly at right angles to the direction of b, a result which has been confirmed by many experiments: but by our analysis we moreover see that when b is sufficiently reduced, the angle γ may be rendered sensible, and the direction of the new polarity will then form with that of b the angle $\frac{1}{2}\pi + \gamma$; γ being determined by the equation

$$\sin \gamma = \frac{\beta}{b}.$$

This would be very easily put to the test of experiment by employing a solid sphere of iron.

The values of the forces induced by the rotation of the body, which would be observed in the space exterior to it, may be obtained by differentiating that of V' before given, and will be found to agree with the observations of Mr BARLOW (*Phil. Tran.* 1825), on the supposition of β being very small.

As the experimental investigation of the magnetic phenomena developed by the rotation of bodies, has lately engaged the attention of several distinguished philosophers, it may not be amiss to consider the subject in a more general way, as we shall thus not only confirm the preceding analysis, but be able to show with what rapidity the body approaches that permanent state, which it has been the object of the preceding part of this article to determine.

Let us now, therefore, consider a body A fixed in space, under the influence of electric forces which vary according to any given law; then we might propose to determine the electrical state of the body, after a certain interval of time, from the knowledge of its initial state; supposing a constant coercive force to exist within it. To resolve this in its most general form, it would be necessary to distinguish between those parts of the body where the fluid was at rest, from the forces acting there being less than the coercive force, and those where it would be in motion; moreover these parts would vary at every instant, and the problem therefore become very intricate: were we however to suppose the initial state so chosen, that the total force to move any particle p within A, arising from its electric state and exterior actions, was then just equal to the coercive force β; also, that the alteration in the exterior forces should always be such, that if the electric fluid remained at rest during the next instant, this total force should no where be less than β; the problem would become more easy, and still possess a great degree of generality. For in this case, when the fluid is moveable, the whole force tending to move any particle p within A, will, at every instant, be exactly equal to the coercive force. If therefore x, y, z represent the co-ordinates of p, and V the value of the total potential function at any instant of time t, arising from the electric state of the body and exterior forces, we shall have the equation

$$\beta^2 = \left(\frac{dV}{dx}\right)^2 + \left(\frac{dV}{dy}\right)^2 + \left(\frac{dV}{dz}\right)^2 \text{ .... } (a),$$

whose general integral may be thus constructed:

Take the value of V arbitrarily over any surface whatever S, plane or curved, and suppose three rectangular co-ordinates w, w', w'', whose origin is at a point P on S: the axis of w being a normal to S, and those of w', w'', in its plane tangent. Then the values of $\frac{dV}{dw'}$ and $\frac{dV}{dw''}$ are known at the point P, and the value of $\frac{dV}{dw}$ will be determined by the equation

$$\left(\frac{dV}{dw}\right)^2 + \left(\frac{dV}{dw'}\right)^2 + \left(\frac{dV}{dw''}\right)^2 = \beta^2,$$

which is merely a transformation of the above.

Take now another point $P_{,}$, whose co-ordinates referred to these axes are $\frac{dV}{dw}$, $\frac{dV}{dw'}$ and $\frac{dV}{dw''}$, and draw a right line L

through the points P, P_1, then will the value of V at any point p, on L, be expressed by

$$V_0 + \beta\lambda;$$

λ being the distance Pp, measured along the line L, considered as increasing in the direction PP_1, and V_0, the given value of V at P. For it is very easy to see that the value of V furnished by this construction, satisfies the partial differential equation (a), and is its general integral; moreover the system of lines L, L', L'', &c. belonging to the points P, P', P'', &c. on S, are evidently those along which the electric fluid tends to move, and will move during the following instant.

Let now $V + DV$ represent what V becomes at the end of the time $t + dt$; substituting this for V in (a) we obtain

$$0 = \frac{dV}{dx} \cdot \frac{dDV}{dx} + \frac{dV}{dy} \cdot \frac{dDV}{dy} + \frac{dV}{dz} \cdot \frac{dDV}{dz} \quad \text{.........}(b).$$

Then, if we designate by $D'V$, the augmentation of the potential function, arising from the change which takes place in the exterior forces during the element of time dt,

$$DV - D'V$$

will be the increment of the potential function, due to the corresponding alterations $D\rho$ and $D\rho'$ in the densities of the electric fluid at the surface of A and within it, which may be determined from $DV - D'V$ by Art. 7. But, by the known theory of partial differential equations, the most general value of DV satisfying (b), will be constant along every one of the lines L, L', L'', &c., and may vary arbitrarily in passing from one of them to another: as it is also along these lines the electric fluid moves during the instant dt, it is clear the total quantity of fluid in any infinitely thin needle, formed by them, and terminating in the opposite surfaces of A, will undergo no alteration during this instant. Hence therefore

$$0 = \int D\rho' dv + D\rho d\sigma + D\rho_1 d\sigma_1 \text{..............}(c);$$

dv being an element of the volume of the needle, and $d\sigma$, $d\sigma_1$, the two elements of A's surface by which it is terminated. This condition, combined with the equation (b), will completely determine the value of DV, and we shall thus have the value of the potential function $V+DV$, at the instant of time $t+dt$, when its value V, at the time t, is known.

As an application of this general solution; suppose the body A is a solid of revolution, whose axis is that of the co-ordinate z, and let the two other axes X, Y, situate in its equator, be fixed in space. If now the exterior electric forces are such that they may be reduced to two, one equal to c, acting parallel to z, the other equal to b, directed parallel to a line in the plane (xy), making the variable angle ϕ with X; the value of the potential function arising from the exterior forces, will be

$$-zc - xb\cos\phi - yb\sin\phi\,;$$

where b and c are constant quantities, and ϕ varies with the time so as to be constantly increasing. When the time is equal to t, suppose the value of V to be

$$V = \beta\,(x\cos\varpi + y\sin\varpi):$$

then the system of lines L, L', L'' will make the angle ϖ with the plane (xz), and be perpendicular to another plane whose equation is

$$0 = x\cos\varpi + y\sin\varpi.$$

If during the instant of time dt, ϕ becomes $\phi + D\phi$, the augmentation of the potential function due to the elementary change in the exterior forces, will be

$$D'V = (x\sin\phi - y\cos\phi)\,bD\phi\,;$$

moreover the equation (b) becomes

$$0 = \cos\varpi\,.\,\frac{dDV}{dx} + \sin\varpi\,.\,\frac{dDV}{dy} \;\ldots\ldots\ldots\ldots..(b')$$

and therefore the general value of DV is

$$DV = DF\,(y\cos\varpi - x\sin\varpi\,;\; z)\,;$$

DF being the characteristic of an infinitely small arbitrary function. But, it has been before remarked that the value of DV

will be completely determined, by satisfying the equation (b) and the condition (c). Let us then assume

$$DF(y\cos\varpi - x\sin\varpi;\ z) = hD\phi\,(y\cos\varpi - x\sin\varpi);$$

h being a quantity independent of x, y, z, and see if it be possible to determine h so as to satisfy the condition (c). Now on this supposition

$$\begin{aligned} DV - D'V &= hD\phi\,(y\cos\varpi - x\sin\varpi) - (x\sin\phi - y\cos\phi)\,bd\phi \\ &= D\phi\,\{y\,(h\cos\varpi + b\cos\phi) - x\,(h\sin\varpi + b\cos\phi)\}. \end{aligned}$$

The value of $D\rho'$ corresponding to this potential function is (Art. 7)

$$D\rho' = 0,$$

and on account of the parallelism of the lines L, L', &c. to each other, and to A's equator $d\sigma = d\sigma_1$. The condition (c) thus becomes

$$0 = D\rho + D\rho_1 \quad \ldots\ldots\ldots\ldots\ldots\ldots\ldots \quad (c'):$$

$D\rho$ and $D\rho_1$ being the elementary densities on A's surface at opposite ends of any of the lines L, L', &c. corresponding to the potential function $DV - D'V$. But it is easy to see from the form of this function, that these elementary densities at opposite ends of any line perpendicular to a plane whose equation is

$$0 = y\,(h\cos\varpi + b\cos\phi) - x\,(h\sin\varpi + b\sin\phi),$$

are equal and of contrary signs, and therefore the condition (c) will be satisfied by making this plane coincide with that perpendicular to L, L', &c., whose equation, as before remarked, is

$$0 = x\cos\varpi + y\sin\varpi;$$

that is the condition (c) will be satisfied, if h be determined by the equation

$$\frac{h\cos\varpi + b\cos\phi}{\sin\varpi} = -\frac{h\sin\varpi + b\sin\phi}{\cos\varpi},$$

which by reduction becomes

$$0 = h + b\cos(\phi - \varpi),$$

and consequently

$$V + DV = \beta\,(x\cos\varpi + y\sin\varpi) + hD\phi\,(y\cos\varpi - x\sin\varpi)$$

$$= \beta x\left\{\cos\varpi + \frac{b}{\beta}\sin\varpi\cos(\phi-\varpi)\,D\phi\right\}$$
$$+ \beta y\left\{\sin\varpi - \frac{b}{\beta}\cos\varpi\cos(\phi-\varpi)\,D\phi\right\},$$

$$= \beta x\cos\left\{\varpi - \frac{b}{\beta}\cos(\phi-\varpi)\,D\phi\right\}$$
$$+ \beta y\sin\left\{\varpi - \frac{b}{\beta}\cos(\phi-\varpi)\,D\phi\right\}.$$

When therefore ϕ is augmented by the infinitely small angle $D\phi$, ϖ receives the corresponding increment $-\frac{b}{\beta}\cos(\phi-\varpi)\,D\phi$, and the form of V remains unaltered; the preceding reasoning is consequently applicable to every instant, and the general relation between ϕ and ϖ expressed by

$$0 = D\varpi + \frac{b}{\beta}\cos(\phi-\varpi)\,D\phi:$$

a common differential equation, which by integration gives

$$H\,.\,e^{\phi\,\mathrm{cotan}\,\gamma} = \frac{\sin\left(\frac{3}{4}\pi - \frac{1}{2}\gamma + \frac{1}{2}\varpi - \frac{1}{2}\phi\right)}{\sin\left(\frac{1}{4}\pi + \frac{1}{2}\gamma + \frac{1}{2}\varpi - \frac{1}{2}\phi\right)};$$

H being an arbitrary constant, and γ, as in the former part of this article, the smallest root of

$$0 = b\sin\gamma - \beta.$$

Let ϖ_0 and ϕ_0 be the initial values of ϖ and ϕ; then the total potential function at the next instant, if the electric fluid remained fixed, would be

$$V_{\prime} = \beta\,(x\cos\varpi_0 + y\sin\varpi_0) + (x\sin\phi_0 - \gamma\cos\phi_0)\,bd\phi,$$

and the whole force to move a particle p, whose co-ordinates are x, y, z,

$$\sqrt{\left\{\left(\frac{dV_{\prime}}{dx}\right)^2 + \left(\frac{dV_{\prime}}{dy}\right)^2 + \left(\frac{dV_{\prime}}{dz}\right)^2\right\}} = \beta + d\phi\,.\,b\sin(\phi_0 - \varpi_0),$$

which, in order that our solution may be applicable, must not be less than β, and consequently the angle $\phi_0 - \varpi_0$ must be between 0 and π: when this is the case, ϖ is immediately determined from ϕ by what has preceded. In fact, by finding the value of H from the initial values ϖ_0 and ϕ_0, and making $\zeta = \frac{1}{4}\pi + \frac{1}{2}\gamma + \frac{1}{2}\varpi - \frac{1}{2}\phi$, we obtain

$$\tan \zeta = \frac{\tan \zeta_0}{e^{(\phi - \phi_0)\cot\gamma} + \tan\gamma \tan\zeta_0 \{e^{(\phi-\phi_0)\cot\gamma} - 1\}};$$

ζ_0 being the initial value of ζ.

We have, in the latter part of this article, considered the body A at rest, and the line X', parallel to the direction of b, as revolving round it: but if, as in the former, we now suppose this line immovable and the body to turn the contrary way, so that the relative motion of X' to X may remain unaltered, the electric state of the body referred to the axes X, Y, Z, evidently depending on this relative motion only, will consequently remain the same as before. In order to determine it on the supposition just made, let X' be the axis of x', one of the co-ordinates of p, referred to the rectangular axes X', Y', Z, also y', z, the other two; the direction X' Y', being that in which A revolves. Then, if ϖ' be the angle the system of lines L, L', &c. forms with the plane (x', z), we shall have

$$\varpi + \varpi' = \phi;$$

ϕ, as before stated, being the angle included by the axes X, X'. Moreover the general values of V and ζ will be

$$V = \beta\,(x' \cos\varpi' + y' \sin\varpi') \text{ and } \zeta = \tfrac{1}{4}\pi + \tfrac{1}{2}\gamma - \tfrac{1}{2}\varpi',$$

and the initial condition, in order that our solution may be applicable, will evidently become $\phi_0 - \varpi_0 = \varpi'_0 =$ a quantity betwixt 0 and π.

As an example, let $\tan\gamma = \frac{1}{10}$, since we know by experiment that γ is generally very small; then taking the most unfavourable case, viz. where $\varpi'_0 = 0$, and supposing the body to make one revolution only, the value of ζ, determined from its initial one, $\zeta_0 = \frac{1}{4}\pi + \frac{1}{2}\gamma - \frac{1}{2}\varpi'_0$, will be found extremely small and only equal to a unit in the 27th decimal place. We thus see with what rapidity ζ decreases, and consequently, the body approaches to a permanent state, defined by the equation

$$0 = \zeta = \tfrac{1}{4}\pi + \tfrac{1}{2}\gamma - \tfrac{1}{2}\varpi'.$$

Hence, the polarity induced by the rotation is ultimately directed along a line, making an angle equal to $\frac{1}{2}\pi + \gamma$ with the axis X', which agrees with what was shown in the former part of this article.

The value of V at the body's surface being thus known at any instant whatever, that of the potential function at a point p' exterior to the body, together with the forces acting there, will be immediately determined as before.

APPLICATION OF THE PRELIMINARY RESULTS TO THE THEORY OF MAGNETISM.

(14.) The electric fluid appears to pass freely from one part of a continuous conductor to another, but this is by no means the case with the magnetic fluid, even with respect to those bodies which, from their instantly returning to a natural state the moment the forces inducing a magnetic one are removed, must be considered, in a certain sense, as perfect conductors of magnetism. Coulomb, I believe, was the first who proposed to consider these as formed of an infinite number of particles, each of which conducts the magnetic fluid in its interior with perfect freedom, but which are so constituted that it is impossible there shall be any communication of it from one particle to the next. This hypothesis is now generally adopted by philosophers, and its consequences, as far as they have hitherto been developed, are found to agree with observation; we will therefore admit it in what follows, and endeavour thence to deduce, mathematically, the laws of the distribution of magnetism in bodies of any shape whatever.

Firstly, let us endeavour to determine the value of the potential function, arising from the magnetic state induced in a very small body A, by the action of constant forces directed parallel to a given right line; the body being composed of an infinite number of particles, all perfect conductors of magnetism and originally in a natural state. In order to deduce this more immediately from Art. 6, we will conceive these forces to arise from an infinite quantity Q of magnetic fluid, concentrated in a point p on this line, at an infinite distance from A. Then the origin O of the rectangular co-ordinates being anywhere within A, if x, y, z, be those of the point p, and x', y', z', those of any other exterior point p', to which the potential function V arising from A belongs, we shall have (vide *Méc. Cél.* Liv. 3)

$$V = \frac{U^{(0)}}{r'} + \frac{U^{(1)}}{r'^2} + \frac{U^{(2)}}{r'^3} + \&c.;$$

$r' = \sqrt{(x'^2 + y'^2 + z'^2)}$ being the distance Op'.

Moreover, since the total quantity of magnetic fluid in A is equal to zero, $U^{(0)} = 0$. Supposing now r' very great compared with the dimensions of the body, all the terms after $\dfrac{U^{(0)}}{r'^2}$ in the expression just given will be exceedingly small compared with this, by neglecting them, therefore, and substituting for $U^{(1)}$ its most general value, we obtain

$$V = \frac{U^{(1)}}{r'^2} = \frac{Ax' + By' + Cz'}{r'^3};$$

A, B, C, being quantities independent of x', y', z', but which may contain x, y, z.

Now (Art. 6) the value of V will remain unaltered, when we change x, y, z, into x', y', z', and reciprocally. Therefore

$$V = \frac{Ax' + By' + Cz'}{r'^3} = \frac{A'x + B'y + C'z}{r^3};$$

A', B', C', being the same functions of x', y', z', as A, B, C, are of x, y, z. Hence it is easy to see that V must be of the form

$$V = \frac{a''xx' + b''yy' + c''zz' + e''(xy' + yx') + f''(xz' + zx') + g''(yz' + zy')}{r^3 r'^3};$$

a'', b'', c'', e'', f'', g'', being constant quantities.

If X, Y, Z, represent the forces arising from the magnetism concentrated in p, in the directions of x, y, z, positive, we shall have

$$X = \frac{-Qx}{r^3}; \quad Y = \frac{-Qy}{r^3}; \quad Z = \frac{-Qz}{r^3};$$

and therefore V is of the form

$$V=\frac{a'Xx'+b'Yy'+c'Zz'+e'(Xy'+Yx')+f'(Xz'+Zx')+g'(Yz'+Zy')}{r'^3};$$

a', b', &c. being other constant quantities. But it will always be possible to determine the situation of three rectangular axes, so that e, f, and g may each be equal to zero, and consequently V be reduced to the following simple form

$$V=\frac{aXx'+bYy'+cZz'}{r'^3} \quad\ldots\ldots\ldots\ldots\ldots\ldots (a);$$

a, b, and c being three constant quantities.

When A is a sphere, and its magnetic particles are either spherical, or, like the integrant particles of non-crystallized bodies, arranged in a confused manner; it is evident the constant quantities a', b', c', &c. in the general value of V, must be the same for every system of rectangular co-ordinates, and consequently we must have $a'=b'=c'$, $e'=o$, $f'=o$, and $g'=o$, therefore in this case

$$V=\frac{a'(Xx'+Yy'+Zz')}{r'^3} \quad\ldots\ldots\ldots\ldots\ldots\ldots (b);$$

a' being a constant quantity dependant on the magnitude and nature of A.

The formula (a) will give the value of the forces acting on any point p', arising from a mass A of soft iron or other similar matter, whose magnetic state is induced by the influence of the earth's action; supposing the distance Ap' to be great compared with the dimensions of A, and if it be a solid of revolution, one of the rectangular axes, say X, must coincide with the axis ot revolution, and the value of V reduce itself to

$$V=\frac{a'Xx'+b'(Yy'+Zz')}{r'^3};$$

a' and b' being two constant quantities dependant on the form and nature of the body. Moreover the forces acting in the directions of x', y', z', positive, are expressed by

$$-\left(\frac{dV}{dx'}\right), \quad -\left(\frac{dV}{dy'}\right), \quad -\left(\frac{dV}{dz'}\right).$$

We have thus the means of comparing theory with experiment, but these are details into which our limits will not permit us to enter.

The formula (*b*), which is strictly correct for an infinitely small sphere, on the supposition of its magnetic particles being arranged in a confused manner, will, in fact, form the basis of our theory, and although the preceding analysis seems sufficiently general and rigorous, it may not be amiss to give a simpler proof of this particular case. Let, therefore, the origin O of the rectangular co-ordinates be placed at the centre of the infinitely small sphere A, and OB be the direction of the parallel forces acting upon it; then, since the total quantity of magnetic fluid in A is equal to zero, the value of the potential function V, at the point p', arising from A, must evidently be of the form

$$V = \frac{k \cos\theta}{r'^2};$$

r' representing as before the distance Op', and θ the angle formed between the line Op', and another line OD fixed in A. If now f be the magnitude of the force directed along OB, the constant k will evidently be of the form $k = a'f$; a' being a constant quantity. The value of V, just given, holds good for any arrangement, regular or irregular, of the magnetic particles composing A, but on the latter supposition, the value of V would evidently remain unchanged, provided the sphere, and consequently the line OD, revolved round OB as an axis, which could not be the case unless OB and OD coincided. Hence $\theta =$ angle BOp' and

$$V = \frac{a'f \cos\theta}{r'^2}.$$

Let now α, β, γ, be the angles that the line $Op' = r'$ makes with the axes of x, y, z, and α', β', γ', those which OB makes with the same axes; then, substituting for $\cos\theta$ its value

$$\cos\alpha \cos\alpha' + \cos\beta \cos\beta' + \cos\gamma \cos\gamma',$$

we have, since

$$f\cos\alpha = X, \quad f\cos\beta = Y, \quad f\cos\gamma = Z,$$

$$V = \frac{a'(X\cos\alpha + Y\cos\beta + Z\cos\gamma)}{r'^2} \quad \text{.........} \ (b').$$

Which agrees with the equation (b), seeing that

$$\cos\alpha = \frac{x'}{r'}, \quad \cos\beta = \frac{y'}{r'}, \quad \cos\gamma = \frac{z'}{r'}.$$

(15.) Conceive now a body A, of any form, to have a magnetic state induced in its particles by the influence of exterior forces, it is clear that if dv be an element of its volume, the value of the potential function arising from this element, at any point p' whose co-ordinates are x', y', z', must, since the total quantity of magnetic fluid in dv is equal to zero, be of the form

$$\frac{dv\,\{X(x'-x) + Y(y'-y) + Z(z'-z)\}}{r^3} \quad \ldots\ldots\ldots (a);$$

x, y, z, being the co-ordinates of dv, r the distance p', dv and X, Y, Z, three quantities dependant on the magnetic state induced in dv, and serving to define this state. If therefore dv' be an infinitely small volume within the body A and inclosing the point p', the potential function arising from the whole A exterior to dv', will be expressed by

$$\int dx\,dy\,dz\,\frac{X(x'-x) + Y(y'-y) + Z(z'-z)}{r^3};$$

the integral extending over the whole volume of A exterior to dv'.

It is easy to show from this expression that, in general, although dv' be infinitely small, the forces acting in its interior vary in magnitude and direction by passing from one part of it to another; but, when dv' is spherical, these forces are sensibly constant in magnitude and direction, and consequently, in this case, the value of the potential function induced in dv' by their action, may be immediately deduced from the preceding article.

Let ψ' represent the value of the integral just given, when dv' is an infinitely small sphere. The force acting on p' arising from the mass exterior to dv', tending to increase x', will be

$$-\left(\frac{d\psi'}{dx'}\right);$$

the line above the differential coefficient indicating that it is to be obtained by supposing the radius of dv' to vanish after differentiation, and this may differ from the one obtained by first making the radius vanish, and afterwards differentiating the resulting function of x', y', z', which last being represented as usual by $\frac{d\psi'}{dx'}$, we have

$$\overline{\frac{d\psi'}{dx'}} = \frac{d}{dx'}\int dx\,dy\,dz\,\frac{X(x'-x)+Y(y'-y)+Z(z'-z)}{r^3},$$

$$\frac{d\psi'}{dx'} = \frac{d}{dx'}\int dx\,dy\,dz\,\frac{X(x'-x)+Y(y'-y)+Z(z'-z)}{r^3};$$

the first integral being taken over the whole volume of A exterior to dv', and the second over the whole of A including dv'. Hence

$$\frac{d\psi'}{dx'} - \overline{\frac{d\psi'}{dx'}} = \frac{d}{dx'}\int dx\,dy\,dz\,\frac{X(x'-x)+Y(y'-y)+Z(z'-z)}{r^3};$$

the last integral comprehending the volume of the spherical particle dv' only, whose radius a is supposed to vanish after differentiation. In order to effect the integration here indicated, we may remark that X, Y and Z are sensibly constant within dv', and may therefore be replaced by $X_{,}$, $Y_{,}$ and $Z_{,}$, their values at the centre of the sphere dv', whose co-ordinates are $x_{,}$, $y_{,}$, $z_{,}$; the required integral will thus become

$$\int dx\,dy\,dz\,\frac{X_{,}(x'-x)+Y_{,}(y'-y)+Z_{,}(z'-z)}{r^3}.$$

Making for a moment $E = X_{,}x + Y_{,}y + Z_{,}z$, we shall have

$$X_{,} = \frac{dE}{dx}, \quad Y_{,} = \frac{dE}{dy}, \quad Z_{,} = \frac{dE}{dz},$$

and as also

$$\frac{x'-x}{r^3} = \frac{d\frac{1}{r}}{dx}; \quad \frac{y'-y}{r^3} = \frac{d\frac{1}{r}}{dy}; \quad \frac{z'-z}{r^3} = \frac{d\frac{1}{r}}{dz},$$

this integral may be written

$$\int dx\,dy\,dz\left(\frac{dE}{dx}\cdot\frac{d\frac{1}{r}}{dx}+\frac{dE}{dy}\cdot\frac{d\frac{1}{r}}{dy}+\frac{dE}{dz}\cdot\frac{d\frac{1}{r}}{dz}\right),$$

which since $\delta E = 0$, and $\delta\frac{1}{r} = 0$, reduces itself by what is proved in Art. 3, to

$$-\int\frac{d\sigma}{r}\left(\frac{dE}{dw}\right) = (\text{because } dw = -\,da)\int\frac{d\sigma}{r}\frac{dE}{da};$$

the integral extending over the whole surface of the sphere dv', of which $d\sigma$ is an element; r being the distance p', $d\sigma$, and dw measured from the surface towards the interior of dv'. Now $\int\frac{d\sigma}{r}\frac{dE}{da}$ expresses the value of the potential function for a point p', within the sphere, supposing its surface everywhere covered with electricity whose density is $\frac{dE}{da}$, and may very easily be obtained by No. 13, Liv. 3, *Méc. Céleste.* In fact, using for a moment the notation there employed, supposing the origin of the polar co-ordinates at the centre of the sphere, we have

$$E = E_{,} + a\,(X_{,}\cos\theta + Y_{,}\sin\theta\cos\varpi + Z_{,}\sin\theta\sin\varpi);$$

$E_{,}$ being the value of E at the centre of the sphere. Hence

$$\frac{dE}{da} = X_{,}\cos\theta + Y_{,}\sin\theta'\cos\varpi + Z_{,}\sin\theta\sin\varpi,$$

and as this is of the form $U^{(0)}$ (Vide *Méc. Céleste,* Liv. 3), we immediately obtain

$$\int\frac{d\sigma}{r}\frac{dE}{da} = \tfrac{4}{3}\pi r'\,\{X_{,}\cos\theta' + Y_{,}\sin\theta'\cos\varpi' + Z_{,}\sin\theta'\sin\varpi'\},$$

where r', θ', ϖ' are the polar co-ordinates of p'. Or by restoring x', y' and z'

$$\int\frac{d\sigma}{r}\frac{dE}{da} = \tfrac{4}{3}\pi\,\{X_{,}(x'-x_{,}) + Y_{,}(y'-y_{,}) + Z_{,}(z'-z_{,})\}.$$

Hence we deduce successively

$$\frac{d\psi'}{dx'} - \frac{\overline{d\psi'}}{dx'} = \frac{d}{dx}\int dx\,dy\,dz\,\frac{X(x'-x) + Y(y'-y) + Z(z'-z)}{r^3}$$

$$= \frac{d}{dx'}\int \frac{d\sigma}{r}\frac{dE}{da} = \frac{d}{dx'}\tfrac{4}{3}\pi\{X_{,}(x'-x_{,}) + Y_{,}(y'-y_{,})$$

$$+ Z_{,}(z'-z_{,})\} = \tfrac{4}{3}\pi X_{,}.$$

If now we make the radius a vanish, $X_{,}$ must become equal to X', the value of X at the point p', and there will result

$$\frac{d\psi'}{dx'} - \frac{\overline{d\psi'}}{dx'} = \tfrac{4}{3}\pi X', \text{ i.e. } \frac{\overline{d\psi'}}{dx'} = \frac{d\psi'}{dx'} - \tfrac{4}{3}\pi X'.$$

But $-\dfrac{\overline{d\psi'}}{dx'}$ expresses the value of the force acting in the direction of x positive, on a point p' within the infinitely small sphere dv', arising from the whole of A exterior to dv'; substituting now for $\dfrac{\overline{d\psi'}}{dx'}$ its value just found, the expression of this force becomes

$$\tfrac{4}{3}\pi X' - \frac{d\psi'}{dx'}.$$

Supposing V' to represent the value of the potential function at p', arising from the exterior bodies which induce the magnetic state of A, the force due to them acting in the same direction, is

$$-\frac{dV'}{dx'},$$

and therefore the total force in the direction of x' positive, tending to induce a magnetic state in the spherical element dv', is

$$\tfrac{4}{3}\pi X' - \frac{d\psi'}{dx'} - \frac{dV'}{dx'} = \overline{X}.$$

In the same way, the total forces in the directions of y' and z' positive, acting upon dv', are shown to be

$$\tfrac{4}{3}\pi Y' - \frac{d\psi'}{dy'} - \frac{dV'}{dy'} = \overline{Y}, \text{ and, } \tfrac{4}{3}\pi Z' - \frac{d\psi'}{dz'} - \frac{dV'}{dz'} = \overline{Z}.$$

By the equation (b') of the preceding article, we see that when dv' is a perfect conductor of magnetism, and its particles are not regularly arranged, the value of the potential function at any point p'', arising from the magnetic state induced in dv' by the action of the forces $\overline{X}$, $\overline{Y}$, $\overline{Z}$, is of the form

$$\frac{a'(\overline{X}\cos\alpha + \overline{Y}\cos\beta + \overline{Z}\cos\gamma)}{r'^2};$$

r' being the distance p'', dv', and α, β, γ the angles which r' forms with the axes of the rectangular co-ordinates. If then x'', y'', z'' be the co-ordinates of p'', this becomes, by observing that here $a' = kdv'$,

$$\frac{kdv'\{\overline{X}(x''-x') + \overline{Y}(y''-y) + \overline{Z}'(z''-z')\}}{r'^3}$$

k being a constant quantity dependant on the nature of the body. The same potential function will evidently be obtained from the expression (a) of this article, by changing dv, p', and their co-ordinates, into dv', p'', and their co-ordinates; thus we have

$$\frac{dv'\{X'(x''-x') + Y'(y''-y') + Z(z''-z')\}}{r'^3}.$$

Equating these two forms of the same quantity, there results the three following equations:

$$X' = k\overline{X} = \tfrac{4}{3}\pi kX' - k\frac{d\psi'}{dx'} - k\frac{dV'}{dx'},$$

$$Y' = k\overline{Y} = \tfrac{4}{3}\pi kY' - k\frac{d\psi'}{dy'} - k\frac{dV'}{dy'},$$

$$Z' = k\overline{Z} = \tfrac{4}{3}\pi kZ' - k\frac{d\psi'}{dz'} - k\frac{dV'}{dz'},$$

since the quantities x'', y'', z'' are perfectly arbitrary. Multiplying the first of these equations by dx', the second by dy', the third by dz', and taking their sum, we obtain

$$0 = (1 - \tfrac{4}{3}\pi k)(X'dx' + Y'dy' + Z'dz') + k'd\psi' + kdV'.$$

But $d\psi'$ and dV' being perfect differentials, $X'dx' + Y'dy' + Z'dz$ must be so likewise, making therefore

$$d\phi' = X'dx' + Y'dy' + Z'dz',$$

the above, by integration, becomes

$$\text{const.} = (1 - \tfrac{4}{3}\pi k)\,\phi' + k\psi' + kV'.$$

Although the value of k depends wholly on the nature of the body under consideration, and is to be determined for each by experiment, we may yet assign the limits between which it must fall. For we have, in this theory, supposed the body composed of conducting particles, separated by intervals absolutely impervious to the magnetic fluid; it is therefore clear the magnetic state induced in the infinitely small sphere dv', cannot be greater than that which would be induced, supposing it one continuous conducting mass, but may be made less in any proportion, at will, by augmenting the non-conducting intervals.

When dv' is a continuous conductor, it is easy to see the value of the potential function at the point p'', arising from the magnetic state induced in it by the action of the forces $\overline{X}$, $\overline{Y}$, $\overline{Z}$, will be

$$\frac{3dv}{4\pi} \cdot \frac{X(x'' - x') + Y(y'' - y') + Z(z'' - z')}{r'^3},$$

seeing that $\frac{3dv}{4\pi} = a^3$; a representing, as before, the radius of the sphere dv'. By comparing this expression with that before found, when dv' was not a continuous conductor, it is evident k must be between the limits 0 and $\frac{3}{4}\pi$, or, which is the same thing,

$$k = \frac{3g}{4\pi};$$

g being any positive quantity less than 1.

The value of k, just found, being substituted in the equation serving to determine ϕ', there arises

$$\text{const.} = (1 - g)\,\phi' + \frac{3g}{4\pi}\,(\psi' + V').$$

Moreover

$$\psi' = \int dx\, dy\, dz \frac{X(x'-x) + Y(y'-y) + Z(z'-z)}{r^3}$$

$$= \int dx\, dy\, dz \left(\frac{d\phi}{dx} \cdot \frac{d\frac{1}{r}}{dx} + \frac{d\phi}{dy} \cdot \frac{d\frac{1}{r}}{dy} + \frac{d\phi}{dz} \cdot \frac{d\frac{1}{r}}{dz} \right)$$

$$= 4\pi\phi' - \int d\sigma \overline{\phi} \left(\frac{\overline{d\frac{1}{r}}}{dw} \right) \text{ (Art. 3)};$$

the triple integrals extending over the whole volume of A, and that relative to $d\sigma$ over its surface, of which $d\sigma$ is an element; the quantities $\overline{\phi}$ and $\frac{\overline{d\frac{1}{r}}}{dw}$ belonging to this element. We have, therefore, by substitution

$$\text{const.} = (1 + 2g)\, \phi' + \frac{3g}{4\pi} \left\{ V' - \int d\sigma \overline{\phi} \left(\frac{\overline{d\frac{1}{r}}}{dw} \right) \right\}.$$

Now $\delta' V' = 0$, and

$$\delta' \int d\sigma \overline{\phi} \left(\frac{d\frac{1}{r}}{dw} \right) = 0,$$

and consequently $\delta'\phi' = 0$; the symbol δ' referring to x', y', z' the co-ordinates of p'; or, since x', y' and z' are arbitrary, by making them equal to x, y, z respectively, there results

$$0 = \phi,$$

in virtue of which, the value of ψ', by Article 3, becomes

$$\psi' = -\int \frac{d\sigma}{r} \left(\frac{\overline{d\phi}}{dw} \right) \text{ .... } (b);$$

r being the distance p', $d\sigma$, and $\left(\frac{\overline{d\phi}}{dw}\right)$ belonging to $d\sigma$. The former equation serving to determine ϕ' gives, by changing x', y', z' into x, y, z,

$$\text{const.} = (1 - g)\, \phi + \frac{3g}{4\pi} (\psi + V) \text{ .... } (c);$$

ϕ, ψ and V belonging to a point p, within the body, whose co-ordinates are x, y, z. It is moreover evident from what precedes that the functions ϕ, ψ and V satisfy the equations $0 = \delta\phi$, $0 = \delta\psi$ and $0 = \delta V$, and have no singular values in the interior of A.

The equations (b) and (c) serve to determine ϕ and ψ, completely, when the value of V arising from the exterior bodies is known, and therefore they enable us to assign the magnetic state of every part of the body A, seeing that it depends on X, Y, Z, the differential co-efficients of ϕ. It is also evident that ψ', when calculated for any point p', not contained within the body A, is the value of the potential function at this point arising from the magnetic state induced in A, and therefore this function is always given by the equation (b).

The constant quantity g, which enters into our formulæ, depends on the nature of the body solely, and, in a subsequent article, its value is determined for a cylindric wire used by Coulomb. This value differs very little from unity: supposing therefore $g = 1$, the equations (b) and (c) become

$$\psi' = -\int \frac{d\sigma}{r} \left(\frac{\overline{d\phi}}{dw}\right) \qquad \text{.....................} \qquad (b'),$$

$$\text{const.} = \psi + V \qquad \text{.....................} \qquad (c'),$$

evidently the same, in effect, as would be obtained by considering the magnetic fluid at liberty to move from one part of the conducting body to another; the density ρ being here replaced by $-\left(\frac{\overline{d\phi}}{dw}\right)$, and since the value of the potential function for any point exterior to the body is, on either supposition, given by the formula (b), the exterior actions will be precisely the same in both cases. Hence, when we employ iron, nickel, or similar bodies, in which the value of g is nearly equal to 1, the observed phenomena will differ little from those produced on the latter hypothesis, except when one of their dimensions is very small compared with the others, in which case the results of the two hypotheses differ widely, as will be seen in some of the applications which follow.

If the magnetic particles composing the body were not perfect conductors, but indued with a coercive force, it is clear there might always be equilibrium, provided the magnetic state of the element dv' was such as would be induced by the forces $\frac{\overline{d\psi'}}{dx'}+\frac{dV'}{dx'}+A'$, $\frac{\overline{d\psi'}}{dy'}+\frac{dV'}{dy'}+B'$ and $\frac{\overline{d\psi'}}{dz'}+\frac{dV'}{dz'}+C'$, instead of $\frac{\overline{d\psi'}}{dx'}+\frac{dV'}{dx'}$, $\frac{\overline{d\psi'}}{dy'}+\frac{dV'}{dy'}$ and $\frac{\overline{d\psi'}}{dz'}+\frac{dV'}{dz'}$; supposing the resultant of the forces A', B', C' no where exceeds a quantity β, serving to measure the coercive force. This is expressed by the condition

$$A'^2+B'^2+C'^2<\beta^2,$$

the equation (c) would then be replaced by

$$0=(1-g)\,d\phi+\frac{3g}{4\pi}(d\psi+dV+Adx+Bdy+Cdz)\ldots\ldots(c');$$

A, B, C being any functions of x, y, z, as A', B', C' are of x', y', z' subject only to the condition just given.

It would be extremely easy so to modify the preceding theory, as to adapt it to a body whose magnetic particles are regularly arranged, by using the equation (a) in the place of the equation (b) of the preceding article; but, as observation has not yet offered any thing which would indicate a regular arrangement of magnetic particles, in any body hitherto examined, it seems superfluous to introduce this degree of generality, more particularly as the omission may be so easily supplied.

(16.) As an application of the general theory contained in the preceding article, suppose the body A to be a hollow spherical shell of uniform thickness, the radius of whose inner surface is a, and that of its outer one $a_{\prime}$; and let the forces inducing a magnetic state in A, arise from any bodies whatever, situate at will, within or without the shell. Then since in the interior of A's mass $0=\delta\phi$, and $0=\delta V$, we shall have (*Méc. Cél.* Liv. 3)

$$\phi=\Sigma\phi^{(i)}r^i+\Sigma\phi_{\prime}^{(i)}r^{-i-1} \text{ and } V=\Sigma U^{(i)}r^i+\Sigma U_{\prime}^{(i)}r^{-i-1};$$

r being the distance of the point p, to which ϕ and V belong, from the shell's centre, $\phi^{(0)}$, $\phi^{(1)}$, &c.,—$U^{(0)}$, $U^{(1)}$, &c. functions of θ and ϖ, the two other polar co-ordinates of p, whose nature has been fully explained by Laplace in the work just cited; the finite integrals extending from $i=0$ to $i=\infty$.

If now, to prevent ambiguity, we enclose the r of equation (*b*) Art. 15 in a parenthesis, it will become

$$\psi = \int \frac{d\sigma}{(r)} \overline{\left(\frac{d\phi}{dw}\right)};$$

(r) representing the distance p, $d\sigma$, and the integral extending over both surfaces of the shell. At the inner surface we have $\overline{\frac{d\phi}{dw}} = \overline{\frac{d\phi}{dr}}$ and $r=a$: hence the part of ψ due to this surface is

$$-\int \frac{d\sigma}{(r)} \overline{\frac{d\phi}{dr}} = -\int \frac{d\sigma}{(r)} \Sigma i\phi^{(i)} a^{i-1} + \int \frac{d\sigma}{(r)} \Sigma (i+1)\, \phi_{,}^{(i)} a^{-i-2};$$

the integrals extending over the whole of the inner surface, and $d\sigma$ being one of its elements. Effecting the integrations by the formulæ of Laplace (*Méc. Céleste*, Liv. 3), we immediately obtain the part ψ, due to the inner surface, viz.

$$\frac{4\pi a^2}{r} \Sigma \frac{a^i}{(2i+1)\, r^i} \{-ia^{i-1}\phi^{(i)} + (i+1)\, \phi_{,}^{(i)} a^{-i-2}\}.$$

In the same way the part of ψ due to the outer surface, by observing that for it $\overline{\frac{d\phi}{dw}} = -\overline{\frac{d\phi}{dr}}$ and $r=a_{,}$, is found to be

$$4\pi a_{,} \Sigma \frac{r^i}{(2i+1)\, a_{,}^i} \{ia_{,}^{i-1}\phi^{(i)} - (i+1)\, \phi^{(i)} a_{,}^{-i-2}\}.$$

The sum of these two expressions is the complete value of ψ, which, together with the values of ϕ and V before given, being substituted in the equation (*c*) Art. 15, we obtain

$$\text{const.} = (1-g)\, \Sigma\phi_{,}^{(i)} r^{-i-1} + (1-g)\, \Sigma\phi^{(i)} r^i + \frac{3g}{4\pi} \Sigma U_{,}^{(i)} r^{-i-1} + \frac{3g}{4\pi} \Sigma U^{(i)} r^i$$

$$+ \frac{3ga^2}{r} \Sigma \frac{a^i}{(2i+1)\, r^i} \{-ia^{i-1}\phi^{(i)} + (i+1)\, \phi_{,}^{(i)} a^{-i-2}\}$$

$$+ 3ga_{,} \Sigma \frac{r^i}{(2i+1)\, a_{,}^i} \{ia_{,}^{i-1}\phi^{(i)} - (i+1)\, \phi_{,}^{(i)} a_{,}^{-i-2}\}.$$

Equating the coefficients of like powers of the variable r, we have generally, whatever i may be,

$$0 = (1-g)\phi_{,}^{(i)} + \frac{3ga^{i+2}}{2i+1}\{-ia^{i-1}\phi^{(i)} + (i+1)\phi_{,}^{(i)}a^{-i-2}\} + \frac{3g}{4\pi}U_{,}^{(i)},$$

$$0 = (1-g)\phi^{(i)} + \frac{3g}{(2i+1)a_{,}^{i-1}}\{ia_{,}^{i-1}\phi^{(i)} - (i+1)\phi_{,}^{(i)}a_{,}^{-i-2}\} + \frac{3g}{4\pi}U^{(i)};$$

neglecting the constant on the right side of the equation in r as superfluous, since it may always be made to enter into $\phi^{(0)}$. If now, for abridgment, we make

$$D = (2i+1)^2(1+g) + (i-1)(i+2)g^2 - 9g^2 i(i+1)\left(\frac{a}{a_{,}}\right)^{2i+1},$$

we shall obtain by elimination

$$\phi^{(i)} = -\frac{3g}{4\pi}U^{(i)}\frac{\{2i+1)(2i+1+(i+2)g\}}{D} - \frac{3g}{4\pi}U_{,}^{(i)}\frac{3g(i+1)(2i+1)a_{,}^{-2i-1}}{D},$$

$$\phi_{,}^{(i)} = -\frac{3g}{4\pi}U^{(i)}\frac{3gi(2i+1)a^{2i+1}}{D} - \frac{3g}{4\pi}U_{,}^{(i)}\frac{(2i+1)\{2i+1+(i-1)g\}}{D}.$$

These values substituted in the expression

$$\phi = \Sigma\phi^{(i)}r^i + \Sigma\phi_{,}^{(i)}r^{-i-1},$$

give the general value of ϕ in a series of the powers of r, when the potential function due to the bodies inducing a magnetic state in the shell is known, and thence we may determine the value of the potential function ψ arising from the shell itself, for any point whatever, either within or without it.

When all the bodies are situate in the space exterior to the shell, we may obtain the total actions exerted on a magnetic particle in its exterior, by the following simple method, applicable to hollow shells of any shape and thickness.

The equation (*c*) Art. 15 becomes, by neglecting the superfluous constant,

$$0 = (1-g)\phi + \frac{3g}{4\pi}(\psi + V).$$

If now (ϕ) represent the value of the potential function, corresponding to ϕ the value of ϕ at the inner surface of the shell, each of the functions (ϕ), ψ and V, will satisfy the equations

$0=\delta(\phi)$, $0=\delta\psi$ and $0=\delta V$, and moreover, have no singular values in the space within the shell; the same may therefore be said of the function

$$(1-g)(\phi)+\frac{3g}{4\pi}(\psi+V),$$

and as this function is equal to zero at the inner surface, it follows (Art. 5) that it is so for any point p of the interior space. Hence

$$0=(1-g)(\phi)+\frac{3g}{4\pi}(\psi+V).$$

But $\psi+V$ is the value of the total potential function at the point p, arising from the exterior bodies and shell itself: this function will therefore be expressed by

$$-\frac{4\pi(1-g)}{3g}(\phi).$$

In precisely the same way, the value of the total potential function at any point p', exterior to the shell, when the inducing bodies are all within it, is shown to be

$$-\frac{4\pi(1-g)}{3g}.(\phi');$$

(ϕ') being the potential function corresponding to the value of ϕ at the exterior surface of the shell. Having thus the total potential functions, the total action exerted on a magnetic particle in any direction, is immediately given by differentiation.

To apply this general solution to our spherical shell, the inducing bodies being all exterior to it, we must first determine $\bar{\phi}$, the value of ϕ at its inner surface, making $0=\Sigma U_{,}^{(i)}r^{-i-1}$ since there are no interior bodies, and thence deduce the value of (ϕ). Substituting for $\phi^{(i)}$ and $\phi_{,}^{(i)}$ their values before given, making $U_{,}^{(i)}=0$ and $r=a$, we obtain

$$\bar{\phi}=\frac{-3g}{4\pi}(1+2g)\,\Sigma U^{(i)}\frac{(2i+1)^2a^i}{D},$$

and the corresponding value of (ϕ) is (*Méc. Cél.* Liv. 3)

$$(\phi)=\frac{-3g}{4\pi}(1+2g)\,\Sigma U^{(i)}\frac{(2i+1)^2r^i}{D}.$$

The value of the total potential function at any point p within the shell, whose polar co-ordinates are r, θ, ϖ, is

$$-\frac{4\pi}{3g}(1-g)(\phi) = (1-g)(1+2g)\,\Sigma U^{(i)}\frac{(2i+1)^2 r^i}{D}.$$

In a similar way, the value of the same function at a point p' exterior to the shell, all the inducing bodies being within it, is found to be

$$(1-g)(1-2g)\,\Sigma U_{,}^{(i)}\frac{(2i+1)^2}{D\,.\,r^{i+1}}$$

r, θ and ϖ in this expression representing the polar co-ordinates of p'.

To give a very simple example of the use of the first of these formulæ, suppose it were required to determine the total action exerted in the interior of a hollow spherical shell, by the magnetic influence of the earth; then making the axis of x to coincide with the direction of the dipping needle, and designating by f, the constant force tending to impel a particle of positive fluid in the direction of x positive, the potential function V, due to the exterior bodies, will here become

$$V = -f\,.\,x = -f\cos\theta\,.\,r = U^{(i)}\,.\,r.$$

The finite integrals expressing the value of V reduce themselves therefore, in this case, to a single term, in which $i=1$, and the corresponding value of D being

$$9\left(1+g-2g^2\frac{a^3}{a_{,}^3}\right),$$

the total potential function within the shell is

$$-(1-g)(1+2g)\,U^{(1)}\frac{r}{1+g-2g^2\dfrac{a^3}{a_{,}^3}} = -\frac{1+g-2g^2}{1+g-2g^2\dfrac{a^3}{a_{,}^3}}f\,.\,x.$$

We therefore see that the effect produced by the intervening shell, is to reduce the directive force which would act on a very small magnetic needle,

$$\text{from } f, \text{ to } \frac{1+g-2g^2}{1+g-2g^2\dfrac{a^3}{a_{,}^3}}f.$$

In iron and other similar bodies, g is very nearly equal to 1, and therefore the directive force in the interior of a hollow spherical shell is greatly diminished, except when its thickness is very small compared with its radius, in which case, as is evident from the formula, it approaches towards the original value f, and becomes equal to it when this thickness is infinitely small.

To give an example of the use of the second formula, let it be proposed to determine the total action upon a point p, situate on one side of an infinitely extended plate of uniform thickness, when another point P, containing a unit of positive fluid, is placed on the other side of the same plate considering it as a perfect conductor of magnetism. For this, let fall the perpendicular PQ upon the side of the plate next P, on PQ prolonged, demit the perpendicular pq, and make $PQ = b$, $Pq = u$, $pq = v$, and $t =$ the thickness of the plate; then, since its action is evidently equal to that of an infinite sphere of the same thickness, whose centre is upon the line QP at an infinite distance from P, we shall have the required value of the total potential function at p by supposing $a_{,} = a + t$, a infinite, and the line PQ prolonged to be the axis from which the angle θ is measured. Now in the present case

$$V = \frac{1}{Pp} = \frac{1}{\sqrt{\{r^2 - 2r(a-b)\cos\theta + (a-b)^2\}}} = \Sigma U_{,}^{(i)} r^{-i-1},$$

and the value of the potential function, as before determined, is

$$(1-g)(1-2g)\,\Sigma \frac{(2i+1)^2}{D} U_{,}^{(i)} r^{-i-1}.$$

From the first expression we see that the general term $U_{,}^{(i)} r^{-i-1}$ is a quantity of the order $(a-b)^i r^{-i-1}$. Moreover, by substituting for r its value in u,

$$(a-b)^i r^{-i-1} = (a-b)^i (a-b+u)^{-i-1} = \frac{1}{a} e^{-\frac{iu}{a}};$$

neglecting such quantities as are of the order $\frac{1}{a}$ compared with those retained. The general term $U_{,}^{(i)} r^{-i-1}$, and consequently $U_{,}^{(i)}$, ought therefore to be considered as functions of $\frac{i}{a} = \gamma$.

In the finite integrals just given, the increment of i is 1, and the corresponding increment of γ is $\frac{1}{a} = d\gamma$ (because a is infinite), the finite integrals thus change themselves into ordinary integrals or fluents. In fact (*Méc. Cél.* Liv. 3), $U_{\prime}^{(i)}$ always satisfies the equation

$$\frac{d^2 U_{\prime}^{(i)}}{d\theta^2} + \frac{\cos\theta}{\sin\theta}\frac{dU_{\prime}^{(i)}}{d\theta} + i(i+1)\,U_{\prime}^{(i)} = 0,$$

and as θ is infinitely small whenever V has a sensible value, we may eliminate it from the above by means of the equation $a\theta = v$, and we obtain by neglecting infinitesimals of higher orders than those retained, since $\frac{i}{a} = \gamma$,

$$\frac{d^2 U_{\prime}^{(i)}}{dv^2} + \frac{dU_{\prime}^{(i)}}{vdv} + \gamma^2 U_{\prime}^{(i)} = 0.$$

Hence the value $U_{\prime}^{(i)}$ is of the form

$$U_{\prime}^{(i)} = A\int_0^1 \frac{d\beta}{\sqrt{(1-\beta^2)}}\cos(\beta\gamma v);$$

seeing that the remaining part of the general integral becomes infinite when v vanishes, and ought therefore to be rejected. It now only remains to determine the value of the arbitrary constant A. Making, for this purpose, $\theta = 0$, i.e. $v = 0$, we have

$$U_{\prime}^{(i)} = (a-b)^i \text{ and } \int_0^1 \frac{d\beta}{\sqrt{(1-\beta^2)}} = \tfrac{1}{2}\pi: \text{ hence } (a-b)^i = \tfrac{1}{2}(A\pi),$$

$$\text{i.e. } A = \frac{2}{\pi}(a-b)^i.$$

By substituting for A and r their values, there results

$$U_{\prime}^{(i)} r^{-i-1} = \frac{\pi}{2}(a-b)^i(a-b+u)^{-i-1}\int_0^1 \frac{d\beta}{\sqrt{(1-\beta^2)}}\cos(\beta\gamma v)$$

$$= \frac{2d\gamma}{\pi}e^{-\gamma u}\int_0^1 \frac{d\beta}{\sqrt{(1-\beta^2)}}\cos(\beta\gamma v);$$

because $\frac{i}{a}=\gamma$ and $\frac{1}{a}=d\gamma$. Writing now in the place of i its value $a\gamma$, and neglecting infinitesimal quantities, we have

$$\frac{(2i+1)^2}{D}=\frac{4}{4+4g+g^2-9g^2e^{-2\gamma t}}.$$

Hence the value of the total potential function becomes

$$\frac{8}{\pi}(1-g)(1+2g)\int_0^\infty \frac{d\gamma\,.\,e^{-\gamma u}}{4+4g+g^2-9g^2e^{-2\gamma t}}\int_0^1 \frac{d\beta}{\sqrt{(1-\beta^2)}}\cos(\beta\gamma v);$$

where the integral relative to γ is taken from $\gamma=0$ to $\gamma=\infty$, to correspond with the limits 0 and ∞ of i, seeing that $i=a\gamma$.

The preceding solution is immediately applicable to the imaginary case only, in which the inducing bodies reduce themselves to a single point P, but by the following simple artifice we may give it a much greater degree of generality:

Conceive another point P', on the line PQ, at an arbitrary distance c from P, and suppose the unit of positive fluid concentrated in P' instead of P; then if we make $r'=Pp$, and $\theta'=\angle pPQ$, we shall have $u=r'\cos\theta'$, $v=r'\sin\theta'$, and the value of the potential function arising from P' will be

$$\frac{1}{P'p}=\frac{1}{\sqrt{(r'^2-2r'c\cos\theta'+c^2)}}=Q^{(0)}\frac{1}{r'}+Q^{(1)}\frac{c}{r'^2}+Q^{(2)}\frac{c^2}{r'^3}+\text{etc.}$$

Moreover, the value of the total potential function at p due to this, arising from P' and the plate itself, will evidently be obtained by changing u into $u-c$ in that before given, and is therefore

$$\frac{8}{\pi}(1-g)(1+2g)\int_0^\infty \frac{e^{\gamma c}\,d\gamma e^{-\gamma u}}{(2+g)^2-9g^2e^{-2\gamma t}}\int_0^1 \frac{d\beta}{\sqrt{(1-\beta^2)}}\cos(\beta\gamma v).$$

Expanding this function in an ascending series of the powers of c, the term multiplied by c^t is

$$\frac{8}{\pi}(1-g)(1+2g)\int_0^\infty \frac{\frac{\gamma^t c^t}{1.2.3\ldots n}\,d\gamma e^{-\gamma u}}{(2+g)^2-9g^2e^{-2\gamma t}}\int_0^1 \frac{d\beta}{\sqrt{(1-\beta^2)}}\cos(\beta\gamma v),$$

which, as c is perfectly arbitrary, must be the part due to the term $Q^{(i)} \frac{c_i}{r^{i+1}}$ in the potential function arising from the inducing bodies. If then this function had been

$$Q^{(0)} \frac{k_0}{r'} + Q^{(1)} \frac{k_1}{r'^2} + Q^{(2)} \frac{k_2}{r'^3} + Q^{(3)} \frac{k_3}{r'^4} + \&c.;$$

where the successive powers c^0, c^1, c^2, &c. of c are replaced by the arbitrary constant quantities k_0, k_1, k_2, &c., the corresponding value of the total potential function will be given by making a like change in that due to P'. Hence if, for abridgement, we make

$$\phi(\gamma) = k_0 + \frac{k_1}{1}\gamma + \frac{k_2}{1.2}\gamma^2 + \frac{k_3}{1.2.3}\gamma^3 + \&c.,$$

the value of this function at the point p will be

$$\frac{8}{\pi}(1-g)(1+2g)\int_0^\infty \frac{\phi(3)\, d\gamma e^{-\gamma u}}{(2+g)^2 - 9g^2 e^{-2\gamma t}} \int_0^1 \frac{d\beta}{\sqrt{(1-\beta^2)}} \cos(\beta\gamma v).$$

Now, if the original one due to the point P be called F, it is clear the expression just given may be written

$$\phi\left(\frac{-d}{du}\right).F;$$

where the symbols of operation are separated from those of quantity, according to ARBOGAST'S method; thus all the difficulty is reduced to the determination of F.

Resuming therefore the original supposition of the plate's magnetic state being induced by a particle of positive fluid concentrated in P, the value of the total potential function at p will be

$$F = \frac{8}{\pi}(1-g)(1+2g)\int_0^\infty \frac{d\gamma e^{-u\gamma}}{(2+g)^2 - 9g^2 e^{-2\gamma t}} \int_0^1 \frac{d\beta}{\sqrt{(1-\beta^2)}} \cos(\beta\gamma v),$$

as was before shown.

[Let $\dfrac{3g}{2+g} = m$: we shall have

$$F = \frac{2}{\pi}(1 - m^2)\int_0^1 \frac{d\beta}{(1-\beta^2)^{\frac{1}{2}}}\int_0^\infty \frac{d\gamma e^{-u\gamma}}{1 - m^2 e^{-2\gamma t}}\cos(\beta\gamma v)$$

$$= \frac{2}{\pi}(1 - m^2)\int_0^1 \frac{d\beta}{(1-\beta^2)^{\frac{1}{2}}}\int_0^\infty d\gamma e^{-u\gamma}(1 + m^2 e^{-2\gamma t} + m^4 e^{-4\gamma t} + \&c.)\cos(\beta\gamma v)$$

$$= \frac{2}{\pi}(1 - m^2)\int_0^1 \frac{d\beta}{(1-\beta^2)^{\frac{1}{2}}}\left\{\frac{u}{u^2+\beta^2 v^2} + \frac{m^2(u+2t)}{(u+2t)^2+\beta^2 v^2} + \frac{m^4(u+4t)}{(u+4t)^2+\beta^2 v^2} + \&c.\right\}$$

$$= \frac{2}{\pi}(1 - m^2)\,\Sigma\int_0^{\frac{1}{2}\pi}\frac{m^{2i}u_i\,d\theta}{u_i^2 + v^2\sin^2\theta}, \quad \text{where } u_i = u + 2it,$$

$$= (1 - m^2)\,\Sigma\,\frac{m^{2i}}{(u_i^2 + v^2)^{\frac{1}{2}}}\Bigg].$$

Writing now $e^{-v\beta\gamma\sqrt{-1}}$ in the place of $\cos(\beta\gamma v)$, we obtain

$$F = \frac{8}{\pi}(1-g)(1+2g)\int_0^1\frac{d\beta}{\sqrt{(1-\beta^2)}}\int_0^\infty\frac{d\gamma e^{-\gamma(u+\beta v\sqrt{-1})}}{(2+g)^2 - 9g^2 e^{-2\gamma t}},$$

provided we reject the imaginary quantities which may arise. In order to transform this double integral let

$$z = \frac{3g}{2+g}e^{-\gamma t},$$

and we shall have

$$F = \frac{8(1-g)(1+2g)}{9\pi g^2 t}\left(\frac{2+g}{3g}\right)^{\frac{u}{t}-2}\int_0^1\frac{d\beta}{\sqrt{(1-\beta^2)}}\left(\frac{2+g}{3g}\right)^{\frac{\beta u\sqrt{-1}}{t}}\int\frac{dz\,.\,z^{\frac{u}{t}-1+\frac{\beta v\sqrt{-1}}{t}}}{1-z^2}$$

the integral relative to z being taken from $z = 0$ to $z = \dfrac{3g}{2+g}$.

The value of $1 - g$, for iron and other similar bodies, is very small; neglecting therefore quantities which are of the order $(1 - g)$ compared with those retained, there results

$$F = \frac{8(1-g)}{3\pi t}\int_0^1\frac{d\beta}{\sqrt{(1-\beta^2)}}\int_0^1\frac{dz}{1-z^2}\,z^{\frac{u}{t}-1+\frac{\beta v}{t}\sqrt{-1}} \quad \ldots\ldots\ldots (a);$$

where u and v may have any values whatever provided they are not very great and of the order $\frac{t}{1-g}$. If F_1 represents what F becomes by changing u into $u+2t$, we have

$$F_1 = \frac{8(1-g)}{3\pi t}\int_0^1 \frac{d\beta}{\sqrt{(1-\beta^2)}}\int_0^1 \frac{z^2 dz}{1-z^2}\, z^{\frac{u}{t}-1+\frac{\beta v}{t}\sqrt{-1}};$$

and consequently

$$F - F_1 = \frac{8(1-g)}{3\pi t}\int_0^1 \frac{d\beta}{\sqrt{(1-\beta^2)}}\int_0^1 dz\,.\, z^{\frac{u}{t}-1+\frac{\beta v}{t}\sqrt{-1}},$$

which, by effecting the integrations and rejecting the imaginary quantities, becomes

$$F - F_1 = \frac{4(1-g)}{3\sqrt{(u^2+v^2)}} = \frac{4(1-g)}{3r'}.$$

Suppose now pO is a perpendicular falling from the point p upon the surface of the plate, and on this line, indefinitely extended in the direction Op, take the points p_1, p_2, p_3, &c., at the distances $2t$, $4t$, $6t$, &c. from p; then F_1, F_2, F_3, &c. being the values of F, calculated for the points p_1, p_2, p_3, &c. by the formula (a) of this article, and r'_1, r'_2, r'_3, &c. the corresponding values of r', we shall equally have

$$F_1 - F_2 = \frac{4(1-g)}{3r'_1};$$

$$F_2 - F_3 = \frac{4(1-g)}{3r'_2}, \text{ \&c.};$$

and consequently

$$F = \frac{4}{3}(1-g)\left(\frac{1}{r'} + \frac{1}{r'_1} + \frac{1}{r'_2} + \text{\&c. in infinitum}\right);$$

seeing that $F_\infty = 0$.

From this value of F, it is evident the total action exerted upon the point p, in any given direction pn, is equal to the sum of the actions which would be exerted without the interposition of the plate, on each of the points p, p_1, p_2, &c. in infinitum, in the directions pn, p_1n_1, p_2n_2, &c. multiplied by the constant factor $\frac{4}{3}(1-g)$: the lines pn, p_1n_1, p_2n_2, &c. being all parallel. Moreover, as this is the case wherever the inducing point P

may be situate, the same will hold good when, instead of P, we substitute a body of any figure whatever magnetized at will. The only condition to be observed, is, that the distance between p and every part of the inducing body be not a very great quantity of the order $\frac{t}{1-g}$.

On the contrary, when the distance between p and the inducing body is great enough to render $\frac{(1-g)r'}{t}$ a very considerable quantity, it will be easy to show, by expanding F in a descending series of the powers of r', that the actions exerted upon p are very nearly the same as if no plate were interposed.

We have before remarked (art. 15), that when the dimensions of a body are all quantities of the same order, the results of the true theory differ little from those, which would be obtained by supposing the magnetic like the electric fluid, at liberty to move from one part of a conducting body to another; but when, as in the present example, one of the dimensions is very small compared with the others, the case is widely different; for if we make g rigorously equal to 1 in the preceding formulæ, they will belong to the latter supposition (art. 15), and as F will then vanish, the interposing plate will exactly neutralize the action of any magnetic bodies however they may be situate, provided they are on the side opposite the attracted point. This differs completely from what has been deduced above by employing the correct theory. A like difference between the results of the two suppositions takes place, when we consider the action exerted by the earth on a magnetic particle, placed in the interior of a hollow spherical shell, provided its thickness is very small compared with its radius, as will be evident by making $g=1$ in the formulæ belonging to this case, which are given in a preceding part of the present article.

17. Since COULOMB's experiments on cylindric wires magnetized to saturation are numerous and very accurate, it was thought this little work could not be better terminated, than by directly deducing from theory such consequences as would admit of an immediate comparison with them, and in order to effect this, we

will, in the first place, suppose a cylindric wire whose radius is a and length 2λ, is exposed to the action of a constant force, equal to f, and directed parallel to the axis of the wire, and then endeavour to determine the magnetic state which will thus be induced in it. For this, let r be a perpendicular falling from a point p within the wire upon its axis, and x, the distance of the foot of this perpendicular from the middle of the axis; then f being directed along x positive, we shall have for the value of the potential function due to the exterior forces

$$V = -fx,$$

and the equations (b), (c) (art. 15) become, by omitting the superfluous constant,

$$\psi' = -\int \frac{d\sigma}{(r)} \overline{\left(\frac{d\phi}{dw}\right)}, \qquad (b),$$

$$0 = (1-g)\phi + \frac{3g}{4\pi}\psi - \frac{3gf}{4\pi}x \qquad (c):$$

(r), the distance p', $d\sigma$ being inclosed in a parenthesis to prevent ambiguity, and p' being the point to which ψ' belongs. By the same article we have $0 = \delta\phi$ and $0 = \delta\psi$, and as ϕ and ψ evidently depend on x and r only, these equations being written at length are

$$0 = r^2 \frac{d^2\phi}{dx^2} = \frac{rd}{dr}\left(\frac{rd\phi}{dr}\right),$$

$$0 = r^2 \frac{d^2\psi}{dx^2} + \frac{rd}{dr}\left(\frac{rd\psi}{dr}\right).$$

Since r is always very small compared with the length of the wire, we may expand ϕ in an ascending series of the powers of r, and thus

$$\phi = X + X_1 r + X_2 r^2 + \text{etc.};$$

X, X_1, X_2, etc. being functions of x only. By substituting this value in the equation just given, and comparing the coefficients of like powers r, we obtain

$$\phi = X - \frac{d^2X}{dx^2}\,\frac{r^2}{2^2} + \frac{d^4X}{dx^4}\,\frac{r^4}{2^2.4^2} + \text{etc.}$$

In precisely the same way the value of ψ is found to be

$$\psi = Y - \frac{d^2Y}{dx^2}\frac{r^2}{2^2} + \frac{d^4Y}{dx^4}\frac{r^4}{2^2.4^2} - \text{etc.}$$

It now only remains to find the values of X and Y in functions of x. By supposing p' placed on the axis of the wire, the equation (c) becomes

$$Y' = -\int \frac{d\sigma}{(r)}\left(\frac{d\phi}{dw}\right);$$

the integral being extended over the whole surface of the wire: Y' belonging to the point p', whose co-ordinates will be marked with an accent.

The part of Y' due to the circular plane at the end of the cylinder, where $x = -\lambda$, is

$$= \frac{dX''}{dx}\int_0^a \frac{2\pi r dr}{(r)} = -2\pi \frac{dX''}{dx}\left[\sqrt{\{(\lambda + x')^2 + a^2\}} - \lambda - x'\right],$$

since here $d\sigma = 2\pi r dr$ and $\frac{\overline{d\phi}}{dw} = \frac{dX''}{dx}$, by neglecting quantities of the order a^2 on account of their smallness; X'' representing the value of X when $x = -\lambda$.

At the other end where $x = +\lambda$ we have $d\sigma = 2\pi r dr$,

$$\frac{d\phi}{dw} = -\frac{dX'''}{dx},$$

and consequently the part due to it is

$$\frac{dX'''}{dx}\int_0^a \frac{2\pi r dr}{(r)} = 2\pi \frac{dX'''}{dx}\left[\sqrt{\{(\lambda - x')^2 + a^2\}} - \lambda + x'\right];$$

X''' designating the value of X when $x = +\lambda$.

At the curve surface of the cylinder

$$d\sigma = 2\pi a dx \text{ and } \frac{\overline{d\phi}}{dw} = -\frac{\overline{d\phi}}{dr} = \tfrac{1}{2}a\frac{d^2X}{dx^2},$$

provided we omit quantities of the order a^2 compared with those retained. Hence the remaining part due to this surface is

$$-\pi a^2 \int \frac{dx}{(r)}\frac{d^2X}{dx^2};$$

the integral being taken from $x=-\lambda$ to $x=+\lambda$. The total value of Y' is therefore

$$Y' = 2\pi \frac{dX'''}{dx} [\sqrt{\{(\lambda - x')^2 + a^2\}} + \lambda - x']$$

$$- 2\pi \frac{nX''}{dx} [\sqrt{\{(\lambda + x')^2 + a^2\}} - \lambda - x']$$

$$- \pi a^2 \int \frac{dx}{(r)} \frac{d^2X}{dx^2};$$

the limits of the integral being the same as before. If now we substitute for (r) its value $\sqrt{\{(x-x')^2 + a)^2\}}$ we shall have

$$- \pi a^2 \int \frac{dx}{(r)} \frac{d^2X}{dx^2} = - \pi a^2 \int \frac{dx}{\sqrt{\{(x-x')^2 + a^2\}}} \frac{d^2X}{dx^2};$$

both integrals extending from $x=-\lambda$ to $x=+\lambda$.

On account of the smallness of a, the elements of the last integral where x is nearly equal to x' are very great compared with the others, and therefore the approximate value of the expression just given, will be

$$- \pi a^2 A \frac{d^2X'}{dx'^2} \text{ where } A = \int \frac{dx}{\sqrt{\{(x-x')^2 + a^2\}}} = 2 \log \frac{2\mu}{a}$$

very nearly; the two limits of the integral being $-\mu$ and $+\mu$ and μ so chosen that when p' is situate anywhere on the wire's axis, except in the immediate vicinity of either end, the approximate shall differ very little from the true value, which may in every case be done without difficulty. Having thus, by substitution, a value of Y' free from the sign of integration, the value of Y is given by merely changing x' into x and X' into X; in this way

$$Y = 2\pi \frac{dX'''}{dx} [\sqrt{\{(\lambda - x)^2 + a^2\}} - \lambda + x]$$

$$- 2\pi \frac{dX''}{dx} [\sqrt{\{(\lambda + x)^2 + a^2\}} - \lambda - x]$$

$$- \pi a^2 \frac{d^2X}{dx^2}.$$

The equation (*c*), by making $r = 0$, becomes

$$0 = (1 - g)\,X + \frac{3g}{4\pi}\,Y - \frac{3gf}{4\pi}\,x,$$

or by substituting for Y

$$0 = (1 - g)\,X - \tfrac{3}{4}\,(ga^2A)\,\frac{d^2X}{dx^2} - \frac{3gf}{4\pi}\,x$$

$$+\tfrac{3}{2}g\,\frac{dX'''}{dx}\,[\sqrt{\{(\lambda - x)^2 + a^2\}} - \lambda + x]$$

$$-\tfrac{3}{2}g\,\frac{dX''}{dx}\,[\sqrt{\{(\lambda + x)^2 + a^2\}} - \lambda - x]\,;$$

an equation which ought to hold good, for every value of x, from $x = -\lambda$ to $x = +\lambda$.

In those cases to which our theory will be applied, $1 - g$ is a small quantity of the same order as a^2A, and thus the three terms of the first line of our equation will be of the order a^2AX; making now $x = +\lambda$, $\tfrac{3}{2}g\,\frac{dX'''}{dx}\,a$ is shown to be of the order a^2AX''', and therefore $\frac{dX'''}{dx} \div X'''$ is a small quantity of the order aA; but for any other value of x the function multiplying $\frac{dX'''}{dx}$ becomes of the order a^2, and therefore we may without sensible error neglect the term containing it, and likewise suppose

$$\frac{dX'''}{dx} \div X''' = 0.$$

In the same way by making $x = -\lambda$, it may be shown that the term containing $\frac{dX''}{dx}$ is negligible, and

$$\frac{dX''}{dx} \div X'' = 0.$$

Thus our equation reduces itself to

$$0 = (1 - g)\,X - \tfrac{3}{4}\,(ga^2A)\,\frac{d^2X}{dx^2} - \frac{3gf}{4\pi}\,x,$$

of which the general integral is

$$X = \frac{3gfx}{4\pi(1-g)} + Be^{-\beta x} + Ce^{+\beta x};$$

where $\beta^2 = \frac{4(1-g)}{3ga^2 A}$, B and C being two arbitrary constants. Determining these by the conditions $0 = \frac{dX'''}{dx} \div X'''$ and $0 = \frac{dX''}{dx} \div X''$ we ultimately obtain

$$X = \frac{3gf}{4\pi(1-g)} \left\{ x - \frac{e^{\beta x} - e^{-\beta x}}{\beta(e^{\beta\lambda} + e^{-\beta\lambda})} \right\}.$$

But the density of the fluid at the surface of the wire, which would produce the same effect as the magnetized wire itself, is

$$-\overline{\frac{d\phi}{dw}} = \overline{\frac{d\phi}{dr}} = -\frac{1}{2} a \frac{d^2X}{dx^2} \text{ very nearly,}$$

and therefore the total quantity in an infinitely thin section whose breadth is dx, will be

$$-\pi a^2 \frac{d^2X}{dx^2} dx = \frac{3gf\beta a^2}{4(1-g)} \cdot \frac{e^{\beta x} - e^{-\beta x}}{e^{\beta\lambda} + e^{-\beta\lambda}} dx.$$

As the constant quantity f may represent the coercive force of steel or other similar matter, provided we are allowed to suppose this force the same for every particle of the mass, it is clear that when a wire is magnetized to saturation, the effort it makes to return to a natural state must, in every part, be just equal to f; and therefore, on account of its elongated form, the degree of magnetism retained by it will be equal to that which would be induced in a conducting wire of the same form by the force f, directed along lines parallel to its axis. Hence the preceding formulæ are applicable to magnetized steel wires. But it has been shown by M. BIOT (*Traité de Phy.* Tome 3, Chap. 6) from COULOMB'S experiments, that the apparent quantity of free fluid in any infinitely thin section is represented by

$$A'(\mu'^{-x} - \mu'^{+x})\, dx.$$

This expression agrees precisely with the one before deduced from theory, and gives, for the determination of the constants A' and μ', the equations

$$\beta = -\log \mu'; \quad A' = \frac{3gf\beta a^2}{4(1-g)(e^{\beta\lambda} + e^{-\beta\lambda})}.$$

The chapter in which these experiments are related, contains also a number of results, relative to the forces with which magnetized wires tend to turn towards the meridian, when retained at a given angle from it, and it is easy to prove that this force for a fine wire, whose variable section is s, will be proportional to the quantity

$$\int s dx \frac{d\phi}{dx},$$

where the wire is magnetized in any way either to saturation or otherwise, the integral extending over its whole length. But in a cylindric wire magnetized to saturation, we have, by neglecting quantities of the order a^2,

$$\frac{d\phi}{dx} = \frac{dX}{dx} = \frac{3gf}{4\pi(1-g)}\left\{1 - \frac{e^{\beta x} + e^{-\beta x}}{e^{\beta\lambda} + e^{-\beta\lambda}}\right\} \text{ and } s = \pi a^2,$$

and therefore for this wire the force in question is proportional to

$$\frac{3gfa^2}{4(1-g)}\left\{2\lambda - \frac{2(e^{\beta\lambda} - e^{-\beta\lambda})}{\beta(e^{\beta\lambda} - e^{-\beta\lambda})}\right\}.$$

The value of g, dependent on the nature of the substance of which the needles are formed, being supposed given as it ought to be, we have only to determine β in order to compare this result with observation. But β depends upon $A = 2 \log \frac{\mu}{a}$, and on account of the smallness of a, A undergoes but little alteration for very considerable variations in μ, so that we shall be able in every case to judge with sufficient accuracy what value of μ ought to be employed: nevertheless, as it is always desirable to avoid every thing at all vague, it will be better to determine A by the condition, that the sum of the squares of the errors committed by employing, as we have done, $A \frac{d^2X'}{dx'^2}$ for the approximate value of $\int_{-\lambda}^{+\lambda} \frac{dx}{\sqrt{\{(x-x')^2+a^2\}}}$, shall be a minimum for the whole length of the wire. In this way I find when λ is so great that quantities of the order $\frac{1}{\beta\lambda}$ may be neglected,

$$A = .231863 - 2 \log a\beta + 2a\beta;$$

where $.231863$ &c. $= 2 \log 2 - 2(A)$; (A) being the quantity represented by A in LACROIX' *Traité du Cal. Diff.* Tome 3, p. 521. Substituting the value of A just found in the equation

$$\beta^2 = \frac{4(1-g)}{3ga^2 A}$$

before given, we obtain

$$(a) \quad \frac{4(1-g)}{3g \,.\, a^2\beta^2} = .231863 - 2 \log a\beta + 2a\beta.$$

We hence see that when the nature of the substance of which the wires are formed remains unchanged, the quantity $a\beta$ is constant, and therefore β varies in the inverse ratio of a. This agrees with what M. BIOT has found by experiment in the chapter before cited, as will be evident by recollecting that

$$\beta = -\log \mu'.$$

From an experiment made with extreme care by COULOMB, on a magnetized wire whose radius was $\frac{1}{12}$ inch, M. BIOT has found the value of μ' to be .517948 (*Traité de Phy.* Tome 3, p. 78). Hence we have in this case

$$a\beta = \frac{-1}{12} \log \mu' = .0548235,$$

which, according to a remark just made, ought to serve for all steel wires. Substituting this value in the equation (a) of the present article, we obtain

$$g = .986636.$$

With this value of g we may calculate the forces with which different lengths of a steel wire whose radius is $\frac{1}{12}$ inch, tend to turn towards the meridian, in order to compare the results with the table of COULOMB's observations, given by M. BIOT (*Traité de Phy.* Tome 3, p. 84). Now we have before proved that this force for any wire may be represented by

$$K\left(\beta\lambda - \frac{e^{\beta\lambda} - e^{-\beta\lambda}}{e^{\beta\lambda} + e^{-\beta\lambda}}\right) = K\left(\beta\lambda - \frac{1 - e^{-2\beta\lambda}}{1 + e^{-2\beta\lambda}}\right);$$

where, for abridgment, we have supposed

$$K = \frac{3gfa^2}{2\beta(1-g)}.$$

It has also been shown that for any steel wire

$$a\beta = .0548235,$$

the French inch being the unit of space, and as in the present case $a = \frac{1}{12}$, there results $\beta = .657882$. It only remains therefore to determine K from one observation, the first for example, from which we obtain $K = 58^{\circ}.5$ very nearly; the forces being measured by their equivalent torsions. With this value of K we have calculated the last column of the following table:

Length 2λ.	Observed Torsion.	Calculated Torsion.
18 in.	288°	287°.9
12	172	172 .1
9	115	115 .3
6	59	59 .3
4.5	34	33 .9
3	13	13 .5

The last three observations have been purposely omitted, because the approximate equation (a) does not hold good for very short wires.

The very small difference existing between the observed and calculated results will appear the more remarkable, if we reflect that the value of β was determined from an experiment of quite a different kind to any of the present series, and that only one of these has been employed for the determination of the constant quantity K, which depends on f, the measure of the coercive force.

The table page 87 of the volume just cited, contains another set of observed torsions, for different lengths of a much finer wire whose radius $a = \frac{1}{12}\sqrt{\frac{38}{865}}$: hence we find the corresponding value of $\beta = 3{,}13880$, and the first observation in the table gives $K = {}^{0}.6448$. With these values the last column of the following table has been calculated as before:

Length 2λ.	Observed Torsion.	Calculated Torsion.
12 in.	11°.50	11 .50
9	8 .50	8 .46
6	5 .30	5 .43
3	2 .30	2 .39
2	1 .30	1 .38
1	.35	.42
.5	.07	.084
.25	.02	.012

Here also the differences between the observed and calculated values are extremely small, and as the wire is a very fine one, our formula is applicable to much shorter pieces than in the former case. In general, when the length of the wire exceeds 10 or 15 times its diameter, we may employ it without hesitation.

MATHEMATICAL INVESTIGATIONS

CONCERNING THE

LAWS OF THE EQUILIBRIUM OF FLUIDS

ANALOGOUS TO THE ELECTRIC FLUID,

WITH

OTHER SIMILAR RESEARCHES*.

* From the *Transactions of the Cambridge Philosophical Society*, 1833.
[Read *Nov.* 12, 1832.]

MATHEMATICAL INVESTIGATIONS CONCERNING THE LAWS OF THE EQUILIBRIUM OF FLUIDS ANALOGOUS TO THE ELECTRIC FLUID, WITH OTHER SIMILAR RESEARCHES.

Amongst the various subjects which have at different times occupied the attention of Mathematicians, there are probably few more interesting in themselves, or which offer greater difficulties in their investigation, than those in which it is required to determine mathematically the laws of the equilibrium or motion of a system composed of an infinite number of free particles all acting upon each other mutually, and according to some given law. When we conceive, moreover, the law of the mutual action of the particles to be such that the forces which emanate from them may become insensible at sensible distances, the researches to which the consideration of these forces lead will be greatly simplified by the limitation thus introduced, and may be regarded as forming a class distinct from the rest. Indeed they then for the most part terminate in the resolution of equations between the values of certain functions at any point taken at will in the interior of the system, and the values of the partial differentials of these functions at the same point. When on the contrary the forces in question continue sensible at every finite distance, the researches dependent upon them become far more complicated, and often require all the resources of the modern analysis for their successful prosecution. It would be easy so to exhibit the theories of the equilibrium and motion of ordinary fluids, as to offer instances of researches appertaining to the former class, whilst the mathematical investigations to which the theories of Electricity and Magnetism have given rise may be considered as interesting examples of such as belong to the latter class.

It is not my chief design in this paper to determine mathematically the density of the electric fluid in bodies under given circumstances, having elsewhere* given some general methods by which this may be effected, and applied these methods to a variety of cases not before submitted to calculation. My present object will be to determine the laws of the equilibrium of an hypothetical fluid analogous to the electric fluid, but of which the law of the repulsion of the particles, instead of being inversely as the square of the distance, shall be inversely as any power n of the distance; and I shall have more particularly in view the determination of the density of this fluid in the interior of conducting spheres when in equilibrium, and acted upon by any exterior bodies whatever, though since the general method by which this is effected will be equally applicable to circular plates and ellipsoids. I shall present a sketch of these applications also.

It is well known that in enquiries of a nature similar to the one about to engage our attention, it is always advantageous to avoid the direct consideration of the various forces acting upon any particle p of the fluid in the system, by introducing a particular function V of the co-ordinates of this particle, from the differentials of which the values of all these forces may be immediately deduced†. We have, therefore, in the present paper endeavoured, in the first place, to find the value of V, where the density of the fluid in the interior of a sphere is given by means of a very simple consideration, which in a great measure obviates the difficulties usually attendant on researches of this kind, have been able to determine the value V, where ρ, the density of the fluid in any element dv of the sphere's volume, is equal to the product of two factors, one of which is a very simple function

* *Essay on the Application of Mathematical Analysis to the Theories of Electricity and Magnetism.*

† This function in the present case will be obtained by taking the sum of all the molecules of a fluid acting upon p, divided by the $(n-1)^{\text{th}}$ power of their respective distances from p; and indeed the function which Laplace has represented by V in the third book of the *Mécanique Céleste,* is only a particular value of our more general one produced by writing 2 in the place of the general exponent n.

containing an arbitrary exponent β, and the remaining one f is equal to any rational and entire function whatever of the rectangular co-ordinates of the element dv, and afterwards by a proper determination of the exponent β, have reduced the resulting quantity V to a rational and entire function of the rectangular co-ordinates of the particle p, of the same degree as the function f. This being done, it is easy to perceive that the resolution of the inverse problem may readily be effected, because the coefficients of the required factor f will then be determined from the given coefficients of the rational and entire function V, by means of linear algebraic equations.

The method alluded to in what precedes, and which is exposed in the two first articles of the following paper, will enable us to assign generally the value of the induced density ρ for any ellipsoid, whatever its axes may be, provided the inducing forces are given explicitly in functions of the co-ordinates of p; but when by supposing these axes equal we reduce the ellipsoid to a sphere, it is natural to expect that as the form of the solid has become more simple, a corresponding degree of simplicity will be introduced into the results; and accordingly, as will be seen in the fourth and fifth articles, the complete solutions both of the direct and inverse problems, considered under their most general point of view, are such that the required quantities are there always expressed by simple and explicit functions of the known ones, independent of the resolution of any equations whatever.

The first five articles of the present paper being entirely analytical, serve to exhibit the relations which exist between the density ρ of our hypothetical fluid, and its dependent function V; but in the following ones our principal object has been to point out some particular applications of these general relations.

In the seventh article, for example, the law of the density of our fluid when in equilibrium in the interior of a conductory sphere, has been investigated, and the analytical value of ρ there found admits of the following simple enunciation.

The density ρ of free fluid at any point p within a conducting sphere A, of which O is the centre, is always proportional to the $(n-4)^{\text{th}}$ power of the radius of the circle formed by the intersection of a plane perpendicular to the ray Op with the surface of the sphere itself, provided n is greater than 2. When on the contrary n is less than 2, this law requires a certain modification; the nature of which has been fully investigated in the article just named, and the one immediately following.

It has before been remarked, that the generality of our analysis will enable us to assign the density of the free fluid which would be induced in a sphere by the action of exterior forces, supposing these forces are given explicitly in functions of the rectangular co-ordinates of the point of space to which they belong. But, as in the particular case in which our formulæ admit of an application to natural phenomena, the forces in question arise from electric fluid diffused in the inducing bodies, we have in the ninth article considered more especially the case of a conducting sphere acted upon by the fluid contained in any exterior bodies whatever, and have ultimately been able to exhibit the value of the induced density under a very simple form, whatever the given density of the fluid in these bodies may be.

The tenth and last article contains an application of the general method to circular planes, from which results, analogous to those formed for spheres in some of the preceding ones, are deduced; and towards the latter part, a very simple formula is given, which serves to express the value of the density of the free fluid in an infinitely thin plate, supposing it acted upon by other fluid, distributed according to any given law in its own plane. Now it is clear, that if to the general exponent n we assign the particular value 2, all our results will become applicable to electrical phenomena. In this way the density of the electric fluid on an infinitely thin circular plate, when under the influence of any electrified bodies whatever, situated in its own plane, will become known. The analytical expression which serves to represent the value of this density, is remarkable for its simplicity; and by suppressing the term due to the exterior bodies, immediately gives the density of the electric fluid on a circular conducting plate, when quite free from all extraneous action. Fortunately, the manner in which the electric fluid distributes itself in the latter case, has long since been determined experimentally by Coulomb. We have thus had the advantage of comparing our theoretical results with those of a very accurate observer, and the differences between them are not greater than may be supposed due to the unavoidable errors of experiment, and to that which would necessarily be produced by employing plates of a finite thickness, whilst the theory supposes this thickness infinitely small. Moreover, the errors are all of the same kind with regard to sign, as would arise from the latter cause.

1. If we conceive a fluid analogous to the electric fluid, but of which the law of the repulsion of the particles instead of being inversely as the square of the distance is inversely as some power n of the distance, and suppose ρ to represent the density of this fluid, so that dv being an element of the volume of a body A through which it is diffused, ρdv may represent the quantity contained in this element, and if afterwards we write g for the distance between dv and any particle p under consideration, and these form the quantity

$$V = \int \frac{\rho dv}{g^{n-1}};$$

the integral extending over the whole volume of A, it is well known that the force with which a particle p of this fluid situate in any point of space is impelled in the direction of any line q and tending to increase this line will always be represented by

$$\frac{1}{1-n}\left(\frac{dV}{dq}\right) \text{......................} (1);$$

V being regarded as a function of three rectangular co-ordinates of p, one of which co-ordinates coincides with the line q, and $\left(\frac{dV}{dq}\right)$ being the partial differential of V, relative to this last co-ordinate.

In order now to make known the principal artifices on which the success of our general method for determining the function V mainly depends, it will be convenient to begin with a very simple example.

Let us therefore suppose that the body A is a sphere, whose centre is at the origin O of the co-ordinates, the radius being 1; and ρ is such a function of x', y', z', that where we substitute for x', y', z' their values in polar co-ordinates

$$x' = r' \cos\theta', \quad y' = r' \sin\theta' \cos\varpi', \quad z' = r' \sin\theta' \sin\varpi',$$

it shall reduce itself to the form

$$\rho = (1 - r'^2)^\beta . f(r'^2);$$

f being the characteristic of any rational and entire function whatever: which is in fact equivalent to supposing

$$\rho = (1 - x'^2 - y'^2 - z'^2)^\beta . f(x'^2 + y'^2 + z'^2).$$

Now, when as in the present case, ρ can be expanded in a series of the entire powers of the quantities x', y', z', and of the various products of these powers, the function V will always admit of a similar expansion in the entire powers and products of the quantities x, y, z, provided the point p continues within the body A*, and as moreover V evidently depends on the distance $Op = r$ and is independent of θ and ϖ, the two other polar co-ordinates of p, it is easy to see that the quantity V, when we substitute for x, y, z these values

$$x = r\cos\theta, \quad y = r\sin\theta\cos\varpi, \quad z = r\sin\theta\sin\varpi,$$

will become a function of r, only containing none but the even powers of this variable.

But since we have

$$dv = r'^2\, dr'\, d\theta'\, d\varpi' \sin\theta', \quad \text{and} \quad \rho = (1 - r'^2)^\beta . f(r'^2),$$

the value of V becomes

$$V = \int \frac{\rho dv}{g^{n-1}} = \int r'^2\, dr'\, d\theta'\, d\varpi' \sin\theta' \,(1 - r'^2)^\beta f(r'^2) . g^{1-n};$$

* The truth of this assertion will become tolerably clear, if we recollect that V may be regarded as the sum of every element ρdv of the body's mass divided by the $(n-1)^{\text{th}}$ power of the distance of each element from the point p, supposing the density of the body A to be expressed by ρ, a continuous function of x', y', z'. For then the quantity V is represented by a continuous function, so long as p remains within A; but there is in general a violation of the law of continuity whenever the point p passes from the interior to the exterior space. This truth, however, as enunciated in the text, is demonstrable, but since the present paper is a long one, I have suppressed the demonstrations to save room.

the integrals being taken from $\varpi'=0$ to $\varpi'=2\pi$, from $\theta'=0$ to $\theta'=\pi$, and from $r'=0$ to $r'=1$.

Now V may be considered as composed of two parts, one V' due to the sphere B whose centre is at the origin O, and surface passes through the point p, and another V'' due to the shell S exterior to B. In order to obtain the first part, we must expand the quantity g^{1-n} in an ascending series of the powers of $\frac{r'}{r}$. In this way we get

$$g^{1-n}=[r^2-2rr'\{\cos\theta\cos\theta'+\sin\theta\sin\theta'\cos(\varpi'-\varpi)\}+r'^2]^{\frac{1-n}{2}}$$
$$=r^{1-n}.\left(Q_0+Q_1\frac{r'}{r}+Q_2\frac{r'^2}{r^2}+Q_3\frac{r'^3}{r^3}+\&c.\right).$$

If then we substitute this series for g^{1-n} in the value of V', and after having expanded the quantity $(1-r'^2)^\beta$, we effect the integrations relative to r', θ', and ϖ', we shall have a result of the form

$$V'=r^{4-n}\{A+Br^2+Cr^4+\&c.\}$$

seeing that in obtaining the part of V before represented by V', the integral relative to r' ought to be taken from $r'=0$ to $r'=r$ only.

To obtain the value of V'', we must expand the quantity g^{1-n} in an ascending series of the powers of $\frac{r}{r'}$, and we shall thus have

$$g^{1-n}=(r^2-2rr'[\cos\theta\cos\theta'+\sin\theta\sin\theta'\cos(\varpi-\varpi')]+r'^2)^{\frac{1-n}{2}}$$
$$=r'^{1-n}.\{Q_0+Q_1\frac{r}{r'}+Q_2\frac{r^2}{r'^2}+Q_3\frac{r^3}{r'^3}+\&c.\};$$

the coefficients Q_0, Q_1, Q_2, &c. being the same as before.

The expansion here given being substituted in V'', there will arise a series of the form

$$V''=T_0+T_1+T_2+T_3+\&c.$$

of which the general term T_s is

$$T_s=\int d\theta'\,d\varpi'\sin\theta'\,Q_s\int r'^2dr'\frac{r^s}{r'^{s+n-1}}(1-r'^2)^\beta.f(r'^2);$$

the integrals being taken from $r' = r$ to $r' = 1$, from $\theta' = 0$ to $\theta' = \pi$, and from $\varpi' = 0$ to $\varpi' = 2\pi$. This will be evident by recollecting that the triple integral by which the value of V'' is expressed, is the same as the one before given for V, except that the integration relative to r', instead of extending from $r' = 0$ to $r' = 1$, ought only to extend from $r' = r$ to $r = 1$.

But the general term in the function $f(r'^2)$ being represented by $A_t r'^{2t}$, the part of T_s dependent on this term will evidently be

$$A_t r \int d\theta' \, d\varpi' \sin\theta' . Q_s \int r'^{2t+3-s-n} \, dr' \, (1 - r'^2)^\beta \ldots\ldots\ldots\ldots (2);$$

the limits of the integrals being the same as before.

We thus see that the value of T_s and consequently of V'' would immediately be obtained, provided we had the value of the general integral

$$\int_r^1 r'^b \, dr' \, (1 - r'^2)^\beta,$$

which being expanded and integrated becomes

$$\left.\begin{array}{l} \dfrac{1}{b+1} - \dfrac{\beta}{1} \cdot \dfrac{1}{b+3} + \dfrac{\beta(\beta-1)}{1.2} \cdot \dfrac{1}{b+5} - \text{\&c.} \\ - \dfrac{r^{b+1}}{b+1} + \dfrac{\beta}{1} \cdot \dfrac{r^{b+3}}{b+3} - \dfrac{\beta(\beta-1)}{1.2} \cdot \dfrac{r^{b+5}}{b+5} + \text{\&c.} \end{array}\right\}$$

but since the first line of this expression is the well-known expansion of

$$\left(\frac{p}{q}\right) \text{ or } \frac{\Gamma\left(\frac{p}{n}\right) \Gamma\left(\frac{q}{n}\right)}{n\Gamma\left(\frac{p+q}{n}\right)},$$

when $n = 2 . p = b + 1$ and $q = 2(\beta + 1)$ we have ultimately,

$$\int_r^1 r'^b \, dr' \, (1 - r'^2)^\beta$$

$$= \frac{\Gamma\left(\frac{b+1}{2}\right) \Gamma(\beta + 1)}{2\Gamma\left(\frac{b+3}{2} + \beta\right)} - 1 \times \frac{r^{b+1}}{b+1} + \frac{\beta}{1} \times \frac{r^{b+3}}{b+3} - \text{\&c.} \ldots\ldots (3).$$

By means of the result here obtained, we shall readily find the value of the expression (2), which will evidently contain one term multiplied by r^s and an infinite number of others, in all of which the quantity r is affected with the exponent n. But, as in the case under consideration, n may represent any number whatever, fractionary or irrational, it is clear that none of the terms last mentioned can enter into V, seeing that it ought to contain the even powers of r only, thence the terms of this kind entering into V'' must necessarily be destroyed by corresponding ones in V'. By rejecting them, therefore, the formula (2) will become

$$\frac{\Gamma\left(t+2-\frac{s+n}{2}\right)\Gamma(\beta+1)}{2\Gamma\left(t+\beta+3-\frac{s+n}{2}\right)} A_t r^s . \int d\theta' \, d\varpi' \sin\theta' \, Q_s \ldots\ldots\ldots (2').$$

But as V ought to contain the even powers of r only, those terms in which the exponent s is an odd number, will vanish of themselves after all the integrations have been effected, and consequently the only terms which can appear in V, are of the form

$$\frac{\Gamma\left(t+2-s'-\frac{n}{2}\right)\Gamma(\beta+1)}{2\Gamma\left(t+\beta+3-s'-\frac{n}{2}\right)} A_t r^{2s'} \int d\theta' \, d\varpi' \sin\theta' \, Q_{2s'} \ldots\ldots (4);$$

where, since s is an even number, we have written $2s'$ in the place of s, and as $Q_{2s'}$ is always a rational and entire function of $\cos\theta'$, $\sin\theta'\cos\varpi'$, and $\sin\theta'\sin\varpi'$, the remaining integrations may immediately be effected.

Having thus the part of $T'_{2s'}$ due to any term $A_t r'^{2t}$ of the function $f(r'^2)$ we have immediately the value of T' and consequently of V'', since

$$V'' = U' + T_0' + T_2' + T_4' + T_6' + \&c.;$$

U' representing the sum of all the terms in V'' which have been rejected on account of their form, and $T_0'\ T_1'\ T_2'$ the value of $T_0\ T_1\ T_2$, &c. obtained by employing the truncated formula (2) in the place of the complete one (2).

But $-V = V' + V'' = V' + U' + T_0' + T_2' + T_4' + T_6' + \&c.$ or by transposition,

$$V - T_0' - T_2' - T_4' - T_6' - \&c. = V' + U';$$

and as in this equation, the function on the left side contains none but the even powers of the indeterminate quantity r, whilst that on the right does not contain any of the even powers of r, it is clear that each of its sides ought to be equated separately to zero. In this way the left side gives

$$V = T_0' + T_2' + T_4' + T_9' + \&c. \dots\dots\dots (5).$$

Hitherto the value of the exponent β has remained quite arbitrary, but the known properties of the function Γ will enable us so to determine β, that the series just given shall contain a finite number of terms only. We shall thus greatly simplify the value of V, and reduce it in fact to a rational and entire function of r^2.

For this purpose, we may remark that

$$\Gamma(0) = \infty, \ \Gamma(-1) = \infty, \ \Gamma(-2) = \infty, \ \textit{in infinitum}.$$

If therefore we make $-\frac{n}{2} + \beta =$ any whole number positive or negative, the denominator of the function (4) will become infinite, and consequently the function itself will vanish when s' is so great that $-\frac{n}{2} + \beta + t + 3 - s'$ is equal to zero or any negative number, and as the value of t never exceeds a certain number, seeing that $f(r'^2)$ is a rational and entire function, it is clear that the series (4) will terminate of itself, and V become a rational and entire function of r^2.

2. The method that has been employed in the preceding article, where the function by which the density is expressed is of the particular form

$$\rho = (1 - r'^2)^\beta . f(r'^2),$$

may, by means of a very slight modification, be applied to the far more general value

$$\rho = (1 - r'^2)^\beta f(x', y', z') = (1 - x'^2 - y'^2 - z'^2)^\beta f(x', y', z'),$$

where f is the characteristic of any rational and entire function whatever: and the same value of β which reduces V to a rational and entire function of r^2 in the first case, reduces it in the second to a similar function of x, y, z and the rectangular co-ordinates of p.

To prove this, we may remark that the corresponding value V will become

$$V = \int r'^2 dr' d\theta' d\varpi' \sin\theta' \, (1 - r'^2)^\beta f(x', y', z') \, g^{1-n};$$

the integral being conceived to comprehend the whole volume of the sphere.

Let now the function f be divided into two parts, so that

$$f(x', y', z') = f_1(x', y', z') + f_2(x', y', z');$$

f_1 containing all the terms of the function f, in which the sum of the exponents of x', y', z' is an odd number; and f_2 the remaining terms, or those where the same sum is an even number. In this way we get

$$V = V_1 + V_2;$$

the functions V_1 and V_2 corresponding to f_1 and f_2, being

$$V_1 = \int r'^2 dr' d\theta' d\varpi' \sin\theta' \, (1 - r'^2)^\beta f_1(x', y', z') \, g^{1-n},$$

$$V_2 = \int r'^2 dr' d\theta' d\varpi' \sin\theta' \, (1 - r'^2)^\beta f_2(x', y', z') \, g^{1-n}.$$

We will in the first place endeavour to determine the value V_1; and for this purpose, by writing for x', y', z' their values before given in r', θ', ϖ', we get

$$f_1(x', y', z') = r' \psi(r'^2);$$

the coefficients of the various powers of r'^2 in $\psi(r'^2)$ being evidently rational and entire functions of $\cos\theta'$, $\sin\theta' \cos\varpi'$, and $\sin\theta' \sin\varpi'_2$. Thus

$$V_1 = \int r'^2 dr' d\theta' d\varpi' \sin\theta' \, (1 - r'^2)^\beta r' \psi(r'^2) \, g^{1-n};$$

this integral, like the foregoing, comprehending the whole volume of the sphere.

Now as the density corresponding to the function V_1 is

$$\rho_1 = (1 - x'^2 - y'^2 - z'^2)^\beta f_1(x', y', z'),$$

it is clear that it may be expanded in an ascending series of the entire powers of x', y', z', and the various products of these powers consequently, as was before remarked (Art. 1), V_1 ad-

9

mits of an analogous expansion in entire powers and products of x, y, z. Moreover, as the density ρ_1 retains the same numerical value, and merely changes its sign when we pass from the element dv to a point diametrically opposite, where the co-ordinates x', y', z' are replaced by $-x'$, $-y'$, $-z'$: it is easy to see that the function V_1, depending upon ρ_1, possesses a similar property, and merely changes its sign when x, y, z, the co-ordinates of p, are changed into $-x$, $-y$, $-z$. Hence the nature of the function V_1 is such that it can contain none but the odd powers of r, when we substitute for the rectangular co-ordinates x, y, z, their values in the polar co-ordinates r, θ, ϖ.

Having premised these remarks, let us now suppose V_1 is divided into two parts, one V_1' due to the sphere B which passes through the particle p, and the other V_1'' due to the exterior shell S. Then it is evident by proceeding, as in the case where $\rho = (1 - r'^2)^\beta f(r^2)$, that V_1' will be of the form

$$V_1' = r^{5-n} \{A + Br^2 + Cr^4 + \&c.\};$$

the coefficients A, B, C, &c. being quantities independent of the variable r.

In like manner we have also

$$V_1'' = \int r'^2 dr' d\theta' d\varpi' \sin\theta' (1 - r'^2)^\beta . r'\psi(r'^2)\, g^{1-n};$$

the integrals being taken from $r' = r$ to $r = 1$, from $\theta' = 0$ to $\theta' = \pi$, and from $\varpi' = 0$ to $\varpi' = 2\pi$.

By substituting now the second expansion of g^{1-n} before used (Art. 1), the last expression will become

$$V_1'' = T_0 + T_1 + T_2 + T_3 + \&c.$$

of which series the general term is

$$T_s = \int d\theta' d\varpi' \sin\theta'\, Q_s \int r'^{4-n} dr' (1 - r'^2)^\beta \frac{r^s}{r'^s} \psi(r'^2).$$

Moreover, the general term of the function $\psi(r'^2)$ being represented by $A_t r'^{2t}$, the portion of T_s due to this term will be

$$r^s \int d\theta' d\varpi' \sin\theta'\, Q_s A^t \int r'^{4-n+2t-s} dr' (1 - r'^2)^\beta \;\ldots\ldots\ldots (a);$$

the limits of the integrals being the same as before.

If now we effect the integrations relative to r' by means of the formula (3), Art. 1, and reject as before those powers of the variable r, in which it is affected, with the exponent n, since these ought not to enter into the function V_1, the last formula will become

$$\frac{\Gamma\left(\frac{5-n+2t-s}{2}\right)\Gamma(\beta+1)}{2\Gamma\left(\frac{7+2\beta-n+2t-s}{2}\right)} r^s \int d\theta' d\varpi' \sin\theta' \, Q_s A_t \, \ldots\ldots\ldots (a'),$$

and as V_1 ought to contain none but the odd powers of r, we may make $s = 2s' + 1$, and disregard all those terms in which s is an even number, since they will necessarily vanish after all the operations have been effected. Thus the only remaining terms will be of the form

$$\frac{\Gamma\left(\frac{4-n+2t-2s'}{2}\right)\Gamma(\beta+1)}{2 . \Gamma\left(\frac{6+2\beta-n+2t-2s'}{2}\right)} r^{2s'+1} \int d\theta' d\varpi' \sin\theta' \; Q_{2s'+1} A_t;$$

where, as A_t and $Q_{2s'+1}$ are both rational and entire functions of $\cos\theta'$, $\sin\theta' \cos\varpi'$, $\sin\theta' \sin\varpi'$, the remaining integrations from $\theta' = 0$ to $\theta' = \pi$, and $\varpi' = 0$ to $\varpi' = 2\pi$, may easily be effected in the ordinary way.

If now we follow the process employed in the preceding article, and suppose T_0', T_1', T_2', &c. are what T_0, T_1, T_2, &c. become when we use the truncated formula (a') instead of the complete one (a), we shall readily get

$$V_1 = T_1' + T_3' + T_5' + T_7' + \&c.$$

In like manner, from the value of V_2 before given, we get

$$V_2'' = \int r'^2 dr' d\theta' d\varpi' \sin\theta' \, (1 - r'^2)^\beta \, \phi(r'^2) \, g^{1-n};$$

the integrals being taken from $r' = r$ to $r = 1$, from $\theta' = 0$ to $\theta' = \pi$, and from $\varpi = 0$ to $\varpi = 2\pi$.

Expanding now g^{1-n} as before, we have

$$V_2'' = U_0 + U_1 + U_2 + U_3 + \&c.$$

where

$$U_s = \int d\theta' d\varpi' \sin\varpi' Q_s \int_r^1 r'^{3-n} dr' (1 - r'^2)^\beta \frac{r^s}{r'^s} \phi(r'^2),$$

and the part of U_s due to the general term $B_t r'^{2t}$ in $\phi(r'^2)$, will be

$$r \int d\theta' d\varpi' \sin\theta' Q_s B_t \int_r^1 r'^{3-n+2t-s} dr' (1 - r'^2)^\beta \ldots\ldots\ldots (b);$$

which, by employing the formula (3') Art. 1, and rejecting the inadmissible terms, gives for truncated formula

$$\frac{\Gamma\left(\frac{4-n+2t-s}{2}\right)\Gamma(\beta+1)}{2\Gamma\left(\frac{6-n+2\beta+2t-2s}{3}\right)} r^s \int d\theta' d\varpi' \sin\theta' Q_s B_t \ldots\ldots\ldots (b').$$

By continuing to follow exactly the same process as was before employed in finding the value of V_1, we shall see that s must always be an even number, say $2s'$; and thus the expression immediately preceding will become

$$\frac{\Gamma\left(\frac{4-n+2t-2s'}{2}\right)\Gamma(\beta+1)}{2\Gamma\left(\frac{6-n+2\beta+2t-2s'}{2}\right)} r^{2s'} \int d\theta' d\varpi' \sin\theta' Q_{2s'} B_t.$$

Moreover, the value of V_2 will be

$$V_2 = U_0' + U_2' + U_4' + U_6' + \&c.;$$

U_0', U_1', U_2', U_3', &c. being what U_0, U_1, U_2, &c. become when we use the formula (b') instead of the complete one (b).

The value of V answering to the density

$$\rho = \rho_1 + \rho_2 = (1 - r'^2)^\beta . f(x', y', z'),$$

by adding together the two parts into which it was originally divided, therefore, becomes

$$V = V_1 + V_2 = T_1' + T_3' + T_5' + T_7' + \&c.$$
$$+ U_0' + U_2' + U_4' + U_6' + \&c.$$

When β is taken arbitrarily, the two series entering into V extend *in infinitum*, but by supposing as before, Art. 1,

$$\frac{-n}{2} + \beta = \omega;$$

ω representing any whole number, positive or negative, it is clear from the form of the quantities entering into $T_{2s'+1}$ and $U_{2s'}$, and from the known properties of the function Γ, that both these series will terminate of themselves, and the value of V be expressed in a finite form; which, by what has preceded, must necessarily reduce itself to a rational and entire function of the rectangular co-ordinates x, y, z. It seems needless, after what has before been advanced (Art. 1), to offer any proof of this: we will, therefore, only remark that if γ represents the degree of the function $f(x', y', z')$, the highest degree to which V can ascend will be

$$\gamma + 2\omega + 4.$$

In what immediately precedes, ω may represent any whole number whatever, positive or negative; but if we make $\omega = -2$, and consequently, $\beta = \frac{n-4}{2}$, the degree of the function V is the same as that of the factor

$$f(x', y', z'),$$

comprised in ρ. This factor then being supposed the most general of its kind, contains as many arbitrary constant quantities as there are terms in the resulting function V. If, therefore, the form of the rational and entire function V be taken at will, the arbitrary quantities contained in $f(x', y', z')$ will in case $\omega = -2$ always enable us to assign the corresponding value of ρ, and the resulting value of $f(x', y', z')$ will be a rational and entire function of the same degree as V. Therefore, in the case now under consideration, we shall not only be able to determine the value of V when ρ is given, but shall also have the means of solving the inverse problem, or of determining ρ when V is given; and this determination will depend upon the resolution of a certain number of algebraical equations, all of the first degree.

3. The object of the preceding sketch has not been to point out the most convenient way of finding the value of the function V, but merely to make known the spirit of the method; and to show on what its success depends. Moreover, when presented in this simple form, it has the advantage of being, with a very slight modification, as applicable to any ellipsoid whatever as to the sphere itself. But when spheres only are to be considered, the resulting formulæ, as we shall afterwards show, will be much more simple if we expand the density ρ in a series of functions similar to those used by Laplace (*Mec. Cel.* Liv. iii.): it will however be advantageous previously to demonstrate a general property of functions of this kind, which will not only serve to simplify the determination of V, but also admit of various other applications of $d\sigma$.

Suppose, therefore, $Y^{(i)}$ is a function of θ and ϖ, of the form considered by Laplace (*Mec. Cel.* Liv. iii.), r, θ, ϖ being the polar co-ordinates referred to the axes X, Y, Z, fixed in space, so that

$$x = r\cos\theta, \quad y = r\sin\theta\cos\varpi, \quad z = r\sin\theta\sin\varpi;$$

then, if we conceive three other fixed axes X_1, Y_1, Z_1, having the same origin but different directions, $Y^{(i)}$ will become a function of θ_1 and ϖ_1, and may therefore be expanded in a series of the form

$$Y^{(r)} = Y_1^{(0)} + Y_1^{(1)} + Y_1^{(2)} + Y_1^{(3)} + \&c. \;\ldots\ldots\ldots\ldots (6).$$

Suppose now we take any other point p' and mark its various co-ordinates with an accent, in order to distinguish them from those of p; then, if we designate the distance pp' by (p, p'), we shall have

$$\frac{1}{(p,p')} = [r^2 - 2rr'\{\cos\theta\cos\theta' + \sin\theta\sin\theta'\cos(\varpi - \varpi')\} + r'^2]^{-\frac{1}{2}}$$

$$= \frac{1}{r}\left(Q^{(0} + Q^{(1)}.\frac{r'}{r} + Q^{(2)}\frac{r'^2}{r^2} + Q^{(3)}\frac{r'^3}{r^3} + \&c.\right),$$

as has been shown by Laplace in the third book of the *Mec. Cel.*, where the nature of the different functions here employed is completely explained.

In like manner, if the same quantity is expressed in the polar co-ordinates belonging to the new system of axes X_1, Y_1, Z_1, we have, since the quantities r and r' are evidently the same for both systems,

$$\frac{1}{(p, p')} = \frac{1}{r}\left(Q_1^{(0)} + Q_1^{(1)}\frac{r'}{r} + Q_1^{(2)}\frac{r'^2}{r^2} + Q_1^{(3)}\frac{r'^3}{r^3} + \&c.\right);$$

and it is also evident from the form of the radical quantity of which the series just given are expansions, that whatever number i may represent, $Q_1^{(i)}$ will be immediately deduced from $Q^{(i)}$ by changing θ, ϖ, θ', ϖ' into θ_1, ϖ_1, θ_1', ϖ_1'. But since the quantity $\frac{r'}{r}$ is indeterminate, and may be taken at will, we get, by equating the two values of $\frac{1}{(p, p')}$ and comparing the like powers of the indeterminate quantity $\frac{r'}{r}$,

$$Q^{(i)} = Q_1^{(i)}.$$

If now we multiply the equation (6) by the element of a spherical surface whose radius is unity, and then by $Q^{(h)} = Q_1^{(h)}$, we shall have, by integrating and extending the integration over the whole of this spherical surface,

$$\int d\mu d\varpi\, Q^{(h)}\, Y^{(i)} = \int d\mu_1 d\varpi_1\, Q_1^{(h)}\, \{Y_1^{(0)} + Y_1^{(1)} + Y_1^{(2)} + \&c.\}.$$

Which equation, by the known properties of the functions $Q^{(h)}$ and $Y^{(i)}$, reduces itself to

$$0 = \int d\mu_1 d\varpi_1\, Q_1^{(h)}\, Y_1^{(h)},$$

when h and i represent different whole numbers. But by means of a formula given by Laplace (*Mec. Cel.* Liv. iii. No. 17) we may immediately effect the integration here indicated, and there will thus result

$$0 = \frac{4\pi}{2h+1}\, Y_1'^{(h)};$$

$Y_1'^{(h)}$ being what $Y_1^{(h)}$ becomes by changing θ_1, ϖ_1 into θ_1', ϖ_1', and as the values of these last co-ordinates, which belong to p', may be taken arbitrarily like the first, we shall have generally

$Y_1^{(h)}$, except when $h = i$. Hence, the expansion (6) reduces itself to a single term, and becomes

$$Y^{(i)} = Y_1^{(i)}.$$

We thus see that the function $Y^{(i)}$ continues of the same form even when referred to any other system of axes X_1, Y_1, Z_1, having the same origin O with the first.

This being established, let us conceive a spherical surface whose center is at the origin O of the co-ordinates and radius r', covered with fluid, of which the density $\rho = Y'^{(i)}$; then, if $d\sigma'$ represent any element of this surface, and we afterwards form the quantity

$$V = \int \rho d\sigma' \psi(g^2) = \int Y'^{(i)} d\sigma' \psi(g^2);$$

the integral extending over the whole spherical surface, g being the distance p, $d\sigma'$ and ψ the characteristic of any function whatever. I say, the resulting value of V will be of the form

$$V = Y^{(i)} R;$$

R being a function of r, the distance Op only and $Y^{(i)}$ what $Y'^{(i)}$ becomes by changing θ', ϖ', the polar co-ordinates, into θ, ϖ, the like co-ordinates of the point p.

To justify this assertion, let there be taken three new axes X_1, Y_1, Z_1, so that the point p may be upon the axis X_1; then, the new polar co-ordinates of $d\sigma'$ may be written r', θ', ϖ', those of p being r, 0, ϖ, and consequently, the distance will become

$$g = \sqrt{(r'^2 - 2rr' \cos \theta_1' + r^2)};$$

and as $d\sigma' = r'^2 d\theta_1' d\varpi_1' \sin \theta_1'$, we immediately obtain

$$V = \int Y'^{(2)} r'^2 d\theta_1 d\varpi_1 \sin \theta_1 \psi (r^2 - 2rr' \cos \theta_1' + r'^2)$$

$$= r'^2 \int_0^\pi d\theta_1' \sin \theta_1' \psi (r^2 - 2rr' \cos \theta_1' + r'^2) \int_0^{2\pi} d\varpi_1' Y'^{(i)}.$$

Let us here consider more particularly the nature of the integral

$$\int_0^{2\pi} d\varpi_1' Y'^{(i)}.$$

In the preceding part of the present article, it has been

shown that the value of $Y'^{(i)}$, when expressed in the new co-ordinates, will be of the form $Y_1'^{(i)}$; but all functions of this form (Vide *Mec. Cel.* Liv. iii.) may be expanded in a finite series containing $2i+1$ terms, of which the first is independent of the angle ϖ_1', and each of the others has for a factor a sine or cosine of some entire multiple of this same angle. Hence, the integration relative to ϖ_1' will cause all the last mentioned terms to vanish, and we shall only have to attend to the first here. But this term is known to be of the form

$$k\left(\mu_1'^i - \frac{i.i-1}{2.2i-1}\mu_1'^{i-2} + \frac{i.i-1.i-2.i-3}{2.4.2i-1.2i-3}m_1'^{i-4} - \&c.\right),$$

and consequently, there will result

$$\int_0^{2\pi} d\varpi_1' Y'^{(i)} = 2\pi k\left(\mu_1'^i - \frac{i.i-1}{2.2i-1}\mu_1'^{i-2} + \frac{i.i-1.i-2.i-3}{2.4.2i-1.2i-3}\mu_1'^{i-4} - \&c.\right);$$

where $\mu_1' = \cos\theta_1'$ and k is a quantity independent of θ_1' and ϖ_1', but which may contain the co-ordinates θ, ϖ, that serve to define the position of the axis X_1 passing through the point p.

It now only remains to find the value of the quantity k, which may be done by making $\theta_1' = 0$, for then the line r coincides with the axis X_1, and $Y^{(i)}$ during the integration remains constantly equal to $Y^{(i)}$, the value of the density at this axis. Thus we have

$$2\pi Y^{(i)} = 2\pi k\left(1 - \frac{i.i-1}{2.2i-1} + \frac{i.i-1.i-2.i-3}{2.4.2i-1.2i-3} - \&c.\right):$$

or, by summing the series within the parenthesis, and supplying the common factor 2π,

$$Y^{(i)} = \frac{1.2.3\ldots\ldots\ldots i}{1.3.5\ldots\ldots\ldots 2i-1}k,$$

and, by substituting the value of k, drawn from this equation in the value of the required integral given above, we ultimately obtain

$$\int_0^{2\pi} d\varpi_1' Y'^{(i)} = 2\pi Y^{(i)}\frac{1.3.5\ldots\ldots 2i-1}{1.2.3\ldots\ldots i}\left(\mu_1'^i - \frac{i.i-1}{2.2i-1}\mu_1'^{i-2} + \&c.\right).$$

If now, for abridgement, we make

$$(i) = \mu_1'^{i} - \frac{i.i-1}{2.2i-1}\mu_1'^{i-2} + \frac{i.i-1.i-2.i-3}{2.4.2i-1.2i-3}\mu_1'^{i-4} - \&c.$$

we shall obtain, by substituting the value of the integral just found in that of V before given,

$$V = Y^{(i)}.2\pi r'^2 \frac{1.3.5......2i-1}{1.2.3...... \quad i} \int_{-1}^{1} d\mu_1' (i) \psi (r^2 - 2rr' \mu_1' + r'^2);$$

which proves the truth of our assertion.

From what has been advanced in the preceding article, it is likewise very easy to see that if the density of the fluid within a sphere of any radius be every where represented by

$$\rho = Y'^{(i)} \phi (r);$$

ϕ being the characteristic of any function whatever; and we afterwards form the quantity

$$V = \int \rho dv \psi (g^2),$$

where dv represents an element of the sphere's volume, and g the distance between dv and any particle p under consideration, the resulting value of V will always be of the form

$$V = Y^{(i)}.R;$$

$Y^{(i)}$ being what $Y'^{(i)}$ becomes by changing $\theta_{\prime}$, ϖ', the polar co-ordinates of the element dv into θ, ϖ, the co-ordinates of the point p; and R being a function of r, the remaining co-ordinate of p, only.

4. Having thus demonstrated a very general property of functions of the form $Y^{(i)}$, let us now endeavour to determine the value of V for a sphere whose radius is unity, and containing fluid of which the density is every where represented by

$$\rho = (1 - x'^2 - y'^2 - z'^2)^{\beta} f(x', y', z');$$

x', y', z' being the rectangular co-ordinates of dv, an element of the sphere's volume, and f, the characteristic of any rational and entire function whatever.

For this purpose we will substitute in the place of the co-ordinates x', y', z' their values

$$x' = r' \cos\theta' : \quad y' = r' \sin\theta' \cos\varpi' : \quad z' = r' \sin\theta' \sin\varpi' ;$$

and afterwards expand the function $f(x', y', z')$ by Laplace's simple method (*Mec. Cel.* Liv. iii. No. 16). Thus,

$$f(x', y', z') = f'^{(0)} + f'^{(1)} + f'^{(2)} + \&c. \ldots\ldots + f'^{(s)} \ldots\ldots\ldots (7);$$

s being the degree of the function $f(x', y', z')$.

It is likewise easy to perceive that any term $f^{(i)}$ of this expansion may be again developed thus,

$$f'^{(i)} = f_0'^{(i)} r'^i + f_1'^{(i)} r'^{i+2} + f_2'^{(i)} r'^{i+4} + \&c.;$$

and as every coefficient of the last developement is of the form $U^{(i)}$, (*Mec. Cel.* Liv. iii.), it is easy to see that the general term $f'^{(i)} r'^{i+2t}$ may always be reduced to a rational and entire function of the original co-ordinates x', y', z'.

If now we can obtain the part of V due to the term

$$f_t'^{(i)} \,.\, r'^{i+2t},$$

we shall immediately have the value of V by summing all the parts corresponding to the various values of which i and t are susceptible. But from what has before been proved (Art. 3), the part of V now under consideration must necessarily be of the form $Y^{(i)}$; representing, therefore, this part by $V_t^{(i)}$, we shall readily get

$$V_t^{(i)} = \int_0^1 r'^{i+2t+2} dr' \,(1 - r'^2)^\beta \int d\varpi' \, d\theta' \sin\theta' f_t'^{(i)} g^{1-n}.$$

Moreover from what has been shown in the same article, it is easy to see that we have generally

$$\int Y'^{(i)} d\varpi' \, d\theta' \sin\theta' \, \psi(g^2)$$
$$= 2\pi Y^{(i)} \frac{1.3.5.\ldots 2i-1}{1.2.3.\ldots i} \int_{-1}^{'} d\mu_1' \,(i)\, \psi(r^2 - 2rr'\mu_1' + r'^2);$$

ψ being the characteristic of any function whatever, and $Y^{(i)}$ what $Y'^{(i)}$ becomes by substituting θ, ϖ the polar co-ordinates of p in the place of θ', ϖ', the analogous co-ordinates of the element *dv*. If therefore in the expression immediately preceding, we make

$$Y'^{(i)} = f_i'^{(i)} \text{ and } \psi(g^2) = g^{n-1} = (g^2)^{\frac{1-n}{2}},$$

and substitute the value of the integral thus obtained for its equal in $V_i^{(i)}$ there will arise

$$V_i^{(i)} = 2\pi f_i^{(i)} \frac{1.3.5\ldots 2i-1}{1.2.3\ldots\ i} \int_0^1 r'^{i+2t+2} dr' (1-r'^2)^\beta \int_{-1}^1 d\mu'_1 (i)$$

$$. (r^2 - 2rr'\mu'_1 + r^2)^{\frac{1-n}{2}} \ldots\ldots\ldots (8);$$

where $f_i^{(i)}$ is deduced from $f_i'^{(i)}$ by changing θ', ϖ' into θ, ϖ, and (i), for abridgement, is written in the place of the function

$$\mu_1'^i - \frac{i.i-1}{2.2i-1}\mu_1'^{i-2} + \frac{i.i-1.i-2.i-3}{2.4.2i-1.2i-3}\mu_1'^{i-4} - \&c.$$

As the integral relative to μ'_1 which enters into the expression on the right side of the equation (8) is a definite one, and depends therefore on the two extreme values of μ'_1 only, it is evident that in the determination of this integral, it is altogether useless to retain the accents by which μ'_1 is affected. But by omitting these superfluous accents, we shall have to calculate the value of the quantity

$$\int_{-1}^1 d\mu\ (i) . (r^2 - 2rr'\mu + r'^2)^{\frac{1-n}{2}};$$

where

$$(i) = \mu^i - \frac{i.i-1}{2.2i-1}\mu^{i-2} + \frac{i.i-1.i-2.i-3}{2.4.2i-1.2i-3}\mu^{i-4} - \&c.$$

The method which first presents itself for determining the value of the integral in question, is to expand the quantity $(r^2 - 2rr'\mu + r'^2)^{\frac{1-n}{2}}$ by means of the Binomial Theorem, to replace the various powers of μ by their values in functions similar to (i) and afterwards to effect the integrations by the formulæ contained in the third Book of the *Mec. Cel.* For this purpose we have the general equation

$$\mu^i = (i) + \frac{i.i-1}{2.2i-1}(1-2) + \frac{i.i-1.i-2.i-3}{2.4.2i-3.2i-5}(i-4)$$

$$+ \frac{i.i-1.i-2.i-3.i-4.i-5}{2.4.6.2i-5.2i-7.2i-9}(i-6) + \&c.\ldots\ldots (9).$$

To remove all doubt of the correctness of this equation, we may multiply each of its sides by μ, and reduce the products on the right by means of the relation

$$\mu\,(i) = (i+1) + \frac{i^2}{2i-1\,.\,2i+1}\,(i-1),$$

which it is very easy to prove exists between functions of the form (i). In this way it will be seen that if the equation (9) holds good for any power μ^i it will do so likewise for the following power μ^{i+1}, and as it is evidently correct when $i=1$, it is therefore necessarily so, whatever whole number i may represent.

Now by means of the Binomial Theorem, we obtain when $r < r'$

$$r'^{n-1}\,.\,(r^2 - 2rr'\mu + r'^2)^{\frac{1-n}{2}} = \left(1 - 2\mu\frac{r}{r'} + \frac{r^2}{r'^2}\right)^{\frac{1-n}{2}}$$

$$= \Sigma_0^\infty \frac{n-1\,.\,n+1\,.\,n+3\ldots\ldots n+2s-3}{2\;.\;4\;.\;6\;\ldots\ldots\;2s}\left(2\mu\frac{r}{r'} - \frac{r^2}{r'^2}\right)^s.$$

If now we conceive the quantity $\left(2\mu\frac{r}{r'} - \frac{r^2}{r'^2}\right)^s$ to be expanded by the same theorem, it is easy to perceive that the term having $\left(\frac{r}{r'}\right)^{i+2t'}$ for factor is

$$\frac{n-1\,.\,n+1\,.\,n+3\ldots\ldots n+2i+4t'-3}{2\;.\;4\;.\;6\;\ldots\ldots\;2i+4t'}\,2^{i+2t'}\mu^{i+2t'}\left(\frac{r}{r'}\right)^{i+2t'}$$

$$-\frac{n-1\,.\,n+1\ldots n+2i+4t'-5}{2\;.\;4\;\ldots\;2i+4t-2}\,.\,2^{i+2t'-2}\mu^{i+2t'-2}\left(\frac{r}{r'}\right)^{i+2t-2}\frac{r^2}{r'^2}\,.\,\frac{i+2t-1}{1}$$

$$+\frac{n-1\,.\,n+1\ldots n+2i+4t'-7}{2\;.\;4\;\ldots\;2i+4t'-4}\,(2\mu)^{i+2t'-4}\left(\frac{r}{r'}\right)^{i+2t'-4}\frac{r^4}{r'^4}\,.\,\frac{i+2t-2\,.\,i+2t'-3}{1\;.\;2}$$

$$-\ \&\text{c}\ldots\ldots\ldots\ldots\&\text{c}\ldots\ldots\ldots\ldots\&\text{c}\ldots\ldots$$

and therefore the coefficient of $\left(\frac{r}{r'}\right)^{i+2t'}$ in the expansion of the function

$$\left(1 - 2\mu\frac{r}{r'} + \frac{r^2}{r'^2}\right)^{\frac{1-n}{2}},$$

will be expressed by

$$\Sigma \frac{n-1 \,.\, n+1 \ldots\ldots n+2i+4t'-2s-3}{2 \;.\; 4 \;\ldots\ldots\; 2i+4t'-2s} (2\mu)^{i+2t+2s} \,.\, (-1)^s$$

$$\cdot \frac{i+2t'-s \,.\, i+2t'-s-1 \ldots\ldots i+2t'-2s-1}{1 \;.\; 2 \;\ldots\ldots\; s} \,.$$

Hence the portion of this coefficient containing the function (i), when the various powers of μ have been replaced by their values in functions of this kind agreeably to the preceding observation will be found, by means of the equation (9), to be

$$(i)\,\Sigma \frac{n-1 \,.\, n+1 \ldots\ldots n+2i+4t'-2s-3}{2 \;.\; 4 \;\ldots\ldots\; 2i+4t'-2s}$$

$$\times \frac{i+2t'-2s \,.\, i+2t'-2s-1 \ldots\ldots i+1}{2 \,.\, 4 \ldots 2t'-2s \times 2i+2t'-2s+1 \,.\, 2i+2t'-2s-1 \ldots 2i+3}$$

$$\ldots\ldots 2^{i+2t'-2s} (-1)^s \times \frac{i+2t'-s \,.\, i+2t'-s-1 \ldots\ldots i+2t'-2s+1}{1 \;.\; 2 \;.\; 3 \;\ldots\ldots\; s}$$

$$= (i)\,\Sigma \frac{n-1 \,.\, n+1 \,.\, n+3 \ldots\ldots n+2i+4t'-2s-3}{2 \;.\; 4 \;.\; 6 \;\ldots\ldots\; 2i+4t'-2s} \,.\, 2^{i+2t'-2s} (-1)^s$$

$$\ldots\ldots \times \frac{i+1 \,.\, i+2 \,.\, i+3 \,.\, i+4 \ldots\ldots i+2t'-s}{1 \,.\, 2 \,.\, 3 \ldots s \times 2 \,.\, 4 \,.\, 6 \ldots 2t'-2s \times 2i+2t'-2s+1 \ldots 2i+3}$$

$$= 2^i \,.\, (i) \,.\, \Sigma \frac{(-1)^s \,.\, n-1 \,.\, n+1 \,.\, n+3 \ldots\ldots n+2i+4t'-2s-3}{2 \,.\, 4 \ldots 2i \times 2 \,.\, 4 \ldots 2s \times 2 \,.\, 4 \ldots 2t'-2s \times 2i+2t'-2s+1 \ldots 2i+3}$$

$$= \frac{3 \,.\, 5 \,.\, 7 \ldots\ldots 2i+1}{1 \,.\, 2 \,.\, 3 \ldots\ldots i} (i) \times \frac{n-1 \,.\, n+1 \,.\, n+3 \ldots\ldots n+2i+2t'-3}{3 \;.\; 5 \;.\; 7 \;\ldots\ldots\; 2i+2t'+1}$$

$$\ldots\ldots \times \Sigma \,.\, (-1)^s \frac{n+2i+2t'-1 \ldots\ldots n+2i+4t'-2s-3}{2 \,.\, 4 \,.\, 6 \;\ldots\ldots\; 2s}$$

$$\times \frac{2i+2t'-2s+3 \ldots\ldots 2i+2t'+1}{2 \,.\, 4 \,.\, 6 \;\ldots\ldots\; 2s},$$

where all the finite integrals may evidently be extended from $s = 0$ to $s = \infty$, and it is clear that the last of these integrals is equal to the coefficient of $x^{t'}$ in the product

$$\{1 + \frac{n+2i+2t'-1}{2} x + \frac{n+2i+2t'-1 \,.\, n+2i+2t'+1}{2 \,.\, 4} x^2 + \text{\&c. } in\ inf.\}$$

$$\times \{1 - \frac{2i+2t'+1}{2} x + \frac{2i+2t'+1 \,.\, 2i+2t'-1}{2 \,.\, 4} x^2 - \text{\&c. } in\ inf.\}$$

If now we write in the place of the series their known values, the preceding product will become

$$(1-x)^{-\frac{n+2i+2t'-1}{2}} \times (1-x)^{\frac{2i+2t'+1}{2}} = (1-x)^{\frac{2-n}{2}},$$

and consequently the value of the required coefficient of $x^{t'}$ is

$$\frac{n-2\,.\,n\,.\,n+2\ldots\ldots\ldots n+2t'-4}{2\ \ .\,4\,.\ \ 6\ \ \ldots\ldots\ldots\ \ \ 2t'}.$$

This quantity being substituted in the place of the last of the finite integrals gives for the value of that portion of the coefficient of

$$\left(\frac{r}{r'}\right)^{i+2t'} \text{ in } \left(1-2\mu\frac{r}{r'}+\frac{r^2}{r'^2}\right)^{\frac{1-n}{2}},$$

which contains the function (i) the expression

$$\frac{3.5.7\ldots 2i+1}{1.2.3\ldots\ \ i} \times \frac{n-1.n+1\ldots n+2i+2t'-3}{3\ \ .\ 5\ \ \ldots\ \ 2i+2t'+1} \times \frac{n-2.n\ldots n+2t'-4}{2\ \ .4\ldots\ \ \ 2t'}\,(i).$$

By multiplying the last expression by $\left(\frac{r}{r'}\right)^{i+2t'}$, and taking the sum of all the resulting values which arise when we make successively

$t' = 0, 1, 2, 3, 4, 5, 6$, &c. *in infinitum*,

we shall obtain the value of the term $Y^{(i)}$ contained in the expression

$$\left(1-2\mu\frac{r'}{r'}+\frac{r^2}{r'^2}\right)^{\frac{1-n}{2}} = Y^{(0} + Y^{(1)} + Y^{(2)} + Y^{(3)} + \&c.$$

Hence

$$Y^{(i)} = \frac{3\,.\,5\ldots 2i+1}{1\,.\,2\ldots\ \ \ i}\,(i)\,\Sigma\,\frac{n-1\,.\,n+1\ldots n+2i+2t'-3}{3\,.\,5\ldots 2i+2t'+1}$$

$$\times\,\frac{n-2\,.\,n\ldots n+2t'-4}{2\,.\,4\ldots 2t'}\left(\frac{r}{r'}\right)^{i+2t'};$$

the finite integral extending from $t'=0$ to $t'=\infty$.

But by the known properties of functions of this kind, we have by substituting for $Y^{(i)}$ its value

$$\int_{-1}^{1} d\mu\,(i)\left(1-2\mu\frac{r}{r'}+\frac{r^2}{r'^2}\right)^{\frac{1-n}{2}}=\int_{-1}^{1} d\mu\,(i)\,.\,Y^{(i)}$$

$$=\frac{3.5.7\ldots 2i+1}{1.2.3\ldots\quad i}\int d\mu\,(i)^2\times\Sigma\,\frac{n-1.n+1\ldots n+2i+2t'-3}{3.5\ldots 2i+2t'+1}$$

$$\times\frac{n-2.n\ldots n+2t'-4}{2.4\ldots 2t'}\left(\frac{r}{r'}\right)^{i+2t'}$$

$$=2\,\frac{1.2.3\ldots\quad i}{1.3.5\ldots 2i-1}\,\Sigma\,\frac{n-1.n+1\ldots n+2i+2t'-3}{3.5.\ldots 2i+2t'+2}$$

$$\times\frac{n-2.n\ldots n+2t'-4}{2.4\ldots 2t'}\left(\frac{r}{r'}\right)^{i+2t'},$$

since by what Laplace has shown (*Mec. Cel.* Liv. iii. No. 17)

$$\int_{-1}^{1} d\mu\,(i)^2=\frac{2}{2i+1}\left(\frac{1.2.3\ldots\quad i}{1.3.5\ldots 2i-1}\right)^2.$$

If now we restore to μ the accents with which it was originally affected, and multiply the resulting quantity by r'^{n-1}, we shall have when $r<r'$

$$\int_{-1}^{1} d\mu'_1\,(i)\,(r^2-2rr'\mu'_1+r'^2)^{\frac{1-n}{2}}$$

$$=r'^{n-1}\int_{-1}^{1} d\mu'_1\,(i)\left(1-2\mu'_1\frac{r}{r'}+\frac{r^2}{r'^2}\right)^{\frac{1-n}{2}}$$

$$=2\,.\,r'^{1-n}\,.\,\frac{1.2.3\ldots\quad i}{1.3.5\ldots 2i-1}\,\Sigma\,\frac{n-1.n+1\ldots n+2i+2t'-3}{3.5\ldots 2i+2t'+1}$$

$$\times\frac{n-2.n\ldots n+2t'-4}{2.4\ldots 2t'}\left(\frac{r}{r'}\right)^{i+2t'}\;\ldots\ldots\ldots\;(10),$$

and in order to deduce the value of the same integral when $r'<r$, we shall only have to change r into r', and reciprocally, in the formula just given.

We may now readily obtain the value of $V_t^{(i)}$ by means of the formula (8). For the density corresponding thereto being

$$f_t'^{(i)}r^{i+2t}(1-r'^2)^{\beta},$$

it follows from what has been observed in the former part of the present article, that $f_t^{(i)}r^{i+2t}$ may always be reduced to a rational and entire function of x', y', z' the rectangular co-ordinates of the element dv, and therefore the density in question will admit of being expanded in a series of the entire powers of x', y', z' and the various products of these powers. Hence (Art. 1) $V_t^{(i)}$ admits of a similar expansion in entire powers &c. of x, y, z the rectangular co-ordinates of the point p, and by following the methods before exposed Art. 1 and 2, we readily get

$$V_t''^{(i)} = 4\pi f^{(i)} . \int_r^1 r'^{i+2t'+3-n} dr' (1-r'^2)^\beta . \Sigma \frac{n-1.n+1 \dots n+2i+2t'-3}{3.5 \dots 2i+2t'+1}$$
$$\times \frac{n-2.n.n+2 \dots n+2t'-4}{2.4.6 \dots 2t'} \left(\frac{r}{r'}\right)^{i+2t'};$$

and thence we have ultimately,

$$V_t^{(i)} = 2\pi f_t^{(i)} \Sigma \frac{n-1.n+1 \dots n+2i+2t'-3}{3.5 \dots 2i+2t'+1}$$
$$\times \frac{n-2.n \dots n+2t'-4}{2.4 \dots 2t'}$$
$$r^{i+2t} . \frac{\Gamma\left(\frac{2t-2t'+4-n}{2}\right)\Gamma(\beta+1)}{\Gamma\left(\frac{2t-2t'+2\beta+6-n}{2}\right)}$$

$$= 2\pi f_t^{(i)} . \frac{\Gamma(\beta+1)\,\Gamma\left(\frac{4-n}{2}\right)}{\Gamma\left(\frac{6+2\beta-n}{2}\right)} r^i \dots\dots$$

$$\dots\dots \Sigma r^{2t'} \frac{4-n.6-n \dots 2t-2t'+2-n}{6+2\beta-n \dots 2t-2t'+2\beta+4-n} \times \frac{n-2.n \dots n+2t'-4}{2.4 \dots 2t'}$$
$$\times \frac{n-1.n+1 \dots n+2i+2t'-3}{3.5 \dots 2i+2t'+1} \dots\dots\dots(11);$$

the finite integrals being taken from $t'=0$ to $t'=\infty$ and Γ being the characteristic of the well known function Gamma, which is introduced when we effect the integrations relative to r' by means of the formula (3), Art. 1.

Having thus $V_t^{(i)}$ or the part of V corresponding to the term $f_t'^{(i)}$, in $f(x', y', z')$ we immediately deduce the complete value of V by giving to i and t the various values of which these numbers are susceptible, and taking the sum of all the parts corresponding to the different terms in the expansion of the function $f(x', y', z')$.

Although in the present Article we have hitherto supposed f to be the characteristic of a rational and entire function, the same process will evidently be applicable, provided $f(x', y', z')$ can be expanded in an infinite series of the entire powers of x', y', z' and the various products of these powers. In the latter case we have as before the development

$$f(x', y', z') = f'^{(0)} + f'^{(1} + f'^{(2)} + f'^{(3)} + \&c.$$

of which any term, as for example $f'^{(i)}$, may be farther expanded as follows,

$$f'^{(i)} = f_0'^{(i)} r'^{i} + f_1'^{(i)} r'^{i+2} + f_2'^{(i)} r'^{i+4} + \&c.$$

and as we have already determined $V_t^{(i)}$ or the part of V corresponding to $f_t'^{(i)} r'^{i+2t}$, we immediately deduce as before the required value of V, the only difference is, that the numbers i and t, instead of being as in the former case confined within certain limits, may here become indefinitely great.

In the foregoing expression (11) β may be taken at will, but if we assign to it such a value that $\frac{2\beta - n}{2}$ may be a whole number, the series contained therein will terminate of itself, and consequently the value of $V_t^{(i)}$ will be exhibited in a finite form, capable by what has been shown at the beginning of the present Article of being converted into a rational and entire function of x, y, z, the rectangular co-ordinates of p. It is moreover evident, that the complete value of V being composed of a finite number of terms of the form $V_t^{(i)}$ will possess the same property, provided the function $f(x', y', z')$ is rational and entire, which agrees with what has been already proved in the second Article by a very different method.

5. We have before remarked (Art. 2), that in the particular case where $\beta = \frac{n-4}{2}$, the arbitrary constants contained in

$f(x', y', z')$ are just sufficient in number to enable us to determine this function, so as to make the resulting value of V equal to any given rational and entire function of x, y, z, the rectangular co-ordinates of p, and have proved that the corresponding functions V and f will be of the same degree. But when this degree is considerable, the method there proposed becomes impracticable, seeing that it requires the resolution of a system of

$$\frac{s+1 \,.\, s+2 \,.\, s+3}{1 \,.\, 2 \,.\, 3}$$

linear equations containing as many unknown quantities; s being the degree of the functions in question. But by the aid of what has been shown in the preceding Article, it will be very easy to determine for this particular value of β the function $f(x', y', z')$ and consequently the density ρ when V is given, and we shall thus be able to exhibit the complete solution of the inverse problem by means of very simple formulæ.

For this purpose, let us suppose agreeably to the preceding remarks, that ρ the density of the fluid in the element dv is of the form

$$\rho = (1 - r'^2)^{\frac{n-4}{2}} f(x', y', z');$$

f being the characteristic of a rational and entire function of the same degree as V, and which we will here endeavour so to determine, that the value of V thence resulting, may be equal to any given rational and entire function of x, y, z of the degree s.

Then by Laplace's simple method (*Mec. Cel.* Liv. III. No. 16) we may always expand V in a series of the form

$$V = V^{(0)} + V^{(1)} + V^{(2)} + \&c.... + V^{(s)}.$$

In like manner as has before been remarked, we shall have the analogous expansion

$$f(x', y', z') = f'^{(0)} + f'^{(1)} + f'^{(2)} + f'^{(3)} + \&c.... + f'^{(s)},$$

of which any term, $f'^{(i)}$ for example, may be farther developed as follows,

$$f'^{(i)}=f_0'^{(i)}r'^i+f_1'^{(i)}r'^{i+2}+f_2'^{(i)}r'^{i+4}+\&c.=r'^i\,(f'^{(i)}+f_1'^{(i)}r'^2+f'^{(i)}r'^4+\&c.)$$

$f_0'^{(i)}$, $f_1'^{(i)}$, $f_2'^{(i)}$, &c. being quantities independent of r' and all of the form $Y'^{(i)}$ (*Mec. Cel.* Liv. III.). Moreover $V_t^{(i)}$ the part of V due to the general term $f_t'^{(i)}r'^{i+2t}$ of the last series, will be obtained by writing $\frac{n-4}{2}$ for β in the equation (11), and afterwards substituting for

$$\Gamma\left(\frac{n-2}{2}\right)\Gamma\left(\frac{4-n}{2}\right) \text{ its value } \frac{\pi}{\sin\left(\frac{n-2}{2}\pi\right)}.$$

In this way we get

$$V_i^{(i)}=\frac{2\pi^2 f_t^{(i)} r^i}{\sin\left(\frac{n-2}{2}\right)\pi}\,\Sigma r^{2t'}\,\frac{4-n.6-n\ldots 2t-2t'+2-n}{2.4\ldots 2t+2i'-1}$$

$$\times\frac{n-2.n\ldots n+2t'-4}{2.4\ldots 2t'}\times\frac{n-1.n+1\ldots n+2i+2t'-3}{3.5\ldots 2i+2t'+1};$$

$f_t^{(i)}$ being what $f_t'^{(i)}$ becomes by changing θ', ϖ' into θ, ϖ, and the finite integral being taken from $t'=0$ to $t'=\infty$.

Let us now for a moment assume.

$$\phi(t')=\frac{n-2.n\ldots n+2t'-4}{2.4\ldots 2t'}\times\frac{n-1.n+1\ldots n+2i+2t'-3}{3.5\ldots 2i+2t'+1};$$

then the expression immediately preceding may be written

$$V_t^{(i)}=\frac{2\pi^2.f_t^{(i)}.r^i}{\sin\left(\frac{n-2}{2}\pi\right)}\,\Sigma\,\frac{4-n.6-n\ldots 2t-2t'+2-n}{2.4\ldots 2t-2t'}\,\phi(t').r^{2t'},$$

and by giving to t the various values 0, 1, 2, 3, &c. of which it is susceptible, and taking the sum of all the resulting values of $V_t^{(i)}$ the quantity thus obtained will be equal to $V^{(i)}$ or that part of V which is of the form $Y^{(i)}$. Thus we get

$$V^{(i)} = \frac{2\pi^2 . r^i}{\sin\left(\frac{n-2}{2}\pi\right)} \times$$

$$\ldots\ldots\ldots\ldots \phi(0) . f_0^{(i)}$$

$$+ \frac{4-n}{2} \phi(0) f_1^{(i)} + \phi(1) f_1^{(i)} . r^2$$

$$+ \frac{4-n . 6-n}{2 . 4} . \phi(0) . f_2^{(i)} + \frac{4-n}{2} \phi(1) f_2^{(i)} . r^2 + \phi(2) f_2^{(i)} . r^4$$

$$+ \frac{4-n . 6-n . 8-n}{2 . 4 . 6} \phi(0) f_3^{(i)} + \frac{4-n . 6-n}{2 . 4} \phi(1) f_3^{(i)} r^2$$

$$+ \frac{4-n}{2} \phi(2) f_3^{(i)} . r^4 + \phi(3) f_3^{(i)} r^6$$

$$+ \&c. \ldots\ldots\ldots\ldots \&c. \ldots\ldots\ldots\ldots \&c. \ldots\ldots\ldots\ldots$$

since all the terms in the preceding value of $V_t^{(i)}$ in which $t' > t$ vanish of themselves in consequence of the factor

$$\frac{4-n . 6-n \ldots 2t-2t'+2-n}{2 . 4 \ldots 2t-2t'}$$

$$= \frac{\Gamma\left(\frac{2t-2t'+4-n}{2}\right)}{\Gamma(t-t'+1)\,\Gamma\left(\frac{4-n}{2}\right)} = 0 \text{ (when } t' > t\text{)}.$$

But $V^{(i)}$ as deduced from the given value of V may be expanded in a series of the form

$$V^{(i)} = r^i . \{V_0^{(i)} + V_1^{(i)} r^2 + V_2^{(i)} . r^4 + V_3^{(i)} r^6 + \&c.\}$$

and if in order to simplify the remaining operations, we make generally

$$V_t^{(i)} = \frac{2\pi^3}{\sin\left(\frac{n-2}{2}\pi\right)} \times \frac{n-2 . n \ldots n+2t-4}{2 . 4 \ldots 2t}$$

$$\times \frac{n-1 . n+1 \ldots n+2i+2t-3}{3 . 5 \ldots 2i+2t+1} U_t^{(i)}$$

$$= \frac{2\pi^2}{\sin\left(\frac{n-2}{2}\pi\right)} \times \phi(t) . U_t^{(i)},$$

the equation immediately preceding will become

$$V^{(i)} = \frac{2\pi^2 . r^i}{\sin\left(\frac{n-2}{2}\pi\right)} \times \{\phi(0).U_0^{(i)} + \phi(1).U_1^{(i)} r^2 + \phi(2) U_2^{(i)} . r^4 + \&c.\}$$

which compared with the foregoing value of $V^{(i)}$, will give by suppressing the factor $\frac{2\pi^2 . r^i}{\sin\left(\frac{n-2}{2}\pi\right)}$, common to both and equating separately the coefficients of the different powers of the indeterminate quantity r the following system of equations

$$U_0^{(i)} = f_0^{(i)} + \frac{4-n}{2} f_1^{(i)} + \frac{4-n . 6-n}{2.4} f_2^{(i)} + \frac{4-n . 6-n . 8-n}{2.4.6} f_3^{(i)} +$$

$$U_1^{(i)} = f_1^{(i)} + \frac{4-n}{2} f_2^{(i)} + \frac{4-n . 6-n}{2.4} f_3^{(i)} + \&c.$$

$$U_2^{(i)} = f_2^{(i)} + \frac{4-n}{2} f_3^{(i)} + \frac{4-n . 6-n}{2.4} f_4^{(i)} + \&c.$$

&c.............................&c.............................&c.

for determining the unknown functions $f_0^{(i)}$, $f_1^{(i)}$, $f_2^{(i)}$, &c. by means of the known ones $U_0^{(i)}$, $U_1^{(i)}$, $U_2^{(i)}$, &c. In fact the last equation of the system gives $U_s^{(i)} = f_s^{(i)}$, and then by ascending successively from the bottom to the top equation, we shall get the values of $f_s^{(i)}$, $f_{s-1}^{(i)}$, $f_{s-2}^{(i)}$, &c. with very little trouble, it will however be simpler still to remark, that the general type of all our equations is

$$U_u^{(i)} = (1+\epsilon)^{\frac{n-4}{2}} f_u^{(i)},$$

where the symbols of operation have been separated from those of quantity and ϵ employed in its usual acceptation, so that

$$\epsilon f_u^{(i)} = f_{u+1}^{(i)}, \quad \epsilon^2 f_u^{(i)} = \epsilon f_{u+1}^{(i)} = f_{u+2}^{(i)}, \ \&c.$$

But it is evident we may satisfy the last equation by making

$$f_u^{(i)} = (1-\epsilon)^{\frac{4-n}{2}} U_u^{(i)}.$$

Expanding now and replacing $\epsilon U_u^{(i)}$, $\epsilon^2 U_u^{(i)}$, &c. by these values $U_{u+1}^{(i)}$, $U_{u+2}^{(i)}$, &c. we get

$$f_u^{(i)} = U_u^{(i)} + \frac{n-4}{2} U_{u+1}^{(i)} + \frac{n-1 \cdot n-2}{2 \cdot 4} U_{u+2}^{(i)}$$
$$+ \frac{n-4 \cdot n-2 \cdot n}{2 \cdot 4 \cdot 6} U_{u+3}^{(i)} + \&c.,$$

from which we may immediately deduce $f_u'^{(i)}$ and thence successively

$$f'^{(i)} = r'^i (f_0'^{(i)} + f_1'^{(i)} r'^2 + f_2'^{(i)} r'^4 + f_3'^{(i)} r'^6 + \&c.)$$

$$f(x', y', z') = f'^{(0)} + f'^{(1)} + f'^{(2)} + \&c... + f'^{(s)},$$

$$\text{and } \rho = (1 - x'^2 - y'^2 - z'^2)^{\frac{n-4}{2}} . f(x', y', z').$$

Application of the general Methods exposed in the preceding Articles to Spherical conducting Bodies.

6. In order to explain the phenomena which electrified bodies present, Philosophers have found it advantageous either to adopt the hypothesis of two fluids, the vitreous and resinous of Dufay for example, or to suppose with Æpinus and others, that the particles of matter when deprived of their natural quantity of electric fluid, possess a mutual repulsive force. It is easy to perceive that the mathematical laws of equilibrium deducible from these two hypotheses, ought not to differ when the quantity of fluid or fluids (according to the hypothesis we choose to adopt) which bodies in their natural state are supposed to contain, is so great, that a complete decomposition shall never be effected by any forces to which they may be exposed, but that in every part of them a farther decomposition shall always be possible by the application of still greater forces. In fact the mathematical theory of electricity merely consists in determining ρ* the analytical value of the fluid's density, so that the

* It may not be improper to remark that ρ is always supposed to represent the density of the free fluid, or that which manifests itself by its repulsive force; and therefore, when the hypothesis of two fluids is employed, the measure of the excess of the qnantity of either fluid which we choose to consider as positive over that of the fluid of opposite name in any element dv of the volume of the body is expressed by ρdv, whereas on the other hypothesis ρdv serves to measure the excess of the quantity of fluid in the element dv over what it would possess in a natural state.

whole of the electrical actions exerted upon any point p, situated at will in the interior of the conducting bodies may exactly destroy each other, and consequently p have no tendency to move in any direction. For the electric fluid itself, the exponent n is equal to 2, and the resulting value of ρ is always such as not to require that a complete decomposition should take place in the body under consideration, but there are certain values of n for which the resulting values of ρ will render $\int \rho dv$ greater than any assignable quantity; for some portions of the body it is therefore evident that how great soever the quantity of the fluid or fluids may be, which in a natural state this body is supposed to possess, it will then become impossible strictly to realize the analytical value of ρ, and therefore some modification at least will be rendered necessary, by the limit fixed to the quantity of fluid or fluids originally contained in the body, and as Dufay's hypothesis appears the more natural of the two, we shall keep this principally in view, when in what follows it may become requisite to introduce either.

7. The foregoing general observations being premised, we will proceed in the present article to determine mathematically the law of the density ρ, when the equilibrium has established itself in the interior of a conducting sphere A, supposing it free from the actions of exterior bodies, and that the particles of fluid contained therein repel each other with forces which vary inversely as the nth power of the distance. For this purpose it may be remarked, that the formula (1), Art. 1, immediately gives the values of the forces acting on any particle p, in virtue of the repulsion exerted by the whole of the fluid contained in A. In this way we get

$$\frac{1}{1-n} \cdot \frac{dV}{dx} = \text{the force directed parallel to the axis } X,$$

$$\frac{1}{1-n} \cdot \frac{dV}{dy} = \text{the force directed parallel to the axis } Y,$$

$$\frac{1}{1-n} \cdot \frac{dV}{dz} = \text{the force directed parallel to the axis } Z.$$

But since, in consequence of the equilibrium, each of these forces is equal to zero, we shall have

$$0 = \frac{dV}{dx}dx + \frac{dV}{dy}dy + \frac{dV}{dz}dz = dV;$$

and therefore, by integration,

$$V = \text{const.}$$

Having thus the value of V at the point p, whose co-ordinates are x, y, z, we immediately deduce, by the method explained in the fifth article,

$$\rho = \frac{\sin\left(\frac{n-2}{2}\pi\right)}{2\pi^2} V.(1-r'^2)^{\frac{n-4}{2}};$$

seeing that in the present case the general expansion of V there given reduces itself to

$$V = V^{(0)}.$$

If moreover Q serve to designate the total quantity of free fluid in the sphere, we shall have, by substituting for

$$\sin\left(\frac{n-2}{2}\pi\right) \text{ its value } \frac{\pi}{\Gamma\left(\frac{n-2}{4}\right)\Gamma\left(\frac{4-n}{2}\right)},$$

$$Q = \int \rho dv = \frac{\sin\left(\frac{n-2}{2}\pi\right)}{2\pi^2} V \int_0^1 4\pi r'^2 dr'(1-r'^2)^{\frac{n-4}{2}} = \frac{\Gamma\left(\frac{3}{2}\right)}{\Gamma\left(\frac{n+1}{2}\right)\Gamma\left(\frac{4-n}{2}\right)} V.$$

See Legendre, *Exer. de Cal. Int. Quatrième Partie.*

In the preceding values, as in the article cited, the radius of the sphere is taken for the unit of space; but the same formulæ may readily be adapted to any other unit by writing $\frac{r'}{a}$ in the place of r', and recollecting that the quantities ρ, V, and Q, are of the dimensions 0, $4-n$, and 3 respectively, with regard to space; a being the number which represents the radius of the sphere when we employ the new unit. In this way we obtain

$$\rho = \frac{\sin\left(\frac{n-2}{2}\pi\right)}{2\pi^2} V(a^2 - r'^2)^{\frac{n-4}{2}}, \text{ and } Q = \frac{\Gamma\left(\frac{3}{2}\right)a^{n-1}}{\Gamma\left(\frac{n+1}{2}\right)\Gamma\left(\frac{4-n}{2}\right)} . V.$$

Hence, when Q, the quantity of redundant fluid originally introduced into the sphere is given, the values of V and of the density ρ are likewise given. In fact, by writing in the preceding equation for

$$\Gamma\left(\frac{3}{2}\right), \text{ and } \sin\left(\frac{n-2}{2}\pi\right),$$

their values, we thence immediately deduce

$$\rho = \frac{1}{\pi\sqrt{\pi}} \frac{\Gamma\left(\frac{n+1}{2}\right)}{\Gamma\left(\frac{n-2}{2}\right)} a^{1-n} Q (a^2 - r'^2)^{\frac{n-4}{2}} \text{ ....(1)}$$

$$\text{and } V = \frac{2\Gamma\left(\frac{n+1}{2}\right)\Gamma\left(\frac{4-n}{2}\right)}{\sqrt{\pi}} a^{1-n} . Q.$$

The foregoing formulæ present no difficulties where $n > 2$, but when $n < 2$, the value of ρ, if extended to the surface of the sphere A, would require an infinite quantity of fluid of one name to have been originally introduced into its interior, and therefore, agreeably to a preceding observation, could not be strictly realized. In order then to determine the modification which in this case ought to be introduced, let us in the first place make $n > 2$, and conceive an inner sphere B whose radius is $a - \delta a$, in which the density of the fluid is still defined by the first of the equations (12); then, supposing afterwards the rest of the fluid in the exterior shell to be considered on A's surface, the portion so condensed, if we neglect quantities of the order δa, compared with those retained, will be

$$\frac{2^{\frac{n}{2}}}{\sqrt{\pi}} . \frac{\Gamma\left(\frac{n+1}{2}\right)}{\Gamma\left(\frac{n}{2}\right)} a^{\frac{2-n}{2}} Q . \delta a^{\frac{n-2}{2}},$$

and since, in the transfer of the fluid to A's surface, its particles move over spaces of the order δa only, the alteration which will thence be produced in V will evidently be of the order

$$\delta a^{\frac{n-2}{2}} \times \delta a = \delta a^{\frac{n}{2}},$$

and consequently the value of V will become

$$V = \frac{2}{\sqrt{\pi}} \Gamma\left(\frac{n+1}{2}\right) \Gamma\left(\frac{4-n}{2}\right) a^{1-n} Q + k \cdot \delta a^{\frac{n}{2}};$$

k being a quantity which remains finite when δa vanishes.

In establishing the preceding results, n has been supposed greater than 2, but ρ the density of the fluid within B and the quantity of it condensed on A's surface being still determined by the same formulæ, the foregoing value of V ought to hold good in virtue of the generality of analysis whatever n may be, and therefore when n is a positive quantity and δa is exceedingly small, we shall have without sensible errors

$$V = \frac{2}{\sqrt{\pi}} \Gamma\left(\frac{n+1}{2}\right) \Gamma\left(\frac{4-n}{2}\right) a^{1-n} . Q.$$

Conceiving now P' to represent the density of the fluid condensed on A's surface, $4\pi a^2 P'$ will be the total quantity thereon contained, which being equated to the value before given, there results

$$4\pi a^2 P' = \frac{2^{\frac{n}{2}}}{\sqrt{\pi}} \cdot \frac{\Gamma\left(\frac{n+1}{2}\right)}{\Gamma\left(\frac{n}{2}\right)} a^{\frac{2-n}{2}} Q \delta a^{\frac{n-2}{2}},$$

and hence we immediately deduce

$$P' = \frac{2^{\frac{n-4}{2}}}{\pi\sqrt{\pi}} \cdot \frac{\Gamma\left(\frac{n+1}{2}\right)}{\Gamma\left(\frac{n}{2}\right)} a^{-\frac{2+n}{2}} . Q . \delta a^{\frac{n-2}{2}},$$

Moreover as Q represents the total quantity of redundant fluid in the entire sphere A, the quantity contained in B is

$$Q - \frac{2^{\frac{n}{2}}}{\sqrt{\pi}} \cdot \frac{\Gamma\left(\frac{n+1}{2}\right)}{\Gamma\left(\frac{n}{2}\right)} a^{\frac{2-n}{2}} . Q . \delta a^{\frac{n-2}{2}},$$

If now when n is supposed less than 2, we adopt an hypothesis similar to Dufay's, and conceive that the quantities of fluid of opposite denominations in the interior of A are exceedingly great when this body is in a natural state, then after having introduced the quantity Q of redundant fluid, we may always by means of the expression just given, determine the value of δa, so that the whole of the fluid of contrary name to Q, may be contained in the inner sphere B, the density in every part of it being determined by the first of the equations (12). If afterwards the whole of the fluid of the same name as Q is condensed upon A's surface, the value of V in the interior of B as before determined will evidently be constant, provided we neglect indefinitely small quantities of the order $\delta a^{\frac{n}{2}}$. Hence all the fluid contained in B will be in equilibrium, and as the shell included between the two concentric spheres A and B is entirely void of fluid, it follows that the whole system must be in equilibrium.

From what has preceded, we see that the first of the formulæ (12) which served to give the density ρ within the sphere A when n is greater than 2, is still sensibly correct when n represents any positive quantity less than 2, provided we do not extend it to the immediate vicinity of A's surface. But as the foregoing solution is only approximative, and supposes the quantities of the two fluids which originally neutralized each other to be exceedingly great, we shall in the following article endeavour to exhibit a rigorous solution of the problem, in case $n > 2$, which will be totally independent of this supposition.

8. Let us here in the first place conceive a spherical surface whose radius is a, covered with fluid of the uniform density P', and suppose it is required to determine the value of the density ρ in the interior of a concentric conducting sphere, the radius of which is taken for the unit of space, so that the fluid therein

contained, may be in equilibrium in virtue of the joint action of that contained in the sphere itself, and on the exterior spherical surface.

If now V' represents the value of V due to the exterior surface, it is clear from what Laplace has shown (*Mec. Cel.* Liv. II. No. 12) that

$$V' = \int \frac{d\sigma P'}{g'^{1-n}} = \frac{2\pi a P'}{(3-n)\, r} \{(a+r)^{3-n} - (a-r)^{3-n}\};$$

$d\sigma$ being an element of this surface, and g' being the distance of this element from the point p to which V' is supposed to belong.

If afterwards we conceive that the function V is due to the fluid within the sphere itself, it is easy to prove as in the last article, that in consequence of the equilibrium we must have

$$V' + V = \text{const.}$$

But V' and consequently V is of the form $Y^{(0)}$, therefore by employing the method before explained (Art. 4), we get

$$f(x', y', z') = f'^{(0)} = f_0'^{(0)} + f_1'^{(0)} . r'^2 + f_2'^{(0)} . r'^4 + \&c. = B_0 + B_1 r'^2 + B_2 r'^4 + \&c.;$$

where, as in the present case, $f_0'^{(0)}$, $f_1'^{(0)}$, $f_2'^{(0)}$, etc. are all constant quantities, they have for the sake of simplicity been replaced by

$$B_0,\ B_1,\ B_2,\ \&c.$$

Hitherto the exponent β has remained quite arbitrary, but by making $\beta = \frac{n-2}{2}$, the formula (11) Art. 4, will become when $i = 0$,

$$V_t^{(0)} = 2\pi B_t \frac{\Gamma\left(\frac{n}{2}\right)\Gamma\left(\frac{4-n}{2}\right)}{\Gamma(2)} \Sigma r^{2t'} \frac{4-n \,.\, 6-n \ldots 2t-2t'+2-n}{4 \quad .\, 6 \ldots 2t-2t'+2} \times \frac{n-2 \,.\, n-1 \ldots n+2t-2t'-3}{2 \,.\, 3 \,.\, 4 \quad \ldots 2t'+1}$$

$$= \frac{(n-2)\pi^2 B_t}{\sin\left(\frac{n-2}{2}\pi\right)} \Sigma . r^{2t'} . \frac{4-n . 6-n \ldots 2t-2t'+2-n}{4 \quad . 6 \ldots 2t-2t'+2} \times \frac{n-2 . n-1 \ldots n+2t'-3}{2 \quad . 3 \ldots 2t'+1}.$$

Giving now to t the successive values 0, 1, 2, 3, &c. and taking the sum of the functions thence resulting, there arises

$$V = V^{(0)} = V_0^{(0)} + V_1^{(0)} + V_2^{(0)} + V_3^{(0)} + \text{etc.} = S\,.\,V_1^{(0)}$$

$$= \frac{(n-2)\pi^2}{\sin\left(\frac{n-2}{2}\pi\right)} SB_t \Sigma r^{2t'} \frac{4-n.6-n\ldots 2t-2t'+2-n}{4\ \ .6.8\ldots 2t-2t'+2} \times \frac{n-2.n-1\ldots n+2t'-3}{2\ \ .3\ldots 2t'+1},$$

where the sign S is referred to the variable t and Σ to t'.

Again, by substituting for V and V' their values in the equation $V' + V = \text{const.}$ and expanding the function V' we obtain

$$\text{const.} = 4\pi P' a^{3-n}.\,\Sigma \frac{r^{2t'}}{a^{2t'}}.\frac{n-2\,.\,n-1\,.\,n\ldots n+2t'-3}{2\quad .\,3.\quad 4\ldots 2t'+1}$$

$$+\frac{(n-2)\pi^2}{\sin\left(\frac{n-2}{2}\pi\right)} S\Sigma B_t r^{2t'} \frac{4-n.6-n\ldots 2t-2t'+2-n}{4.6\ldots 2t-2t'+2} \times \frac{n-2.n-1\ldots n+2t'+3}{2.3.4\ldots 2t'+1},$$

which by equating separately the coefficients of the various powers of the indeterminate quantity r, and reducing, gives generally

$$\frac{2P'a^{3-n-2t'}.\sin\left(\frac{n-2}{2}\pi\right)}{\pi} = S\frac{2-n\,.\,4-n\ldots 2s-n}{2\quad .\,4\ldots\quad 2s} B_{t'+s-1}.$$

Then by assigning to t' its successive values 1, 2, 3, &c., there results for the determination of the quantities B_0, B_1, B_2, &c., the following system of equations,

$$\frac{2P'}{\pi} a^{1-n}.\sin\left(\frac{n-2}{2}\pi\right) = B_0 + \frac{2-n}{2} B_1 + \frac{2-n\,.\,4-n}{2\ \ .\ \ 4} B_2 + \&c.$$

$$\frac{2P'}{\pi} a^{1-n}.\sin\left(\frac{n-2}{2}\pi\right).a^{-2} = B_1 + \frac{2-n}{2} B_2 + \frac{2-n\,.\,4-n}{2\ \ .\ \ 4} B_3 + \&c.$$

$$\frac{2P'}{\pi} a^{1-n}.\sin\left(\frac{n-2}{2}\pi\right).a^{-4} = B_2 + \frac{2-n}{2} B_3 + \frac{2-n\,.\,4-n}{2\ \ .\ \ 4} B_4 + \&c.$$

&c. .... &c. .... &c. .... &c. ......

But it is evident from the form of these equations, that we may satisfy the whole system by making

$$B_1 = B_0 . a^{-2},\ B_2 = B_1 . a^{-2},\ B_3 = B_2 . a^{-2},\ B_4 = B_3 . a^{-2}, \text{ \&c.}$$

provided we determine B_0 by

$$\frac{2P'}{\pi} a^{1-n} \sin\left(\frac{n-2}{2}\pi\right) = B_0\left(1 + \frac{2-n}{2} a^{-2} + \frac{2-n \,.\, 4-n}{2 \,.\, 4} a^{-4} + \text{\&c.}\right)$$

$$= B_0 (1 - a^{-2})^{\frac{n-2}{2}}.$$

Hence as in the present case, $\beta = \frac{n-2}{2}$, we immediately deduce the successive values

$$B_0 = \frac{2P'}{\pi a} \sin\left(\frac{n-2}{2}\pi\right) . (a^2 - 1)^{\frac{2-n}{2}},$$

$$f(x', y', z') = f'^{(0)} = B_0 + B_1 r'^2 + B_2 r'^2 + \text{\&c.} = B_0\left(1 - \frac{r'^2}{a^2}\right)^{-1},$$

$$\text{and } \rho = (1 - r'^2)^{\frac{n-2}{2}} . f(x', y', z') = \frac{2P'a}{\pi} \sin\left(\frac{n-2}{2}\pi\right) . (a^2 - 1)^{\frac{2-n}{2}} \ldots$$

$$\ldots\ldots (a^2 - r'^2)^{-1} (1 - r'^2)^{\frac{n-2}{2}}.$$

In the value of ρ just exhibited, the radius of the sphere is taken as the unit of space, but the same formula may easily be adapted to any other unit by writing $\frac{a}{b}$ and $\frac{r'}{b}$ in the place of a and r' respectively, and recollecting at the same time that in consequence of the equation

$$\text{const.} = V + V' = \int \frac{dv . \rho}{g} + \int \frac{d\sigma P'}{g'},$$

before given, $\frac{\rho}{P'}$, is a quantity of the dimension -1 with regard to space: b being the number which represents the radius of the sphere when we employ the new unit. Hence we obtain for a sphere whose radius is bg, acted upon by an exterior concentric spherical surface of which the radius is a,

$$(\beta) \ldots\ldots \rho = \frac{2P'a \, . \sin\left(\frac{n-2}{2}\pi\right)}{\pi} (a^2 - b^2)^{\frac{2-n}{2}} (a^2 - r'^2)^{-1} (b^2 - r'^2)^{\frac{n-2}{2}};$$

P' being the density of the fluid on the exterior surface.

If now we conceive a conducting sphere A whose radius is a, and determine P' so that all the fluid of one kind, viz. that which is redundant in this sphere, may be condensed on its surface, and afterwards find b the radius of the interior sphere B from the condition that it shall just contain all the fluid of the opposite kind, it is evident that each of the fluids will be in equilibrium within A, and therefore the problem originally proposed is thus accurately solved. The reason for supposing all the fluid of one name to be completely abstracted from B, is that our formulæ may represent the state of *permanent* equilibrium, for the tendency of the forces acting within the void shell included between the surfaces A and B, is to abstract continually the fluid of the same name as that on A's surface from the sphere B.

To prove the truth of what has just been asserted, we will begin with determining the repulsion exerted by the inner sphere itself, on any point p exterior to it, and situate at the distance r from its centre O. But by what Laplace has shown (*Mec. Cel.* Liv. II. No. 12) the repulsion on an exterior point p, arising from a spherical shell of which the radius is r', thickness dr' and centre is at O will be measured by

$$\frac{2\pi r' dr' \rho}{1-n \, . \, 3-n} \cdot \frac{d}{dr} \cdot \frac{(r+r')^{3-n} - (r-r')^{3-n}}{r},$$

the general term of which when expanded in an ascending series of the powers of $\frac{r'}{r}$ is,

$$+ 4\pi \, . \frac{-2+n \times n \, . \, n+1 \, . \, n+2 \ldots n+2s-3 \times n+2s-1}{2 \, . \, 3 \, . \, 4 \, . \, 5 \ldots 2s+1} r^{n-2s} \, . \, r'^{2s+2} \, . \, \rho dr',$$

and the part of the required repulsion due thereto will, by substituting for ρ its value before found, become

$$+\frac{8P'}{a}\sin\left(\frac{n-2}{2}\pi\right).(a^2-b^2)^{\frac{2-n}{2}}\frac{-2+n\times n.n+1\ldots n+2s-3\times n+1s-1}{2.3.4\ldots 2s+1}r^{-n-2s}$$

$$\times\int_0^b\left(1-\frac{r'^2}{a^2}\right)^{-1}(b^2-r'^2)^{\frac{n-2}{2}}r'^{2s+2}\,dr'.$$

It now remains to find the value of the definite integral herein contained. But when $\left(1-\frac{r'^2}{a^2}\right)^{-1}$ is expanded, and the integrations are effected by known formulæ, we obtain

$$\int_0^b\left(1-\frac{r'^2}{a^2}\right)^{-1}(b^2-r'^2)^{\frac{n-2}{2}}r'^{2s+2}\,dr'=\int_0^b\Sigma_0^\infty\frac{r'^{2t}}{a^{2t}}(b^2-r'^2)^{\frac{n-2}{2}}.r'^{2s+2}dr'$$

$$=\tfrac{1}{2}b^{2s+1+n}\Sigma_0^\infty\frac{b^{2t}}{a^{2t}}\times\frac{\Gamma\left(\frac{n}{2}\right)\Gamma\left(s+t+\frac{3}{2}\right)}{\Gamma\left(s+t+\frac{3}{2}+n\right)}=\tfrac{1}{2}b^{2s+1+n}.\frac{\Gamma\left(\frac{n}{2}\right)\Gamma\left(s+\frac{3}{2}\right)}{\Gamma\left(s+\frac{n}{2}+\frac{3}{2}\right)}.$$

$$\left\{1+\frac{2s+3}{2s+3+n}\frac{b^2}{a^2}+\frac{2s+3\,.\,2s+5}{2s+3+n\,.\,2s+5+n}\frac{b^4}{a^4}+\&c.\right\}$$

$$=\tfrac{1}{2}b^{2s+1+n}.\frac{\Gamma\left(\frac{n}{2}\right)\Gamma\left(s+\frac{3}{2}\right)}{\Gamma\left(s+\frac{n}{2}+\frac{3}{2}\right)}\times\frac{(2s+1+n)\,(1-x^2)^{\frac{n-2}{2}}}{x^{2s+1+n}}\int_0\frac{x^{2s+n}\,dx}{(1-x^2)^{\frac{n}{2}}}$$

$$=\frac{\Gamma\left(\frac{n}{2}\right)\Gamma\left(\frac{1}{2}\right)}{2\Gamma\left(\frac{n+1}{2}\right)}.\frac{1\,.\,3\,.\,5\ldots\ldots 2s+1\;b^{2s+1+n}}{1+n\,.\,3+n\,.\,5+n\ldots\ldots 2s-1+n}$$

$$\times\frac{(1-x^2)^{\frac{n-2}{2}}}{x^{2s+1+n}}\int_0\frac{x^{2s+n}\,dx}{(1-x^2)^{\frac{n}{2}}}\ldots\ldots\ldots(14);$$

where after the integrations have been effected, x ought to be made equal to $\frac{b}{a}$.

The value of the integral last found being substituted in the expression immediately preceding, and the finite integral taken relative to s from $s=0$ to $s=\infty$ gives for the repulsion of the inner sphere,

$$-\frac{4\pi P'b}{a}\cdot\frac{\Gamma\left(\frac{1}{2}\right)}{\Gamma\left(\frac{1+n}{2}\right)\Gamma\left(\frac{2-n}{2.}\right)}(a^2-b^2)^{\frac{2-n}{2}}$$

$$\times\Sigma_0^\infty\frac{n-2\,.\,n\,.\,n+2\,......\,n+2s-4}{2\,.\,4\,.\,6\quad......\quad 2s}\left(\frac{b}{r}\right)^{2s+n}\frac{(1-x^2)^{\frac{n-2}{2}}}{x^{2s+1+n}}\int_0\frac{x^{2s+n}dx}{(1-x^2)^{\frac{n}{2}}}$$

$$=\frac{-4\pi\sqrt{\pi}P'a^2r^{-n}}{\Gamma\left(\frac{1+n}{2}\right)\Gamma\left(\frac{2-n}{2}\right)}\Sigma_0^\infty\frac{n-2.n.n+2...n+2s-4}{2.4.6\quad...\quad 2s}\left(\frac{a}{r}\right)^{2s}\int_0 dx\,x^{2s+n}.(1-x^2)^{\frac{-n}{2}};$$

since $\Gamma\left(\frac{1}{2}\right)=\sqrt{\pi}$, $\sin\left(\frac{n-2}{2}\pi\right)=\dfrac{-\pi}{\Gamma\left(\frac{n}{2}\right)\Gamma\left(\frac{2-n}{2}\right)}$,

and as was before observed, $x=\dfrac{b}{a}$.

But we have evidently by means of the binomial theorem,

$$\left(1-\frac{a^2x^2}{r^2}\right)^{\frac{2-n}{2}}=\Sigma_0^\infty\frac{n-2\,.\,n\,.\,n+2\,......\,n+2s-4}{2\,.\,4\,.\,6\quad......\quad 2s}\left(\frac{ax}{r}\right)^{2s};$$

and therefore the preceding quantity becomes

$$-\frac{4\pi\sqrt{\pi}\,.\,P'a^2.\,r^{-n}}{\Gamma\left(\frac{1-n}{2}\right)\Gamma\left(\frac{2-n}{2}\right)}\int_0^{\frac{b}{a}}dx\,x^n\left(1-\frac{a^2x^2}{r^2}\right)^{\frac{2-n}{2}}(1-x^2)^{\frac{-n}{2}}\,......(15).$$

If now we make $x=\dfrac{rx'}{a}$, the same quantity may be written

$$-\frac{4\pi\sqrt{\pi}P'a^{1-n}r}{\Gamma\left(\frac{1+n}{2}\right)\Gamma\left(\frac{2-n}{2}\right)}\int_0^{\frac{b}{r}}x'^n dx'(1-x'^2)^{\frac{2-n}{2}}\left(1-\frac{r^2x'^2}{a^2}\right)^{\frac{-n}{2}}.....(16).$$

Having thus the value of the repulsion due to the inner sphere B on an exterior point p, it remains to determine that due to the fluid on A's surface. But this last is represented by

$$\frac{2\pi aP'}{1-n\,.\,3-n}\,\frac{d.}{dr}\,\frac{(a+r)^{3-n}-(a-r)^{3-n}}{r}..........(17).$$

(*Mec. Cel.* Liv. II. No. 12.) Now by expanding this function there results

$$4\pi P'a^{1-n}r.\frac{2-n}{3}.\left\{1+\frac{n.n+1}{4.5}2\frac{r^2}{a^2}+\frac{n.n+1.n+2.n+3}{4.5.6.7}.3\frac{r^4}{a^4}+\&c.\right\}$$

$$=4\pi P'a^{1-n}r.\frac{2-n}{3}.\Sigma_0^\infty\frac{n.n+1.n+2\ldots\ldots n+2s-1}{4\ .\ 5\ .\ 6\ \ldots\ldots\ 2s+3}(s+1)\frac{r^{2s}}{a^{2s}}.$$

The last of these expressions may readily be exhibited under a finite form, by remarking that

$$\int_0^1 x^n dx(1-x^2)^{\frac{2-n}{2}}\left(1-\frac{r^2x^2}{a^2}\right)^{\frac{-n}{2}}$$

$$=\int_0^1 x^n dx\,(1-x^2)^{\frac{2-n}{n}}\,\Sigma\frac{n.n+2\ldots\ldots n+2s-2}{2.4.6\ldots\ldots\ 2s}.\frac{r^{2s}x^{2s}}{a^{2s}}$$

$$=\Sigma_0^\infty\frac{n.n+2.n+4\ldots\ldots n+2s-2}{2.\ 4\ .\ 6\ \ldots\ldots\ 2s}.\frac{r^{2s}}{a^{2s}}.\frac{\Gamma\left(\frac{2s+n+1}{2}\right)\Gamma\left(\frac{4-n}{2}\right)}{2\Gamma\left(\frac{2s+5}{2}\right)}$$

$$=\frac{\Gamma\left(\frac{2-n}{2}\right)\Gamma\left(\frac{1+n}{2}\right)}{\Gamma\left(\frac{1}{2}\right)}.\frac{2-n}{3}.\Sigma_0^\infty\frac{n.n+1.n+2\ldots n+2s-1}{4.\ 5\ .\ 6\ \ldots\ 2s+3}(s+1)\frac{r^{2s}}{a^{2s}}.$$

Hence, since $\Gamma(\frac{1}{2})=\sqrt{\pi}$, the value of the repulsion arising from A's surface becomes

$$\frac{4\pi\sqrt{\pi}.P'a^{1-n}.r}{\Gamma\left(\frac{1+n}{2}\right)\Gamma\left(\frac{2-n}{2}\right)}\int_0^1 x^n dx\,(1-x^2)^{\frac{2-n}{2}}\left(1-\frac{r^2x^2}{a^2}\right)^{\frac{-n}{2}}.$$

Now by adding the repulsion due to the inner sphere which is given by the formula (16), we obtain, (since it is evidently indifferent what variable enters into a definite integral, provided each of its limits remain unchanged)

$$\frac{4\pi\sqrt{\pi}P'a^{1-n}r}{\Gamma\left(\frac{1+n}{2}\right)\Gamma\left(\frac{2-n}{2}\right)}.\left\{\int_0^1 x^n dx(1-x^2)^{\frac{2-n}{2}}\left(1-\frac{r^2x^2}{a^2}\right)^{\frac{-n}{2}}-\int_0^{\frac{b}{r}}x^n dx(1-x^2)^{\frac{2-n}{2}}.\left(1-\frac{r^2x^2}{a^2}\right)^{\frac{-n}{2}}\right\}$$

$$=\frac{4\pi\sqrt{\pi}.P'a^{1-n}r}{\Gamma\left(\frac{1+n}{2}\right)\Gamma\left(\frac{2-n}{2}\right)}\int_{\frac{b}{r}}^1 x^n dx\,(1-x^2)^{\frac{2-n}{2}}.\left(1-\frac{r^2x^2}{a^2}\right)^{\frac{-n}{2}},$$

for the value of the total repulsion upon a particle p of positive fluid situate within the sphere A and exterior to B. We thus see that when P' is positive the particle p is always impelled by a force which is equal to zero at B's surface, and which continually increases as p recedes farther from it. Hence, if any particle of positive fluid is separated ever so little from B's surface, it has no tendency to return there, but on the contrary, it is continually impelled therefrom by a regularly increasing force; and consequently, as was before observed, the equilibrium can not be permanent until all the positive fluid has been gradually abstracted from B and carried to the surface of A, where it is retained by the non-conducting medium with which the sphere A is conceived to be surrounded.

Let now q represent the total quantity of fluid in the inner sphere, then the repulsion exerted on p by this will evidently be qr^{-n}, when r is supposed infinite. Making therefore r infinite in the expression (15), and equating the value thus obtained to the one just given, there arises

$$q = \frac{-4\pi\sqrt{\pi}\,.\,P'a^2}{\Gamma\left(\frac{1+n}{2}\right)\Gamma\left(\frac{2-n}{2}\right)}\int_0^{\frac{b}{a}} dx\,.\,x^n\,(1-x^2)^{\frac{-n}{2}}.$$

When the equilibrium has become permanent, q is equal to the total quantity of that kind of fluid, which we choose to consider negative, originally introduced into the sphere A; and if now q_1 represent the total quantity of fluid of opposite name contained within A, we shall have, for the determination of the two unknown quantities P' and b, the equations

$$q_1 = 4\pi a^2 . P',$$

and $$\frac{q}{q_1} = \frac{-\sqrt{\pi}}{\Gamma\left(\frac{1+n}{2}\right)\Gamma\left(\frac{2-n}{2}\right)}\int_0^{\frac{b}{a}} dx\, x^n\,(1-x^2)^{\frac{-n}{2}},$$

and hence we are enabled to assign accurately the manner in which the two fluids will distribute themselves in the interior of A; q and q_1, the quantities of the fluids of opposite names originally introduced into A being supposed given.

9. In the two foregoing articles we have determined the

manner in which our hypothetical fluids will distribute themselves in the interior of a conducting sphere A when in equilibrium and free from all exterior actions, but the method employed in the former is equally applicable when the sphere is under the influence of any exterior forces. In fact, if we conceive them all resolved into three X, Y, Z, in the direction of the co-ordinates x, y, z of a point p, and then make, as in Art. 1,

$$V = \int \frac{\rho dv}{g^{n-1}},$$

we shall have, in consequence of the equilibrium,

$$0 = \frac{1}{1-n} \frac{dV}{dx} + X, \quad 0 = \frac{1}{1-n} \frac{dV}{dy} + Y, \quad 0 = \frac{1}{1-n} \frac{dV}{dz} + Z,$$

which, multiplied by dx, dy and dz respectively, and integrated, give

$$\text{const.} = \frac{1}{1-n} V + \int (Xdx + Ydy + Zdz);$$

where $Xdx + Ydy + Zdz$ is always an exact differential.

We thus see that when X, Y, Z are given rational and entire functions V will be so likewise, and we may thence deduce (Art. 5)

$$\rho = (1 - x'^2 - y'^2 - z'^2)^{\frac{n-4}{2}} . f(x', y', z'),$$

where f is the characteristic of a rational and entire function of the same degree as V.

The preceding method is directly applicable when the forces X, Y, Z are given explicitly in functions of x, y, z. But instead of these forces, we may conceive the density of the fluid in the exterior bodies as given, and thence determine the state which its action will induce in the conducting sphere A. For example, we may in the first place suppose the radius of A to be taken as the unit of space, and an exterior concentric spherical surface, of which the radius is a, to be covered with fluid of the density $U''^{(i)}$: $U''^{(i)}$ being a function of the two polar co-ordinates θ'' and ϖ'' of any element of the spherical surface of the same kind as those considered by Laplace (*Mec. Cel.* Liv. III.). Then it is easy to perceive by what has been proved in the article last cited, that the value of the induced density will be of the form

$$\rho = U'^{(i)} r'^{i} (1 - r'^2)^{\frac{n-4}{2}} . f(r'^2);$$

r', θ', ϖ' being the polar co-ordinates of the element dv, and $U'^{(i)}$ what $U''^{(i)}$ becomes by changing θ'', ϖ'' into θ', ϖ'.

Still continuing to follow the methods before explained, (Art. 4 and 5) we get in the present case

$$f(x', y', z') = U'^{(i)} r'^{i} f(r'^2) = f^{(i)},$$

and by expanding $f(r'^2)$, we have

$$f(r'^2) = B_0 + B_1 r'^2 + B_2 r'^4 + B_3 r'^6 + \&c.$$

Hence, $f_t^{(i)} = B_t U'^{(i)}$, and

$$V_t^{(i)} = \frac{2\pi^2 U^{(i)} r^i}{\sin\left(\frac{n-2}{2}\pi\right)} B_t \Sigma_0^{\infty} r^{2t'} \frac{4-n \,.\, 6-n \ldots\ldots 2t-2t'+2-n}{2 \,.\, 4 \,.\, 6 \ldots\ldots 2t-2t'}$$

$$\times \frac{n-2 \,.\, n \ldots\ldots n+2t'-4}{2 \,.\, 4 \ldots\ldots 2t'} \times \frac{n-1 \,.\, n+1 \ldots\ldots n+2i+2t'-3}{3 \,.\, 5 \ldots\ldots 2i+2t'+1}.$$

Then, by giving to t all the values 1, 2, 3, &c. of which it is susceptible, and taking the sum of all the resulting quantities, we shall have, since in the present case V reduces itself to the single term $V^{(i)}$,

$$V = \frac{2\pi^2 U^{(i)} r^i}{\sin\left(\frac{n-2}{2}\pi\right)} S B_t \Sigma r^{2t'} . \frac{4-n \,.\, 6-n \ldots\ldots 2t-2t'+2-n}{2 \,.\, 4 \ldots\ldots 2t-2t'}$$

$$\times \frac{n-2 \,.\, n \ldots\ldots n+2t'-4}{2 \,.\, 4 \ldots\ldots 2t'} \times \frac{n-1 \,.\, n+1 \ldots\ldots n+2i+2t'-3}{3 \,.\, 5 \ldots\ldots 2i+2t'+1};$$

the sign S belonging to the unaccented letter t.

If now V' represents the function analogous to V and due to the fluid on the spherical surface, we shall obtain by what has been proved (Art. 3)

$$V' = U^{(i)} . 2\pi a^2 \frac{1 \,.\, 3 \,.\, 5 \ldots 2i-1}{1 \,.\, 2 \,.\, 3 \ldots i} \int_{-1}^{1} d\mu \,(i)\, (r^2 - 2ar\mu + a^2)^{\frac{1-n}{2}};$$

(i) representing the same function as in the article just cited.

Moreover, it is evident from the equation (10) Art. 4, that

$$\int_{-1}^{1} d\mu\,(i)\,(r^2 - 2ar\mu + a^2)^{\frac{1-n}{2}}$$

$$= 2a^{1-n}\frac{1\,.\,2\,.\,3\,\ldots\,i}{1\,.\,3\,\ldots\,2i-1}\,\Sigma\,\frac{n-1\,.\,n+1\,\ldots\,n+2i+2t'-3}{3\,.\,5\,\ldots\,2i+2t'+1}$$

$$\times\frac{n-2\,.\,n\,\ldots\,n+2t'-4}{2\,.\,4\,\ldots\,2t'}\left(\frac{r}{a}\right)^{i+2t'};$$

and consequently,

$$V' = U^{(i)}\,.\,4\pi a^{3-n}\,.\,\Sigma\,\frac{n-1\,.\,n+1\,\ldots\,n+2i+2t'-3}{3\,.\,5\,\ldots\,2i+2t'+1}$$

$$\times\frac{n-2\,.\,n\,\ldots\,n+2t'-4}{2\,.\,4\,\ldots\,2t'}\left(\frac{r}{a}\right)^{i+2t}\ldots\ldots\ldots\ldots\ldots(19);$$

the finite integrals extending from $t' = 0$ to $t' = \infty$.

Substituting now for V and V' their values in the equation of equilibrium,

$$\text{const.} = V' + V\ldots\ldots\ldots\ldots\ldots\ldots\ldots\ldots(20),$$

we immediately obtain

$$\text{const.} = U^{(i)}\,.\,4\pi a^{3-n}\,.\,\Sigma\,\frac{n-1\,.\,n+1\,\ldots\,n+2i+2t'-3}{3\,.\,5\,\ldots\,2i+2t'+1}$$

$$\times\frac{n-2\,.\,n\,\ldots\,n+2t'-4}{2\,.\,4\,\ldots\,2t'}\left(\frac{r}{a}\right)^{i+2t'}$$

$$+\frac{2\pi^2}{\sin\left(\frac{n-2}{2}\pi\right)}\,U^{(i)}SB_t\Sigma r^{i+2t'}\,.\,\frac{n-1\,.\,n+1\,\ldots\,n+2i+2t'-3}{3\,.\,5\,\ldots\,2i+2t'-1}$$

$$\times\frac{n-2\,.\,n\,\ldots\,n+2t'-4}{2\,.\,4\,\ldots\,2t'}\times\frac{4-n\,.\,6-n\,\ldots\,2t-2t'+2-n}{2\,.\,4\,\ldots\,2t-2t'},$$

the constant on the left side of this equation being equal to zero, except when $i = 0$.

By equating separately the coefficients of the various powers of the indeterminate quantity r, we get the following system of equations:

$$-\frac{2\sin\left(\frac{n-2}{2}\pi\right)}{\pi}a^{3-n-1} = B_0 + B_1\frac{4-n}{2} + B_2\frac{4-n\,.\,6-n}{2\,.\,4} + \&c.$$

$$-\frac{2\sin\left(\frac{n-2}{2}\pi\right)}{\pi}a^{3-n-i-2} = B_1 + B_2\frac{4-n}{2} + B_3\frac{4-n \,.\, 6-n}{2 \,.\, 4} + \&c.$$

$$-\frac{2\sin\left(\frac{n-2}{2}\pi\right)}{\pi}a^{3-n-i-4} = B_2 + B_3\frac{4-n}{2} + B_4\frac{4-n \,.\, 6-n}{2 \,.\, 4} + \&c.$$

&c.................&c...............&c.......................

But it is evident from the form of these equations, that if we make generally $B_{i+1} = a^{-2}B_i$, they will all be satisfied provided the first is, and as by this means the first equation becomes

$$-\frac{2\sin\left(\frac{n-2}{2}\pi\right)}{\pi}a^{3-n-i} = B_0\left(1 + \frac{4-n}{2}a^{-2} + \frac{4-n \,.\, 6-n}{2 \,.\, 4}a^{-4} + \&c.\right)$$

$$= B_0(1-a^{-2})^{\frac{n-4}{2}} = B_0 a^{4-n}(a^2-1)^{\frac{n-4}{2}},$$

there arises

$$B_0 = -\frac{2\sin\left(\frac{n-2}{2}\pi\right)}{\pi}a^{-i-1}(a^2-1)^{\frac{4-n}{2}},\quad B_1 = B_0 . a^{-2},\quad B_2 = B_0 . a^{-4}, \&c.$$

Hence

$$f(r'^2) = B_0 + B_1 r'^2 + B_2 r'^4 + \&c. = B_0\left(1 + \frac{r'^2}{a^2} + \frac{r'^4}{a^4} + \&c.\right)$$

$$= B_0\left(1 - \frac{r'^2}{a^2}\right)^{-1} = B_0 a^2 (a^2 - r'^2)^{-1}$$

$$= -\frac{2\sin\left(\frac{n-2}{2}\pi\right)}{\pi}a^{i+1}(a^2-1)^{\frac{4-n}{2}}(a^2-r'^2)^{-1},$$

and the required value of ρ becomes

$$\rho = U'^{(i)} r'^i (1-r'^2)^{\frac{n-4}{2}} f(r'^2)$$

$$= -\frac{2\sin\left(\frac{n-2}{2}\pi\right)}{\pi}(a^2-1)^{\frac{4-n}{2}} aU'^{(i)}\left(\frac{r'}{a}\right)^i (a^2-r'^2)^{-1}(1-r'^2)^{\frac{n-4}{2}} \ldots (21).$$

But whatever the density P on the inducing spherical surface may be, we can always expand it in a series of the form

$$P = U''^{(0)} + U''^{(1)} + U''^{(2)} + U''^{(3)} + \&c. \textit{ in inf.}$$

and the corresponding value of ρ by what precedes will be

$$\rho = -\frac{2\sin\left(\frac{n-2}{2}\pi\right)}{\pi} a(a^2-1)^{\frac{4-n}{2}} \cdot (a^2-r'^2)^{-1}(1-r'^2)^{\frac{n-4}{2}} \ldots\ldots$$

$$\ldots\ldots \times \left\{U'^{(0)} + U'^{(1)}\frac{r'}{a} + U'^{(2)}\frac{r'^2}{a^2} + U'^{(3)}\frac{r'^3}{a^3} + \&c.\ in\ inf.\right\};$$

$U'^{(0)}$, $U'^{(1)}$, $U'^{(2)}$, &c. being what $U''^{(0)}$, $U''^{(1)}$, $U''^{(2)}$, &c. become by changing θ'', ϖ'' into θ', ϖ', the polar co-ordinates of the element dv. But, since we have generally

$$\int d\theta'' d\varpi'' \sin\theta'' PQ^{(i)} = \int d\theta'' d\varpi'' \sin\theta'' U'^{(i)} Q^{(i)} = \frac{4\pi}{2i+1} U^{(i)},$$

(*Mec. Cel.* Liv. III.) the preceding expression becomes

$$\rho = \frac{-\sin\left(\frac{n-2}{2}\pi\right)}{2\pi^2} a(a^2-1)^{\frac{4-n}{2}}(a^2-r'^2)^{-1}(1-r'^2)^{\frac{n-4}{2}} \int d\theta'' d\varpi'' \sin\theta'' \ldots.$$

$$\ldots\ldots \Sigma^{\infty}(2i+1) PQ^{(i)}\frac{r'^i}{a^i};$$

the integrals being taken from $\theta''=0$ to $\theta''=\pi$, and from ϖ'' to $\varpi''=2\pi$.

In order to find the value of the finite integral entering into the preceding formula, let R represent the distance between the two elements $d\sigma$, dv; then by expanding $\frac{a}{R}$ in an ascending series of the powers of $\frac{r'}{a}$ we shall obtain

$$\frac{a}{R} = \frac{a}{\sqrt{a^2 - 2ar'\{\cos\theta'\cos\theta'' + \sin\theta'\sin\theta''\cos(\varpi'-\varpi'')\} + r'^2}}$$

$$= \Sigma_0^{\infty} Q^{(i)} \cdot \frac{r'^i}{a^i},$$

(*Mec. Cel.* Liv. III.). Hence we immediately deduce

$$\frac{a\sqrt{r'}}{R} = \Sigma_0^{\infty} Q^{(i)} \frac{r'^{i+\frac{1}{2}}}{a^i}, \text{ and } 2\sqrt{r'}\frac{d.}{dr'}\frac{a\sqrt{r'}}{R} = \Sigma_0^{\infty}(2i+1) Q^{(i)}\frac{r'^i}{a^i}.$$

If now we substitute this in the value of ρ before given, and afterwards write $\frac{ds}{a^2}$ and $\frac{a^2-r'^2}{2R^3}$ in the place of their equivalents,

$$d\theta'' d\varpi'' \sin\theta'', \text{ and } \sqrt{r'}\,\frac{d.}{dr'}\,\frac{\sqrt{r'}}{R},$$

we shall obtain

$$\rho = -\frac{\sin\left(\frac{n-2}{2}\pi\right)}{2\pi^2}(a^2-1)^{\frac{4-n}{2}}(1-r'^2)^{\frac{n-4}{2}}\int\frac{d\sigma P}{R^3};$$

the integral relative to $d\sigma$ being extended over the whole spherical surface.

Lastly, if ρ_1 represents the density of the reducing fluid disseminated over the space exterior to A, it is clear that we shall get the corresponding value of ρ by changing P into $\rho_1 da$ in the preceding expression, and then integrating the whole relative to a. Thus,

$$\rho = -\frac{\sin\left(\frac{n-2}{2}\pi\right)}{2\pi^2}(1-r'^2)^{\frac{n-4}{2}}\int(1-a^2)^{\frac{4-n}{2}}\int\frac{d\sigma da\rho_1}{R^3}.$$

But $d\sigma da = dv_1$; dv_1 being an element of the volume of the exterior space, and therefore we ultimately get

$$\rho = -\frac{\sin\left(\frac{n-2}{2}\pi\right)}{2\pi^2}(1-r'^2)^{\frac{n-4}{2}}.\int\rho_1 dv_1\,\frac{(a^2-1)^{\frac{4-n}{2}}}{R^3}\ldots\ldots(22);$$

where the last integral is supposed to extend over all the space exterior to the sphere and R to represent the distance between the two elements dv and dv_1.

It is easy to perceive from what has before been shown (Art. 7), that we may add to any of the preceding values of ρ, a term of the form

$$h(1-r'^2)^{\frac{n-4}{2}};$$

h being an arbitrary constant quantity: for it is clear from the article just cited, that the only alteration which such an addition could produce would be to change the value of the constant on the left side of the general equation of equilibrium; and as this constant is arbitrary, it is evident that the equilibrium will not be at all affected by the change in question. Moreover, it may be observed, that in general the additive term is necessary to enable us to assign the proper value of ρ, when Q, the quantity of redundant fluid originally introduced into the sphere, is given.

In the foregoing expressions the radius of the sphere has been taken as the unit of space, but it is very easy thence to deduce formulæ adapted to any other unit, by recollecting that $\frac{\rho}{\rho_1}$, $\frac{\rho}{P}$, $\frac{\rho}{U'^{(i)}}$ and $\frac{V_1}{U^{(i)}}$, are quantities of the dimensions 0, -1, -1 and $3-n$ respectively with regard to space: for if b represents the sphere's radius, when we employ any other unit we shall only have to write, $\frac{r}{b}$, $\frac{r'}{b}$, $\frac{R}{b}$, $\frac{dv_1}{b^3}$ and $\frac{a}{b}$ in the place of r, r', R, dv_1 and a, and afterwards to multiply the resulting expressions by such powers of b, as will reduce each of them to their proper dimensions.

If we here take the formula (22) of the present article as an example, there will result,

$$\rho = -\frac{\sin\left(\frac{n-2}{2}\pi\right)}{2\pi^2}(b^2-r'^2)^{\frac{n-4}{2}}\int\rho_1 dv_1\frac{(a^2-b^2)^{\frac{4-n}{2}}}{R^3} \quad \ldots\ldots(23),$$

for the value of the density which would be induced in a sphere A, whose radius is b, by the action of any exterior bodies whatever.

When $n>2$, the value of ρ or of the density of the free fluid here given offers no difficulties, but if $n<2$, we shall not be able strictly to realize it, for reasons before assigned (Art. 6 and 7). If however n is positive, and we adopt the hypothesis of two fluids, supposing that the quantities of each contained by bodies in a natural state are exceedingly great, we shall easily perceive by proceeding as in the last of the articles here cited, that the density given by the formula (23) will be sensibly correct except in the immediate vicinity of A's surface, provided we extend it to the surface of a sphere whose radius is $b-\delta b$ only, and afterwards conceive the exterior shell entirely deprived of fluid: the surface of the conducting sphere itself having such a quantity condensed upon it, that its density may every where be represented by

$$P' = -\frac{\sin\left(\frac{n-2}{2}\pi\right)}{2\pi^2}\times\frac{b^{\frac{n-1}{2}}(2\delta b)^{\frac{n-2}{2}}}{n-2}\int\rho_1 dv_1\frac{(a^2-b^2)^{\frac{4-n}{2}}}{R^3},$$

Application of the general Methods to circular conducting Planes, &c.

10. Methods in every way similar to those which have been used for a sphere, are equally applicable to a circular plane, as we shall immediately proceed to show, by endeavouring in the first place to determine the value of V when the density of the fluid on such a plane is of the form

$$\rho = (1 - r'^2)^\beta . f(x', y');$$

f being the characteristic of a rational and entire function of the degree s; x', y' the rectangular co-ordinates of any element $d\sigma$ of the plane's surface, and r', θ' the corresponding polar co-ordinates.

Then we shall readily obtain the formula

$$V = \int \frac{\rho d\sigma}{g^{n-1}} = \iint \frac{r dr' d\theta' (1 - r'^2)^\beta . f(x', y')}{\{r^2 - 2rr' \cos(\theta - \theta') + r'^2\}^{\frac{n-1}{2}}};$$

where r, θ are the polar co-ordinates of p, and the integrals are to be taken from $\theta' = 0$ to $\theta' = 2\pi$, and from $r' = 0$ to $r' = 1$; the radius of the circular plane being for greater simplicity considered as the unit of distance.

Since the function $f(x', y')$ is rational and entire of the degree s, we may always reduce it to the form

$$f(x', y') = A^{(0)} + A^{(1)} \cos\theta' + A^{(2)} \cos 2\theta' + A^{(3)} \cos 3\theta' + \ldots\ldots$$
$$+ B^{(1)} \sin\theta' + B^{(2)} \sin 2\theta' + B^{(3)} \sin 3\theta' + \ldots\ldots (24),$$

the coefficients $A^{(0)}$, $A^{(1)}$, $A^{(2)}$, &c. $B^{(1)}$, $B^{(2)}$, $B^{(3)}$, &c. being functions of r' only of a degree not exceeding s, and such that

$$A^{(0)} = a_0^{(0)} + a_1^{(0)} r'^2 + a_2^{(0)} r'^4 + \&c.; \quad A^{(1)} = (a_0^{(1)} + a_1^{(1)} r^2 + a_2^{(1)} r'^4 +) r';$$
$$B^{(1)} = (b_0^{(1)} + b_1^{(1)} r'^2 + b_2^{(1)} r'^4 + \&c.) r'; \quad B^{(2)} = (b_0^{(2)} + b_1^{(2)} r'^2 + \&c.) r'^2.$$

We will now consider more particularly the part of V due to any of the terms in f as $A^{(i)} \cos i\theta'$ for example. The value of this part will evidently be

$$\iint \frac{r' dr' d\theta' (1 - r'^2)^\beta A^{(i)} \cos i\theta'}{\{r^2 - 2rr' \cos(\theta - \theta') + r'^2\}^{\frac{n-1}{2}}};$$

the limits of the integrals being the same as before. But if we make $\theta' = \theta + \phi$, there will result $d\theta' = d\phi$, and

$$\cos i\theta' = \cos i\theta \cos i\phi - \sin i\theta \sin i\phi,$$

and hence the double integral here given by observing that the term multiplied $\sin i\phi$ vanishes when the integration relative to ϕ is effected, becomes

$$\cos i\theta \int_0^1 A^{(i)} r' dr' (1 - r'^2)^\beta \int_0^{2\pi} \frac{d\phi \cos i\phi}{(r^2 - 2rr' \cos\phi + r'^2)^{\frac{n-1}{2}}}.$$

If now we write $V_t^{(i)}$ for that portion of V which is due to the term $a_t^{(i)} . r'^{i+2t}$ in the coefficient $A^{(i)}$ we shall have

$$V_t^{(i)} = a_t^{(i)} . \cos i\theta \int_0^1 r'^{i+2t+1} dr' (1 - r'^2)^\beta \int_0^{2\pi} \frac{d\phi \cos i\phi}{(r^2 - 2rr' \cos\phi + r'^2)^{\frac{n-1}{2}}}.$$

But by well known methods we readily get

$$\int_0^{2\pi} \frac{d\phi \cos i\phi}{(r^2 - 2rr' \cos\phi + r'^2)^{\frac{n-1}{2}}}$$

$$= 2\pi r^i . r'^{1-n-i} \Sigma_0^\infty \frac{r^{2t'}}{r'^{2t'}} . \frac{n-1.n+1 \ldots n+2t'-3}{2 \;.\; 4 \;\ldots\ldots\; 2t'} \times \frac{n-1.n+1 \ldots n+2i+2t'-3}{2 \;.\; 4 \;\ldots\ldots\; 2i+2t},$$

when $r' > r$, and when $r' < r$, the same expression will still be correct, provided we change r into r' and reciprocally.

This value being substituted in that of $V_t^{(i)}$ we shall readily have by following the processes before explained, (Art. 1 and 2),

$$V_t^{(i)} = 2\pi a_t^{(i)} r^i \cos i\theta \, \Sigma_0^\infty r^{2t'} \frac{n-1 . n+1 \ldots n+2t'-3}{2 \;.\; 4 \;\ldots\; 2t'}$$

$$\times \frac{n-1 . n+1 \ldots n+2i+2t'-3}{2 \;.\; 4 \;\ldots\; 2i+2t'} \times \frac{\Gamma(\beta+1)\,\Gamma\left(\frac{3+2t-2t'-n}{2}\right)}{2\Gamma\left(\frac{2\beta+5+2t-2t'}{2}\right)}$$

$$= \pi a_t^{(i)} r^i \cos i\theta . \frac{\Gamma(\beta+1)\,\Gamma\left(\frac{3-n}{2}\right)}{\Gamma\left(\frac{2\beta+5-n}{2}\right)}.$$

$$\Sigma_0^\infty r^{2t'} \frac{n-1.n+1 \ldots n+2t'-3}{2\ .\ 4\ \ldots\ 2t'} \times \frac{n-1.n+1 \ldots n+2i+2t'-3}{2\ .\ 4\ \ldots\ 2i+2t'}$$

$$\times \frac{3-n.5-n \ldots\ldots 1+2t-2t'-n}{2\beta+5-n \ldots\ldots 2\beta+3+2t+2t'-n};$$

the sign of integration Σ belonging to the variable t'.

Having thus the part of V due to the term $a_t^{(i)} \cos i\theta'$ in the expansion of $f(x', y')$ it is clear that we may thence deduce the part due to the analogous term $b_t^{(i)} \sin i\theta'$ by simply changing $a_t^{(i)} \cos i\theta$ into $b_t^{(i)} \sin i\theta$, and consequently we shall have the total value of V itself, by taking the sum of the various parts due to all the different terms which enter into the complete expansion of $f(x', y')$.

If now we make $\beta = \frac{n-3}{2}$ and recollect that

$$\Gamma\left(\frac{n-1}{2}\right) \Gamma\left(\frac{3-n}{2}\right) = \frac{\pi}{\sin\left(\frac{n-1}{2}\pi\right)},$$

the foregoing expression will undergo simplifications analogous to those before noticed (Art. 5). Thus we shall obtain

$$V_t^{(i)} = \frac{\pi^2 a_t^{(i)}}{\sin\left(\frac{n-1}{2}\pi\right)} r^i \cos i\theta \,.\, \Sigma r^{2t'} . \frac{n-1.n+1 \ldots n+2t'-3}{2\ .\ 4\ \ldots\ 2t'}$$

$$\times \frac{n-1.n+1 \ldots n+2i+2t'-3}{2\ .\ 4\ \ldots\ 2i+2t'} \times \frac{3-n.5-n \ldots 1+2t-2t'-n}{2\ .\ 4\ \ldots\ 2t-2t'},$$

or by writing for abridgment

$$\phi(i, t') = \frac{n-1.n+1 \ldots n+2t'-3}{2\ .\ 4\ \ldots\ 2t'} \times \frac{n-1.n+1 \ldots n+2i+2t'-3}{2\ .\ 4\ \ldots\ 2i+2t'},$$

there will result this particular value of β

$$V_t^{(i)} = \frac{\pi^2 a_t^{(i)}}{\sin\left(\frac{n-1}{2}\pi\right)} r^i \cos i\theta \,.\, \Sigma r^{2t'} . \frac{3-n.5-n \ldots 1+2t-2t'-n}{2\ .\ 4\ .\ 6\ \ldots\ 2t-2t'} . \phi(i; t'),$$

and afterwards by making

$$V^{(i)} = V_0^{(i)} + V_1^{(i)} + V_2^{(i)} + V_3^{(i)} + V_4^{(i)} + \&c.$$

we shall have

$$V^{(i)} = \frac{\pi^2}{\sin\left(\frac{n-1}{2}\pi\right)} r^i \cos i\theta \text{ into} \times \ldots\ldots$$

$$a_0^{(i)} . 1 . \phi(i; 0)$$

$$+ a_1^{(i)} . \frac{3-n}{2} . \phi(i; 0) + a_1^{(i)} . 1 . \phi(i; 1) . r^2$$

$$+ a_2^{(i)} . \frac{3-n . 5-n}{2 . 4} . \phi(i; 0) + a_2^{(i)} . \frac{3-n}{2} . \phi(i; 1) . r^2 + a_2^{(i)} \phi(i; 2) . r^4$$

$$+ a_3^{(i)} . \frac{3-n . 5-n . 7-n}{2 . 4 . 6} . \phi(i; 0) + a_3^{(i)} \frac{3-n . 5-n}{2 . 4} . \phi(i; 1) . r^2$$

$$+ a_3^{(i)} . \frac{3-n}{2} . \phi(i; 2) . r^4 + a_3^{(i)} . 1 . \phi(i; 3) . r^6$$

$$+ \&c. \ldots\ldots\ldots + \&c. \ldots\ldots\ldots + \&c. \ldots\ldots\ldots + \&c. \ldots\ldots\ldots$$

Conceiving in the next place that V is a given rational and entire function of x, y, the rectangular co-ordinates of p, we shall have since $x = r\cos\theta$, $y = r\sin\theta$,

$$V = C^{(0)} + C^{(1)}\cos\theta + C^{(2)}\cos 2\theta + C^{(3)}\cos 3\theta + \&c.$$
$$+ E^{(1)}\sin\theta + E^{(2)}\sin 2\theta + E^{(3)}\sin 3\theta + \&c. \ldots\ldots (25),$$

of which expansion any coefficient as $C^{(i)}$ for example, may be still farther developed in the form

$$C^{(i)} = \frac{\pi^2 . r^i}{\sin\left(\frac{n-1}{2}\pi\right)} \{c_0^{(i)} . \phi(i; 0) + c_1^{(i)} . \phi(i; 1) . r^2 + c_2^{(i)} . \phi(i; 2) . r^4 + \&c.\}.$$

Now it is clear that the term $C^{(i)}\cos i\theta$ in the developement (25) corresponds to that part of V which we have designated by $V^{(i)}$, and hence by equating these two forms of the same quantity, we get

$$V^{(i)} = C^{(i)}\cos i\theta,$$

which by substituting for $V^{(i)}$ and $C^{(i)}$ their values before exhibited, and comparing like powers of the indeterminate quantity r gives

$$c_0^{(i)} = 1 . a_0^{(i)} + \frac{3-n}{2} a_1^{(i)} + \frac{3-n . 5-n}{2 . 4} a_2^{(i)} + \frac{3-n . 5-n . 7-n}{2 . 4 . 6} a_3^{(i)} + \&c.$$

$$c_1^{(i)} = 1 . a_1^{(i)} + \frac{3-n}{2} a_2^{(i)} + \frac{3-n . 5-n}{2 . 4} a_3^{(i)} + \&c.$$

$$c_2^{(i)} = 1 . a_2^{(i)} + \frac{3-n}{2} a_3^{(i)} + \&c.$$

&c.= .... &c.......... &c.

of which system the general type is

$$c_u^{(i)} = (1-\epsilon)^{\frac{n-3}{2}} . a_u^{(i)} ;$$

the symbols of operation being here separated from those of quantity, and ϵ being used in its ordinary acceptation with reference to the lower index u, so that we shall have generally

$$\epsilon^m . a_u^{(i)} = a_{u+m}^{(i)} .$$

The general equation between $a_u^{(i)}$ and $c_u^{(i)}$ being resolved, evidently gives by expanding the binomial and writing in the place of $\epsilon c_u^{(i)}$, $\epsilon^2 c_u^{(i)}$, $\epsilon^3 c_u^{(i)}$, &c. their values $c_{u+1}^{(i)}$, $c_{u+2}^{(i)}$, $c_{u+3}^{(i)}$, &c.

$$a_u^{(i)} = (1-\epsilon)^{\frac{3-n}{2}} c_u^{(i)} = c_u^{(i)} + \frac{n-3}{2} c_{u+1}^{(i)} + \frac{n-3 . n-1}{2 . 4} c_{u+2}^{(i)}$$

$$+ \frac{n-3 . n-1 . n+1}{2 . 4 . 6} c_{u+3}^{(i)} + \&c. .... (26).$$

Having thus the value of $a_u^{(i)}$ we thence immediately deduce the value of $A^{(i)}$ and this quantity being known, the first line of the expansion (25) evidently becomes known.

In like manner when we suppose that the quantity $E_u^{(i)}$ is expanded in a series of the form

$$E^{(i)} = \frac{\pi^2 r^i}{\sin\left(\frac{n-1}{2}\pi\right)} \{e_0^{(i)} . \phi(i; 0) + e_1^{(i)} \phi(i; 1) . r^2 + e_2^{(i)} \phi(i; 2) . r^4 + \&c.\}$$

we shall readily deduce

$$b_u^{(i)} = (1-\epsilon)^{\frac{3-n}{2}} e_u^{(i)} = e_u^{(i)} + \frac{n-3}{2} e_{u+1}^{(i)} + \frac{n-3 \,.\, n-1}{2 \,.\, 4} e_{u+2}^{(i)} + \&c.,$$

and $b_u^{(i)}$ being thus given, $B^{(i)}$ and consequently the second line of the expansion (25) are also given.

From what has preceded, it is clear that when V is given equal to any rational and entire function whatever of x and y, the value of $f(x', y')$ entering into the expression

$$\rho = (1 - r'^2)^{\frac{n-3}{2}} . f(x', y'),$$

will immediately be determined by means of the most simple formulæ.

The preceding results being quite independent of the degree s of the function $f(x', y')$ will be equally applicable when s is infinite, or wherever this function can be expanded in a series of the entire powers of x', y', and the various products of these powers.

We will now endeavour to determine the manner in which one fluid will distribute itself on the circular conducting plane A when acted upon by fluid distributed in any way in its own plane.

For this purpose, let us in the first place conceive a quantity q of fluid concentrated in a point P, where $r = a$ and $\theta = 0$, to act upon a conducting plate whose radius is unity. Then the value of V due to this fluid will evidently be

$$\frac{q}{(a^2 - 2ar\cos\theta + r^2)^{\frac{n-1}{2}}} = V',$$

and consequently the equation of equilibrium analogous to the one marked (20) Art. 10, will be

$$\text{const.} = \frac{q}{(a^2 - 2ar\cos\theta + r^2)^{\frac{n-1}{2}}} + V \ldots\ldots\ldots\ldots\ldots (27);$$

V being due to the fluid on the conducting plate only.

If now we expand the value of V deduced from this equation, and then compare it with the formulæ (25) of the present article, we shall have generally $E^{(i)}=0$, and

$$C^{(i)}=-2qa^{-n}\frac{r^i}{a^i}\cdot\left\{\phi(i;0)+\phi(i;1)\frac{r^2}{a^2}+\phi(i;2)\frac{r^4}{a^4}+\phi(i;3)\frac{r^6}{a^6}+\&c.\right\},$$

except when $i=0$, in which case we must take only half the quantity furnished by this expression in order to have the correct value of $C^{(0)}$. Hence whatever u may be,

$$e_u^{(i)}=0,\quad\text{and}\quad c_u^{(i)}=-\frac{2\sin\left(\dfrac{n-1}{2}\pi\right)}{\pi^2}qa^{1-n-i-2u};$$

the particular value $i=0$ being excepted, for in this case we have agreeably to the preceding remark

$$c_u^{(0)}=-\frac{\sin\left(\dfrac{n-1}{2}\pi\right)}{\pi^2}q\,.\,a^{1-n-2u},$$

and then the only remaining exception is that due to the constant quantity on the left side of the equation (27). But it will be more simple to avoid considering this last exception here, and to afterwards add to the final result the term which arises from the constant quantity thus neglected.

The equation (26) of the present article gives by substituting for $c_u^{(i)}$ its value just found,

$$a_u^{(i)}=-\frac{2\sin\left(\dfrac{n-1}{2}\pi\right)}{\pi^2}qa^{1-n-i-2u}\,.\left\{1+\frac{n-3}{2}\,.\,a^{-2}\right.$$
$$\left.+\frac{n-3\,.\,n-1}{2\,.\,4}\,.\,a^{-4}+\frac{n-3\,.\,n-1\,.\,n-1}{2\,.\,4\,.\,6}\,.\,a^{-6}+\&c.\right\}$$
$$=-\frac{2\sin\left(\dfrac{n-1}{2}\pi\right)}{\pi^2}qa^{1-n-i-2u}\left(1-a^{-2}\right)^{\frac{3-n}{2}}$$
$$=-\frac{2\sin\left(\dfrac{n-1}{2}\pi\right)}{\pi^2}qa^{-2-i-2u}\,.\,(a^2-1)^{\frac{3-n}{2}},$$

and consequently,

$$A^{(i)} = \{a_0^{(i)} + a_1^{(i)} r'^2 + a_2^{(i)} r'^4 + \&c.\}\, r'^i$$

$$= -\frac{2\sin\left(\frac{n-1}{2}\pi\right)}{\pi^2}\, qa^{-2-i}(a^2-1)^{\frac{3-n}{2}} r'^i . \left\{1 + \frac{r'^2}{a^2} + \frac{r'^4}{a^2} + \&c.\right\}$$

$$= -\frac{2\sin\left(\frac{n-1}{2}\pi\right)}{\pi^2}\, qa^{-2-i}(a^2-1)^{\frac{3-n}{2}} r'^i \left(1 - \frac{r'^2}{a^2}\right)^{-1}$$

$$= -\frac{2\sin\left(\frac{n-2}{2}\pi\right)}{\pi^2}\, q(a^2-1)^{\frac{3-n}{2}} (a^2 - r'^2)^{-1} . \frac{r'^i}{a^i};$$

the particular value $A^{(0)}$ being one half only of what would result from making $i = 0$ in this general formula.

But $e_u^{(i)} = 0$ evidently gives $E^{(i)} = 0$, and therefore the expansion of $f(x', y')$ before given becomes

$$f(x', y') = A^{(0)} + A^{(1)}\cos\theta' + A^{(2)}\cos 2\theta' + A^{(3)}\cos 3\theta' + \&c.$$

$$= -\frac{2\sin\left(\frac{n-1}{2}\pi\right)}{\pi^2}\, q(a^2-1)^{\frac{3-n}{2}} (a^2 - r'^2)^{-1} . \left\{\frac{1}{2} + \frac{r'}{a}\cos\theta' + \frac{r'^2}{a^2}\cos 2\theta' + \&c.\right\}$$

or by summing the series included between the braces,

$$f(x', y') = -\frac{\sin\left(\frac{n-1}{2}\pi\right)}{\pi^2}\, q\, \frac{(a^2-1)^{\frac{3-n}{2}}}{a^2 - 2ar'\cos\theta' + r'^2}$$

$$= -\frac{\sin\left(\frac{n-1}{2}\pi\right)}{\pi^2}\, q\, \frac{(a^2-1)^{\frac{3-n}{2}}}{R^2};$$

R being the distance between P, the point in which the quantity of fluid q is concentrated, and that to which the density ρ is supposed to belong.

Having thus the value of $f(x', y')$ we thence deduce

$$\rho = (1-r'^2)^{\frac{n-3}{2}} f(x', y') = -\frac{\sin\left(\frac{n-1}{2}\pi\right)}{\pi^2} (1-r'^2)^{\frac{n-3}{2}} q \frac{(a^2-1)^{\frac{3-n}{2}}}{R^2}.$$

The value of ρ here given being expressed in quantities perfectly independent of the situation of the axis from which the angle θ' is measured, is evidently applicable when the point P is not situated upon this axis, and in order to have the complete value of ρ, it will now only be requisite to add the term due to the arbitrary constant quantity on the left side of the equation (26), and as it is clear from what has preceded, that the term in question is of the form

$$\text{const.} \times (1-r'^2)^{\frac{n-3}{2}},$$

we shall therefore have generally, wherever P may be placed,

$$\rho = (1-r'^2)^{\frac{n-3}{2}} . \left\{\text{const.} - \frac{\sin\left(\frac{n-1}{2}\pi\right)}{\pi^2} q . \frac{(a^2-1)^{\frac{3-n}{2}}}{R^2}\right\}.$$

The transition from this particular case to the more general one, originally proposed is almost immediate: for if ρ represents the density of the inducing fluid on any element $d\sigma_1$ of the plane coinciding with that of the plate, $\rho_1 d\sigma_1$ will be the quantity of fluid contained in this element, and the density induced thereby will be had from the last formula, by changing q into $\rho_1 d\sigma_1$. If then we integrate the expression thus obtained, and extend the integral over all the fluid acting on the plate, we shall have for the required value of ρ

$$\rho = (1-r'^2)^{\frac{n-3}{2}} . \left\{\text{const.} - \frac{\sin\left(\frac{n-1}{2}\pi\right)}{\pi^2} \int \rho_1 d\sigma_1 \frac{(a^2-1)^{\frac{3-n}{2}}}{R^2}\right\};$$

R being the distance of the element $d\sigma_1$ from the point to which ρ belongs, and a the distance between $d\sigma_1$ and the center of the conducting plate.

Hitherto the radius of the circular plate has been taken as the unit of distance, but if we employ any other unit, and suppose that b is the measure of the same radius, in this case we shall only have to write $\frac{a}{b}$, $\frac{r'}{b}$, $\frac{d\sigma_1}{b^2}$ and $\frac{R}{b}$ in the place of a, r', $d\sigma_1$ and R respectively, recollecting that $\frac{\rho}{\rho_1}$ is a quantity of the dimension 0 with regard to space, by so doing the resulting value of ρ is

$$\rho = (b^2 - r'^2)^{\frac{n-3}{2}} \cdot \left\{ \text{const.} - \frac{\sin\left(\frac{n-1}{2}\pi\right)}{\pi^2} \int \rho_1 d\sigma_1 \frac{(a^2 - b^2)^{\frac{3-n}{2}}}{R^2} \right\} \ldots\ldots (28).$$

By supposing $n = 2$, the preceding investigation will be applicable to the electric fluid, and the value of the density induced upon an infinitely thin conducting plate by the action of a quantity of this fluid, distributed in any way at will in the plane of the plate itself will be immediately given. In fact, when $n = 2$, the foregoing value of ρ becomes

$$\rho = \frac{1}{\sqrt{b^2 - r'^2}} \left\{ \text{const.} - \frac{1}{\pi^2} \int \rho_1 d\sigma_1 \frac{\sqrt{a^2 - b^2}}{R^2} \right\}.$$

If we suppose the plate free from all extraneous action, we shall simply have to make $\rho_1 = 0$ in the preceding formula; and thus

$$\rho = \frac{\text{const.}}{\sqrt{b^2 - r'^2}} \ldots\ldots\ldots\ldots\ldots\ldots\ldots\ldots\ldots\ldots (29).$$

Biot (*Traité de Physique*, Tom. II. p. 277), has related the results of some experiments made by Coulomb on the distribution of the electric fluid when in equilibrium upon a plate of copper 10 inches in diameter, but of which the thickness is not specified. If we conceive this thickness to be very small compared with the diameter of the plate, which was undoubtedly the case, the formula just found ought to be applicable to it, provided we except those parts of the plate which are in the immediate vicinity of its exterior edge. As the comparison of any results mathematically deduced from the received theory of electricity with those of the experiments of so accurate an observer as Coulomb must always be interesting, we will here give a table of the values of the density at different points on the surface of the plate, calculated by means of the formula (29), together with the corresponding values found from experiment.

Distances from the Plate's edge.	Observed densities.	Calculated densities.
5 in.......	1,	1,
4	1,001	1,020
3	1,005	1,090
2	1,17	1,250
1	1,52	1,667
,5	2,07	2,294
0	2,90	infinite.

We thus see that the differences between the calculated and observed densities are trifling; and moreover, that the observed are all something smaller than the calculated ones, which it is evident ought to be the case, since the latter have been determined by considering the thickness of the plate as infinitely small, and consequently they will be somewhat greater than when this thickness is a finite quantity, as it necessarily was in Coulomb's experiments.

It has already been remarked that the method given in the second article is applicable to any ellipsoid whatever, whose axes are a, b, c. In fact, if we suppose that x, y, z are the co-ordinates of a point p within it, and x', y', z' those of any element dv of its volume, and afterwards make

$$x = a \,.\, \cos\theta,\quad y = b \,.\, \sin\theta \cos\varpi,\quad z = c \,.\, \sin\theta \sin\varpi,$$

$$x' = a \,.\, \cos\theta',\quad y' = b \,.\, \sin\theta' \cos\varpi',\quad z' = c \,.\, \sin\theta' \sin\varpi',$$

we shall readily obtain by substitution,

$$V = abc \int \rho \,.\, r'^2 dr' d\theta' d\varpi' \sin\theta' \,.\, (\lambda r^2 - 2\mu r r' + \nu r'^2)^{\frac{1-n}{2}};$$

the limits of the integrals being the same as before (Art. 2), and

$$\lambda = a^2 \cos\theta^2 + b^2 \sin\theta^2 \cos\varpi^2 + c^2 \sin\theta^2 \sin\varpi^2,$$

$$\mu = a^2 \cos\theta \cos\theta' + b^2 \sin\theta \sin\theta' \cos\varpi \cos\varpi' + c^2 \sin\theta \sin\theta' \sin\varpi \sin\varpi',$$

$$\nu = a^2 \cos\theta'^2 + b^2 \sin\theta'^2 \cos\varpi'^2 + c^2 \sin\theta'^2 \sin\varpi'^2.$$

Under the present form it is clear the determination of V can offer no difficulties after what has been shown (Art. 2). I shall not therefore insist upon it here more particularly, as it is my intention in a future paper to give a general and purely analytical method of finding the value of V, whether p is situated within the ellipsoid or not. I shall therefore only observe, that for the particular value

$$\rho = k\left(1 - \frac{x'^2}{a^2} - \frac{y'^2}{b^2} - \frac{z'^2}{c^2}\right)^{\frac{n-4}{2}} = k\,(1 - r'^2)^{\frac{n-4}{2}} \text{........} (30),$$

the series $U_0' + U_2' + U_4' + \&c.$ (Art. 2) will reduce itself to the single term U_0', and we shall ultimately get

$$V = \frac{\pi kabc}{2\sin\left(\frac{n-2}{2}\pi\right)} \int_0^{\pi} d\theta' \sin\theta' \int_0^{2\pi} d\varpi' \left(a^2\cos\theta'^2 + b^2\sin\theta'^2\cos\varpi'^2 + c^2\sin\theta'^2\sin\varpi'^2\right)^{\frac{1-n}{2}},$$

which is evidently a constant quantity. Hence it follows that the expression (30) gives the value of ρ when the fluid is in equilibrium within the ellipsoid, and free from all extraneous action. Moreover, this value is subject, when $n < 2$, to modifications similar to those of the analogous value for the sphere (Art. 7).

ON THE DETERMINATION

OF THE

EXTERIOR AND INTERIOR

ATTRACTIONS OF ELLIPSOIDS

OF

VARIABLE DENSITIES.

* From the *Transactions of the Cambridge Philosophical Society*, 1835.
[Read *May* 6, 1833.]

ON THE DETERMINATION OF THE EXTERIOR AND INTERIOR ATTRACTIONS OF ELLIPSOIDS OF VARIABLE DENSITIES.

THE determination of the attractions of ellipsoids, even on the hypothesis of a uniform density, has, on account of the utility and difficulty of the problem, engaged the attention of the greatest mathematicians. Its solution, first attempted by Newton, has been improved by the successive labours of Maclaurin, d'Alembert, Lagrange, Legendre, Laplace, and Ivory. Before presenting a new solution of such a problem, it will naturally be expected that I should explain in some degree the nature of the method to be employed for that end, in the following paper; and this explanation will be the more requisite, because, from a fear of encroaching too much upon the Society's time, some very comprehensive analytical theorems have been in the first instance given in all their generality.

It is well known, that when the attracted point p is situated within the ellipsoid, the solution of the problem is comparatively easy, but that from a breach of the law of continuity in the values of the attractions when p passes from the interior of the ellipsoid into the exterior space, the functions by which these attractions are given in the former case will not apply to the latter. As however this violation of the law of continuity may always be avoided by simply adding a positive quantity, u^2 for instance, to that under the radical signs in the original integrals, it seemed probable that some advantage might thus be obtained, and the attractions in both cases, deduced from one common formula which would only require the auxiliary variable u to become evanescent in the final result. The principal advantage however which arises from the introduction of the new variable u, depends on the property which a certain function V* then possesses of satisfying a partial differential equation, whenever the law of the attraction is inversely as any power n of the distance. For by a proper application of this equation we may avoid all the difficulty usually presented by the integrations, and at the same time find the required attractions when the density ρ' is expressed by the product of two factors, one of which is a simple algebraic quantity, and the remaining one any rational and entire function of the rectangular co-ordinates of the element to which ρ' belongs.

The original problem being thus brought completely within the pale of analysis, is no longer confined as it were to the three dimensions of space. In fact, ρ' may represent a function of any number s, of independent variables, each of which may be marked with an accent, in order to distinguish this first system from another system of s analogous and unaccented variables, to be afterwards noticed, and V may represent the value of a multiple integral of s dimensions, of which every element is expressed by a fraction having for numerator the continued product of ρ' into the elements of all the accented variables, and for denominator a quantity containing the whole of these, with the unaccented ones also formed exactly on the model of the corresponding one in the value of V belonging to the original problem. Supposing now the auxiliary variable u is introduced, and the s integrations are effected, then will the resulting value of V be a funtion of u and of the s unaccented variables to

* This function in its original form is given by

$$V = \int \frac{\rho' dx' dy' dz'}{\{(x'-x)^2 + (y'-y)^2 + (z'-z)^2\}^{\frac{n-1}{2}}},$$

where $dx'dy'dz'$ represents the volume of any element of the attracting body of which ρ' is the density and x', y', z' are the rectangular co-ordinates; x, y, z being the co-ordinates of the attracted point p. But when we introduce the auxiliary variable u which is to be made equal to zero in the final result,

$$V = \int \frac{\rho' dx' dy' dz'}{\{(x'-x)^2 + (y'-y)^2 + (z'-z)^2 + u^2\}^{\frac{n-1}{2}}};$$

both integrals being supposed to extend over the whole volume of the attracting body.

be determined. But after the introduction of u, the function V has the property of satisfying a partial differential equation of the second order, and by an application of the Calculus of Variations it will be proved in the sequel that the required value of V may always be obtained by merely satisfying this equation, and certain other simple conditions when ρ' is equal to the product of two factors, one of which may be any rational and entire function of the s accented variables, the remaining one being a simple algebraic function whose form continues unchanged, whatever that of the first factor may be.

The chief object of the present paper is to resolve the problem in the more extended signification which we have endeavoured to explain in the preceding paragraph, and, as is by no means unusual, the simplicity of the conclusions corresponds with the generality of the method employed in obtaining them. For when we introduce other variables connected with the original ones by the most simple relations, the rational and entire factor in ρ' still remains rational and entire of the same degree, and may under its altered form be expanded in a series of a finite number of similar quantities, to each of which there corresponds a term in V, expressed by the product of two factors; the first being a rational and entire function of s of the new variables entering into V, and the second a function of the remaining new variable h, whose differential coefficient is an algebraic quantity. Moreover the first is immediately deducible from the corresponding part of ρ' without calculation.

The solution of the problem in its extended signification being thus completed, no difficulties can arise in applying it to pai icular cases. We have therefore on the present occasion given two applications only. In the first, which relates to the attractions of ellipsoids, both the interior and exterior ones are comprised in a common formula agreeably to a preceding observation, and the discontinuity before noticed falls upon one of the independent variables, in functions of which both these attractions are expressed; this variable being constantly equal to zero so long as the attracted point p remains within the ellipsoid, but becoming equal to a determinate function of the co-ordinates of p, when p is situated in the exterior space. Instead too of seeking directly the value of V, all its differentials have first been deduced, and thence the value of V obtained by integration. This slight modification has been given to our method, both because it renders the determination of V in the case considered more easy, and may likewise be usefully employed in the more general one before mentioned. The other application is remarkable both on account of the simplicity of the results to which it leads, and of their analogy with those obtained by Laplace. (*Méc. Cél.* Liv. III. Chap. 2). In fact, it would be easy to shew that these last are only particular cases of the more general ones contained in the article now under notice.

The general solution of the partial differential equation of the second order, deducible from the seventh and three following articles of this paper, and in which the principal variable V is a function of $s+1$ independent variables, is capable of being applied with advantage to various interesting physico-mathematical enquiries. Indeed the law of the distribution of heat in a body of ellipsoidal figure, and that of the motion of a non-elastic fluid over a solid obstacle of similar form, may be thence almost immediately deduced; but the length of our paper entirely precludes any thing more than an allusion to these applications on the present occasion.

---

1. The object of the present paper will be to exhibit certain general analytical formulæ, from which may be deduced as a very particular case the values of the attractions exerted by ellipsoids upon any exterior or interior point, supposing their densities to be represented by functions of great generality.

Let us therefore begin with considering ρ' as a function of the s independent variables

$$x_1', \quad x_2', \quad x_3' \ldots\ldots x_s',$$

and let us afterwards form the function

$$V = \int \frac{dx_1' dx_2' dx_3' \ldots\ldots dx_s' . \rho'}{\{(x_1 - x_1')^2 + (x_2 - x_2')^2 + \ldots + (x_s - x_s')^2 + u^2\}^{\frac{n-1}{2}}} \ldots\ldots (1),$$

the sign $\int$ serving to indicate s integrations relative to the variables x_1', x_2', x_3', ... x_s', and similar to the double and triple ones employed in the solution of geometrical and mechanical problems. Then it is easy to perceive that the function V will satisfy the partial differential equation

$$0 = \frac{d^2V}{dx_1^2} + \frac{d^2V}{dx_2^2} + \dots + \frac{d^2V}{dx_s^2} + \frac{d^2V}{du^2} + \frac{n-s}{u}\frac{dV}{du} \dots\dots (2),$$

seeing that in consequence of the denominator of the expression (1), every one of its elements satisfies for V to the equation (2).

To give an example of the manner in which the multiple integral is to be taken, we may conceive it to comprise all the real values both positive and negative of the variables $x_1' x_2', \dots x_s'$, which satisfy the condition

$$\frac{x_1'^2}{a_1'^2} + \frac{x_2'^2}{a_2'^2} + \frac{x_3'^2}{a_3'^2} + \dots + \frac{x_s'^2}{a_s'^2} < 1 \dots\dots\dots\dots (a),$$

the symbol $<$, as is the case also in what follows, not excluding equality.

2. In order to avoid the difficulties usually attendant on integrations like those of the formula (1), it will here be convenient to notice two or three very simple properties of the function V.

In the first place, then, it is clear that the denominator of the formula (1) may always be expanded in an ascending series of the entire powers of the increments of the variables $x_1, x_2, \dots x_s$, u, and their various products by means of Taylor's Theorem, unless we have simultaneously

$$x_1 = x_1', \quad x_2 = x_2', \dots\dots x_s = x_s' \text{ and } u = 0;$$

and therefore V may always be expanded in a series of like form, unless the $s+1$ equations immediately preceding are all satisfied for one at least of the elements of V. It is thus evident that the function V possesses the property in question, except only when the two conditions

$$\frac{x_1^2}{a_1'^2} + \frac{x_2^2}{a_2'^2} + \frac{x_3^2}{a_3'^2} + \dots + \frac{x_s^2}{a_s^2} < 1 \text{ and } u = 0 \dots\dots\dots\dots (3)$$

are satisfied simultaneously, considering as we shall in what follows the limits of the multiple integral (1) to be determined by the condition (*a*)*.

In like manner it is clear that when

$$\frac{x_1^2}{a_1'^2}+\frac{x_2^2}{a_2'^2}+\ldots+\frac{x_s^2}{a_s'^2}>1 \ldots\ldots\ldots\ldots\ldots\ldots(4),$$

the expansion of V in powers of u will contain none but the even powers of this variable.

Again, it is quite evident from the form of the function V that when any one of the $s+1$ independent variables therein contained becomes infinite, this function will vanish of itself.

3. The three foregoing properties of V combined with the equation (2) will furnish some useful results. In fact, let us consider the quantity

$$\int dx_1 dx_2 \ldots dx_s du\, u^{n-s}.\left\{\left(\frac{dV}{dx_1}\right)^2+\left(\frac{dV}{dx_2}\right)^2+\ldots+\left(\frac{dV}{dx_s}\right)^2+\left(\frac{dV}{du}\right)^2\right\}\ldots(5)$$

where the multiple integral comprises all the real values whether positive or negative of $x_1, x_2, \ldots x_s$, with all the real and positive values of u which satisfy the condition

$$\frac{x_1^2}{a_1^2}+\frac{x_2^2}{a_2^2}+\ldots+\frac{x_s^2}{a_s^2}+\frac{u^2}{h^2}<1 \ldots\ldots\ldots\ldots\ldots\ldots(6),$$

$a_1, a_2, \ldots a_s$ and h being positive constant quantities; and such that we may have generally

$$a_r > a_r'.$$

In this case the multiple integral (5) will have two extreme limits, viz. one in which the conditions

$$\frac{x_1^2}{a_1^2}+\frac{x_2^2}{a_2^2}+\ldots+\frac{x_s^2}{a_s^2}+\frac{u^2}{h^2}=1 \text{ and } u=\text{a positive quantity}\ldots(7)$$

* The necessity of this first property does not explicitly appear in what follows, but it must be understood in order to place the application of the method of integration by parts, in Nos. 3, 4, and 5, beyond the reach of objection. In fact, when V possesses this property, the theorems demonstrated in these Nos. are certainly correct: but they are not necessarily so for every form of the function V, as will be evident from what has been shewn in the third article of my Essay *On the Application of Mathematical Analysis to the Theories of Electricity and Magnetism.* [See pp. 23—27.]

are satisfied; and another defined by

$$\frac{x_1^2}{a_1^2}+\frac{x_2^2}{a_2^2}+\ldots+\frac{x_s^2}{a_s^2}<1 \text{ and } u=0.$$

Moreover, for greater distinctness, we shall mark the quantities belonging to the former with two accents, and those belonging to the latter with one only.

Let us now suppose that V'' is completely given, and likewise V_1' or that portion of V' in which the condition (3) is satisfied; then if we regard V_2' or the rest of V' as quite arbitrary, and afterwards endeavour to make the quantity (5) a minimum, we shall get in the usual way, by applying the Calculus of Variations,

$$0=-\int dx_1\,dx_2\ldots dx_s\,du\,u^{n-s}\,\delta V\left\{\Sigma_1^{s+1}\frac{d^2V}{dx_r^2}+\frac{d^2V}{du^2}+\frac{n-s}{u}\frac{dV}{du}\right\}$$

$$-\int dx_1\,dx_2\ldots dx\,u'^{n-s}\,\delta V_2'\frac{dV_2'}{du}\ldots\ldots\ldots\ldots\ldots(8),$$

seeing that $\delta V''=0$ and $\delta V_1'=0$, because the quantities V'' and V_1' are supposed given.

The first line of the expression immediately preceding gives generally

$$0=\Sigma_1^{s+1}\frac{d^2V}{dx_r^2}+\frac{d^2V}{du^2}+\frac{n-s}{u}\frac{dV}{du}\ldots\ldots\ldots\ldots\ldots(2'),$$

which is identical with the equation (2) No. 1, and the second line gives

$$0=u'^{n-s}\frac{dV_2'}{du}\ (u' \text{ being evanescent})\ldots\ldots\ldots\ldots(9).$$

From the nature of the question *de minimo* just resolved, there can be little doubt but that the equations (2′) and (9) will suffice for the complete determination of V, where V'' and V_1' are both given. But as the truth of this will be of consequence in what follows, we will, before proceeding farther, give a demonstration of it; and the more willingly because it is simple and very general.

4. Now since in the expression (5) u is always positive,

every one of the elements of this expression will therefore be positive; and as moreover V'' and V_1' are given, there must necessarily exist a function V_0 which will render the quantity (5) a proper minimum. But it follows, from the principles of the Calculus of Variations, that this function V_0, whatever it may be, must moreover satisfy the equations (2') and (9). If then there exists any other function V_1, which satisfies the last-named equations, and the given values of V'' and V_1', it is easy to perceive that the function

$$V = V_0 + A\,(V_1 - V_0)$$

will do so likewise, whatever the value of the arbitrary constant quantity A may be. Suppose therefore that A originally equal to zero is augmented successively by the infinitely small increments δA, then the corresponding increment of V will be

$$\delta V = (V_1 - V_0)\,\delta A,$$

and the quantity (5) will remain constantly equal to its minimum value, however great A may become, seeing that by what precedes the variation of this quantity must be equal to zero whatever the variation of V may be, provided the foregoing conditions are all satisfied. If then, besides V_0 there exists another function V_1 satisfying them all, we might give to the partial differentials of V, any values however great, by augmenting the quantity A sufficiently, and thus cause the quantity (5) to exceed any finite positive one, contrary to what has just been proved. Hence no such value as V_1 exists.

We thus see that when V'' and V_1' are both given, there is one and only one way of satisfying simultaneously the partial differential equation (2), and the condition (9).

5. Again, it is clear that the condition (4) is satisfied for the whole of V_2'; and it has before been observed (No. 2) that when V is determined by the formula (1), it may always be expanded in a series of the form

$$V = A + Bu^2 + Cu^4 + \&c.$$

Hence the right side of the equation (9) is a quantity of the order u'^{n-s+1}; and u' being evanescent, this equation will then evidently be satisfied, provided we suppose, as we shall in what follows, that

$$n-s+1 \text{ is positive.}$$

If now we could by any means determine the values of V'' and V_1' belonging to the expression (1), the value of V would be had without integration by simply satisfying (2′) and (9), as is evident from what precedes. But by supposing all the constant quantities $a_1, a_2, a_3 \ldots\ldots a_s$ and h infinite, it is clear that we shall have

$$0 = V'',$$

and then we have only to find V_1', and thence deduce the general value of V.

6. For this purpose let us consider the quantity

$$\int dx_1 dx_2 \ldots dx_s du u^{n-s} \left\{ \frac{dV}{dx_1}\frac{dU}{dx_1} + \frac{dV}{dx_2}\frac{dU}{dx_2} + \ldots \right.$$
$$\left. + \frac{dV}{dx_s}\frac{dU}{dx_s} + \frac{dV}{du}\frac{dU}{du} \right\} \ldots\ldots\ldots\ldots (10);$$

the limits of the multiple integral being the same as those of the expression (5), and U being a function of $x_1, x_2, \ldots\ldots x_s$ and u, satisfying the condition $0 = U''$ when $a_1, a_2, \ldots\ldots a_s$ and h are infinite.

But the method of integration by parts reduces the quantity (10) to

$$-\int dx_1 dx_2 \ldots\ldots dx_s \frac{dU'}{du} u'^{n-s} . V'$$

$$-\int dx_1 dx_2 \ldots dx_s du u^{n-s} V \left\{ \Sigma_1^{s+1} \frac{d^2U}{dx_r^2} + \frac{d^2U}{du^2} + \frac{n-s}{u}\frac{dU}{du} \right\} \ldots\ldots (11),$$

since $0 = V''$; and as we have likewise $0 = U''$, the same quantity (10) may also be put under the form

$$-\int dx_1 dx_2 \ldots\ldots dx_s \frac{dV'}{du} u'^{n-s} . U'$$

$$-\int dx_1 dx_2 \ldots dx_s du u^{n-s} . U \left\{ \Sigma_1^{s+1} \frac{d^2V}{dx_r^2} + \frac{d^2V}{du^2} + \frac{n-s}{u}\frac{dV}{du} \right\} \ldots (12).$$

Supposing therefore that U like V also satisfies the equation (2′), each of the expressions (11) and (12) will be reduced to its upper line, and we shall get by equating these two forms of the same quantity:

$$\int dx_1 dx_2 \ldots dx_s \frac{dU'}{du} u'^{n-s} V' = \int dx_1 dx_2 \ldots dx_s \frac{dV'}{du} u'^{n-s} U';$$

the quantities bearing an accent belonging, as was before explained, to one of the extreme limits.

Because V satisfies the condition (9), the equation immediately preceding may be written

$$\int dx_1 dx_2 \ldots dx_s \frac{dU'}{du} u'^{n-s} V' = \int dx_1 dx_2 \ldots dx_s \frac{dV_1'}{du} u'^{n-s} U_1'.$$

If now we give to the general function U the particular value

$$U = \{(x_1 - x_1'')^2 + (x_2 - x_2'')^2 + \ldots + (x_s - x_s'')^2 + u^2\}^{\frac{1-n}{2}},$$

which is admissible, since it satisfies for V to the equation (2), and gives $U'' = 0$, the last formula will become

$$\int \frac{dx_1 dx_2 \ldots dx_s u'^{n-s} \dfrac{dV_1'}{du}}{\{(x_1 - x_1'')^2 + (x_2 - x_2'')^2 + \ldots + (x_s - x_s'')^2 + u'^2\}^{\frac{n-1}{2}}}$$

$$= \int \frac{dx_1 dx_2 \ldots dx_s . (1-n)\, u'^{n-s+1} V'}{\{(x_1 - x_1'')^2 + (x_2 - x_2'')^2 + \ldots + (x_s - x_s'')^2 + u'^2\}^{\frac{n+1}{2}}} \ldots\ldots(13),$$

in which expression u' must be regarded as an evanescent positive quantity.

In order now to effect the integrations indicated in the second member of this equation, let us make

$$x_1 - x_1'' = u'\rho \cos\theta_1; \quad x_2 - x_2'' = u'\rho \sin\theta_1 \cos\theta_2;$$
$$x_3 - x_3'' = u'\rho \sin\theta_1 \sin\theta_2 \cos\theta_3, \text{ \&c.}$$

until we arrive at the two last, viz.,

$$x_{s-1} - x''_{s-1} = u'\rho \sin\theta_1 \sin\theta_2 \ldots \sin\theta_{s-2} \cos\theta_{s-1},$$
$$x_s \;-\; x_s'' = u'\rho \sin\theta_1 \sin\theta_2 \ldots \sin\theta_{s-2} \sin\theta_{s-1}$$

u' being, as before, a vanishing quantity.

Then by the ordinary formulæ for the transformation of multiple integrals we get

$$dx_1 dx_2 \dots dx_s = u'^s \rho^{s-1} \sin\theta_1^{s-2} \sin\theta_2^{s-3} \dots \sin\theta^1_{s-2} d\rho\, d\theta_1\, d\theta_2 \dots d\theta_{s-1},$$

and the second number of the equation (13) by substitution will become

$$\int \frac{d\rho\, d\theta_1\, d\theta_2 \dots d\theta_{s-1}\, \rho^{s-1} \sin\theta_1^{s-2} \sin\theta_2^{s-3} \dots \sin\theta_{s-2} . (1-n) V'}{(1+\rho^2)^{\frac{n+1}{2}}} \dots (14).$$

But since u' is evanescent, we shall have ρ infinite, whenever $x_1, x_2, \dots x_s$ differ sensibly from $x_1'', x_2'', \dots x_s''$; and as moreover $n-s+1$ is positive, it is easy to perceive that we may neglect all the parts of the last integral for which these differences are sensible. Hence V' may be replaced with the constant value V_0' in which we have generally

$$x_r = x_r''.$$

Again, because the integrals in (14) ought to be taken from $\theta_{s-1} = 0$ to $\theta_{s-1} = 2\pi$, and afterwards from $\theta_r = 0$ to $\theta_r = \pi$, whatever whole number less than $s-1$ may be represented by r, we easily obtain by means of the well known function Gamma:

$$\int \sin\theta_1^{s-2} \sin\theta_2^{s-3} \sin\theta_3^{s-4} \dots\dots \sin\theta_{s-2}\, d\theta_1\, d\theta_2 \dots d\theta_{s-1} = \frac{2\pi^{\frac{s}{2}}}{\Gamma\left(\frac{s}{2}\right)},$$

and as by the aid of the same function we readily get

$$\int_0^\infty \frac{\rho^{s-1}\, d\rho}{(1+\rho^2)^{\frac{n+1}{2}}} = \frac{\Gamma\left(\frac{s}{2}\right)\Gamma\left(\frac{n-s+1}{2}\right)}{2\Gamma\left(\frac{n+1}{2}\right)},$$

the integral (14) will in consequence become

$$\frac{-2\pi^{\frac{s}{2}} . \Gamma\left(\frac{n-s+1}{2}\right)}{\Gamma\left(\frac{n-1}{2}\right)} V_0',$$

and thus the equation (13) will take the form

$$\int \frac{dx_1 dx_2 \ldots dx_s u'^{n-s} \dfrac{dV_1'}{du}}{\{(x_1 - x_1'')^2 + (x_2 - x_2'')^2 + \ldots + (x_s - x_s'')^2 + u'^2\}^{\frac{n-1}{2}}}$$

$$= \frac{-2\pi^{\frac{s}{2}} . \Gamma\left(\dfrac{n-s+1}{2}\right)}{\Gamma\left(\dfrac{n-1}{2}\right)} V_0'.$$

In this equation V is supposed to be such a function of x_1, $x_2 \ldots x_s$ and u, that the equation (2) and condition (9) are both satisfied. Moreover $V'' = 0$, and V_0' is the particular value of V for which

$$x_1 = x_1''; \quad x_2 = x_2''; \ldots\ldots x_s = x_s'', \text{ and } u = 0.$$

Let us now make, for abridgment,

$$P = u^{n-s} \frac{dV}{du}, \text{ (when } u = 0) \ldots\ldots\ldots\ldots\ldots\ldots (b),$$

and afterwards change x into x', and x'' into x in the expression immediately preceding, there will then result

$$\int \frac{dx_1' dx_2' \ldots\ldots dx_s' P_1'}{\{(x_1' - x_1)^2 + (x_2' - x_2)^2 + \ldots + (x_s' - x_s)^2 + u'^2\}^{\frac{n-1}{2}}}$$

$$= \frac{-2\pi^{\frac{s}{2}} \Gamma\left(\dfrac{n-s+1}{2}\right)}{\Gamma\left(\dfrac{n-1}{2}\right)} V' \ldots\ldots\ldots\ldots\ldots\ldots (15),$$

P' being what P becomes by changing generally x_r into x_r', the unit attached to the foot of P' indicating, as before, that the multiple integral comprises only the values admitted by the condition (a), and V' being what V becomes when we make $u = 0$.

The equation just given supposes u' evanescent; but if we were to replace u' with the general value u in the first member, and make a corresponding change in the second by replacing V' with the general value V, this equation would still be correct, and we should thus have

$$\int \frac{dx_1'dx_2'\ldots\ldots dx_s'P_1'}{\{x_1'-x_1)^2+(x_2'-x_2)^2+\ldots+(x_s'-x_s)^2+u^2\}^{\frac{n-1}{2}}}$$
$$=\frac{-2\pi^{\frac{s}{2}}\Gamma\left(\frac{n-s+1}{2}\right)}{\Gamma\left(\frac{n-1}{2}\right)}V\ldots\ldots\ldots\ldots\ldots\ldots\ldots(16).$$

For under the present form both its members evidently satisfy the equation (2), the condition (9), and give $V''=0$. Moreover, when the condition (3) is satisfied, the same members are equal in consequence of (15). Hence by what has before been proved (No. 4), they are necessarily equal in general.

By comparing the equation (16) with the formula (1), it will become evident, that whenever we can by any means obtain a value of V satisfying the foregoing conditions, we shall always be able to assign a value of ρ' which substituted in (1) shall reproduce this value of V. In fact, by omitting the unit at the foot of P', which only serves to indicate the limits of the integral, we readily see that the required value of ρ' is

$$\rho'=-\frac{\Gamma\left(\frac{n-1}{2}\right)}{2\pi^{\frac{s}{2}}\Gamma\left(\frac{n-s+1}{2}\right)}P'\ldots\ldots\ldots\ldots\ldots\ldots\ldots(c).$$

7. The foregoing results being obtained, it will now be convenient to introduce other independent variables in the place of the original ones, such that

$$x_1=a_1\xi_1,\quad x_2=a_2\xi_2,\ldots x_s=a_s\xi_s,\quad u=hv,$$

a_1, a_2, ... a_s being functions of h, one of the new independent variables, determined by

$$a_1^2=a_1'^2+h^2,\quad a_2^2=a_2'^2+h^2,\ldots a_s^2=a_s'^2+h^2,$$

and v a function of the remaining new variables, ξ_1, ξ_2, ξ_3, ... ξ_s satisfying the equation

$$1=v^2+\xi_1^2+\xi_2^2+\ldots+\xi_s^2;$$

a_1', a_2', a_3', ... a_s' being the same constant quantities as in the equation (a), No. 1. Moreover, a_1, a_2, ... a_s will take the values belonging to the extreme limit before marked with two accents, by simply assigning to h an infinite value.

The easiest way of transforming the equation (2) will be to remark, that it is the general one which presents itself when we apply the Calculus of Variations to the quantity (5), in order to render it a minimum. We have therefore in the first place

$$\left(\frac{dV}{du}\right)^2 + \Sigma_1^{s+1}\left(\frac{dV}{dx_r}\right)^2 = \Sigma_1^{s+1}\left(\frac{dV}{a_r d\xi_r}\right)^2$$

$$+ \left\{\left(\frac{dV}{dh}\right)^2 - \left(\Sigma_1^{s+1}\frac{h\xi_r}{a_r^2}\frac{dV}{d\xi_r}\right)^2\right\}\left(1 - \Sigma_1^{s+1}\frac{a_r'^2\xi_r^2}{a_r^2}\right)^{-1};$$

and by the ordinary formula for the transformation of multiple integrals,

$$dx_1 dx_2 \ldots dx_s du = \frac{a_1 a_2 \ldots a_s}{v}\left(1 - \Sigma_1^{s+1}\frac{a_r'^2\xi_r^2}{a_r^2}\right) d\xi_1 d\xi_2 \ldots d\xi_s dh.$$

But since $1 - \Sigma_1^{s+1}\frac{a_r'^2\xi_r^2}{a^2} = v^2 + h^2\Sigma_1^{s+1}\frac{\xi_r^2}{a_r^2}$,

the expression (5) after substitution will become

$$\int d\xi_1 d\xi_2 \ldots d\xi\, dh\, a_1 a_2 a_3 \ldots a_s h^{n-s} v^{n-s-1} \ldots\ldots$$

$$\left\{\left(v^2 + h^2\Sigma_1^{s+1}\frac{\xi_r^2}{a_r^2}\right)\Sigma_1^{s+1}\left(\frac{dV}{a_r d\xi_r}\right)^2 + \left(\frac{dV}{dh}\right)^2 - h^2\left(\Sigma_1^{s+1}\frac{\xi_r}{a_r^2}\frac{dV}{d\xi_r}\right)^2\right\}.$$

Applying now the method of integration by parts to the variation of this quantity, by reduction, we get for the equivalent of (2) the equation

$$0 = \frac{d^2V}{dh^2} + \left(n - \Sigma\frac{a_r'^2}{a^2}\right)\frac{dV}{h\,dh} + (1 - \Sigma\xi_r^2)\,\Sigma\frac{d^2V}{a_r^2 d\xi^2}$$

$$+ (s - n - 1)\,\Sigma\frac{\xi_r}{a_r^2}\frac{dV}{d\xi_r}$$

$$+ h^2\Sigma\frac{\xi^2}{a_r^2}\times\Sigma\frac{d^2V}{a_r^2 d\xi_r^2} - h^2\Sigma\Sigma\frac{\xi_r\xi_{r'}}{a_r^2 a_{r'}^2}\frac{d^2V}{d\xi_r d\xi_{r'}}\ldots\ldots\ldots(2'')$$

$$+ h^2\Sigma\frac{\xi_r dV}{a_r^4 d\xi_r} - h^2\Sigma\frac{1}{a_r^2}\times\Sigma\frac{\xi_r dV}{a_r^2 d\xi_r};$$

where the finite integrals are all supposed taken from $r = 1$ to $r = s + 1$, and from $r' = 1$ to $r' = s + 1$.

The last equation may be put under the abridged form,

$$0=\frac{d^2V}{dh^2}+\left(n-\Sigma\frac{a_r'^2}{a_r^2}\right)\frac{dV}{hdh}+\nabla V\ldots\ldots\ldots\ldots\ldots(2'''),$$

provided we have generally

$$\text{coefficient of } \frac{d^2V}{d\xi_r^2} \text{ in } \nabla V=\frac{1}{a_r^2}\left(1-\xi_r^2-\Sigma_1^{s+1}\frac{a_r'^2}{a_{r'}}\xi_{r'}^2+\frac{a_r'^2}{a_r^2}\xi_r^2\right),$$

$$\text{coefficient of } \frac{d^2V}{d\xi_r d\xi_{r'}} \text{ in } \nabla V=-\frac{2h^2}{a_r^2 a_{r'}^2}\xi_r\xi_{r'},$$

$$\text{coefficient of } \frac{dV}{d\xi_r} \text{ in } \nabla V=\frac{\xi_r}{a_r^2}\left(-n+\Sigma\frac{a_{r'}'^2}{a_{r'}^2}-\frac{a_r'^2}{a_r^2}\right).$$

Moreover, when we employ the new variables

$$\frac{dV}{du}=-v\left(1-\Sigma\frac{a_r'^2\xi_r^2}{a_r^2}\right)^{-1}.\left\{\Sigma\frac{h\xi_r}{a_r^2}\frac{dV}{d\xi_r}-\frac{dV}{dh}\right\},$$

and therefore the condition (9) in like manner will become

$$0=v^{n-s+1}h^{n-s}\left(1-\Sigma\frac{a_r'^2\xi_r^2}{a_r^2}\right)^{-1}.\left\{\Sigma\frac{h\xi_r}{a_r^2}\frac{dV}{d\xi_r}-\frac{dV}{dh}\right\}\ldots\ldots\ldots(9');$$

where the values of the variables $\xi_1, \xi_2, \ldots \xi_s$ must be such as satisfy the equation $v^2=0$, whatever h may be; and as $n-s+1$ is positive, it is clear that this condition will always be satisfied, provided the partial differentials of V relative to the new variables are all finite.

8. Let us now try whether it is possible to satisfy the equation (2''') by means of a function of the form

$$V=H\phi\ldots\ldots\ldots\ldots\ldots\ldots\ldots\ldots\ldots(\beta);$$

H depending on the variable h only, and ϕ being a rational and entire function of $\xi_1, \xi_2, \ldots \xi_s$ of the degree γ, and quite independent of h.

By substituting this value of V in (2''') and making

$$0=\frac{d^2H}{dh^2}+\left(n-\Sigma\frac{a_r'^2}{a_r^2}\right)\frac{dH}{hdh}+\kappa H\ldots\ldots\ldots(17),$$

we readily get

$$0=\nabla\phi-\kappa\phi\ldots\ldots\ldots\ldots\ldots\ldots\ldots(18);$$

where, in virtue of (17) κ must necessarily be a function of h only; and as the required value of ϕ, if it exist, must be independent of h, we have, by making $h=0$ in the equation immediately preceding,

$$0 = \nabla'\phi - k_0\phi \text{ .... } (19);$$

k_0 being the value κ, and $\nabla'\phi$ that of $\nabla\phi$ when $h=0$.

We shall demonstrate almost immediately that every function ϕ of the form (20), No. 9, which satisfies the equation (19), and which therefore is independent of h, will likewise satisfy the equation (18); and the corresponding value of κ obtained from the latter being substituted in the ordinary differential equation (17), we shall only have to integrate this last in order to have a proper value of V.

9. To satisfy the equation (19) let us assume

$$\phi = F(\xi_1^2, \xi_2^2, \xi_3^2, \ldots \xi_s^2)\, \xi_p, \xi_q, \&c. \text{ .... } (20);$$

F being the characteristic of a rational and entire function of the degree $2\gamma'$, and the most general of its kind, and ξ_p, ξ_q, &c. designating the variables in ϕ which are affected with odd exponents only; so that if their number be ν we shall have

$$\gamma = 2\gamma' + \nu,$$

the remaining variables having none but even exponents. Then it is easy to perceive, that after substitution the second member of the equation (19) will be precisely of the same form as the assumed value of ϕ, and by equating separately to zero the coefficients of the various powers and products of $\xi_1, \xi_2, \ldots \xi_s$, we shall obtain just the same number of linear algebraic equations as there are coefficients in ϕ, and consequently be enabled to determine the ratios of these coefficients together with the constant quantity k_0.

In fact, by writing the foregoing value of ϕ under the form

$$\phi = SA_{m_1, m_2, \ldots m_s}\, \xi_1^{m_1} \xi_2^{m_2} \ldots\ldots \xi_s^{m_s} \text{ .... } (20');$$

and proceeding as above described, the coefficient of

$$\xi_1^{m_1} \xi_2^{m_2} \ldots\ldots \xi_s^{m_s}$$

will give the general equation

$$0 = \left\{\Sigma_1^{s+1} \frac{m_r(m_r - s + n)}{a_r'^2} + k_0\right\} A_{m_1, m_2 \ldots m_s}$$

$$+ \Sigma\Sigma \frac{(m_r + 2)(m_r + 1)}{a_r'^2} A_{m_1, m_2, \ldots m_{2+2} \ldots m_{r'-2}, \ldots m_s} \ldots (21)$$

$$- \Sigma_1^{s+1} \frac{(m_r + 2)(m_r + 1)}{a_r'^2} A_{m_1, m_2, \ldots m_{r+2}, \ldots m_s};$$

the double finite integral comprising all the values of r and r', except those in which $r = r'$, and consequently containing when completely expanded $s(s-1)$ terms.

For the terms of the highest degree γ and of which the number is

$$\frac{\gamma' + 1 . \gamma' + 2 \ldots\ldots\ldots \gamma' + s - 1}{1 . 2 . 3 \ldots\ldots\ldots s - 1} = N,$$

the last line of the expression (21) evidently vanishes, and thus we obtain N distinct linear equations between the coefficients of the degree γ in ϕ and k_0.

Moreover, from the form of these equations it is evident that we may obtain by elimination one equation in k_0 of the degree N, of which each of the N roots will give a distinct value of the function $\phi^{(\gamma)}$, having one arbitrary constant for factor; the homogeneous function $\phi^{(\gamma)}$ being composed of all the terms of the highest degree, γ in ϕ. But the coefficients of $\phi^{(\gamma)}$ and k_0 being known, we may thence easily deduce all the remaining coefficients in ϕ, by means of the formula (21).

Now, since the N linear equations have no terms except those of which the coefficients of $\phi^{(\gamma)}$ are factors, it follows that if k_0 were taken at will, the resulting values of all these coefficients would be equal to zero. If however we obtain the values of $N-1$ of the coefficients in terms of the remaining one A from $N-1$ of the equations, by the ordinary formulæ, and substitute these in the remaining equation, we shall get a result of the form

$$K . A = 0,$$

where K is a function of k_0 of the degree N. We shall thus have only two cases to consider: First, that in which $A=0$, and consequently also all the other coefficients of $\phi^{(\gamma)}$ together with the remaining ones in ϕ, as will be evident from the formula (21). Hence, in this case

$$\phi=0:$$

Secondly, that in which k_0 is one of the N roots of $0=K$, as for instance, k_0' in this case all the coefficients of ϕ will become multiples of A, and we shall have

$$\phi=A\phi_1:$$

ϕ_1 being a determinate function of ξ_1, ξ_2, ... ξ_s.

We thus see that when we consider functions of the form (20) only, the most general solution that the equation

$$0=\nabla'\overline{\phi}-k_0'\overline{\phi} \quad \text{.....................(19')}$$

admits is ....

$$\text{or,}\quad \overline{\phi}=0;\qquad \text{or,}\quad \overline{\phi}=\alpha\phi;$$

α being a quantity independent of ξ_1, ξ_2, ... ξ_s, and ϕ any function which satisfies for $\overline{\phi}$ to the equation (19'). But by affecting both sides of the equation

$$0=\nabla'\phi-k_0'\phi$$

with the symbol ∇, we get

$$0=\nabla\,.\,\nabla'\phi-k_0'\,.\,\nabla\phi;$$

and we shall afterwards prove the operations indicated by ∇ and ∇' to be such, that whatever ϕ may be,

$$\nabla\nabla'\phi=\nabla'\nabla\phi.$$

Hence, the last equation becomes

$$\nabla'(\nabla\phi)-k_0'\nabla\phi;$$

and as $\nabla\phi$ like ϕ is of the form (20), it follows from what has just been shewn, that either

$$0=\nabla\phi,\quad \text{or,}\quad \nabla\phi=\alpha\phi,$$

α being a quantity independent of ξ_1, ξ_2, ... ξ_s.

The first is inadmissible, since it would give $\phi=0$; therefore when ϕ satisfies (19'), we have

$$\nabla\phi'=\alpha\phi,\quad \text{i.e.}\ 0=\nabla\phi-\alpha\phi.$$

But since α is independent of $\xi_1, \xi_2, \ldots\ldots \xi_s$, this last equation is evidently identical with (18), since the equation (18) merely requires that κ should be independent of $\xi_1, \xi_2, \ldots\ldots \xi_s$.

Having thus proved that every function of the form (20) which satisfies (19) will likewise satisfy (18), it will be more simple to determine the remaining coefficients of ϕ from those of $\phi^{(\gamma)}$ by means of the last equation, than to employ the formula (21) for that purpose.

Making therefore h infinite in (18), and writing $\frac{k_1}{h^2}$ in the place of κ, we get

$$0 = \Sigma_1^{s+1} . (1 - \xi_r^2) \frac{d^2\phi}{d\xi_r^2} - 2 (\Sigma\Sigma) \xi_r \xi_{r'} \frac{d^2\phi}{d\xi_r d\xi_{r'}} - n\Sigma_1^{s+1} \xi_r \frac{d\phi}{d\xi_r} - k_1 \phi$$

where $(\Sigma\Sigma)$ comprises the $\frac{s(s-1)}{1.2}$ combinations which can be formed of the s indices taken in pairs.

If now we substitute the value of ϕ before given (20′), and recollect that for the terms of the highest degree we have $\Sigma m_r = \gamma$, we shall readily get

$$0 = (\gamma - \Sigma m_r)(\gamma + \Sigma m_r + n - 1) A_{m_1, m_2, \ldots m_s}$$
$$+ (m_r + 2)(m_r + 1) A_{m_1, m_{r+2}, \ldots m_s} \ldots\ldots (22),$$

from which all the remaining coefficients in ϕ will readily be deduced, when those of the part $\phi^{(\gamma)}$ are known.

10. It now remains, as was before observed, to integrate the ordinary differential equation (17) No. 8. But, by the known theory of linear equations, the integration of (17) will always become more simple when we have a particular value satisfying it, and fortunately in the present case such a value may always be obtained from ϕ by simply changing ξ_r into $\frac{a_r}{\sqrt{(\Sigma a_r'^2)}}$. In fact, if we represent the value thus obtained by H_0 we shall have

$$\frac{dH_0}{dh} = \Sigma_1^{s+1} \frac{d\phi}{d\xi_r} \cdot \frac{h}{a_r \sqrt{(\Sigma a_r'^2)}},$$

and by a second differentiation

$$\frac{d^2H_0}{dh^2}=\Sigma\frac{d\phi}{d\xi_r}\cdot\frac{a_r'^2}{a_r^3\sqrt{(\Sigma a_r'^2)}}+\Sigma\frac{d^2\phi}{d\xi_r^2}\cdot\frac{h^2}{a_r^2\,.\,\Sigma a_r'^2}$$

$$+2\,(\Sigma\Sigma)\,\frac{d^2\phi}{d\xi_r d\xi_{r'}}\cdot\frac{h^2}{a_r a_{r'}\Sigma a_r'^2},$$

$(\Sigma\Sigma)$ as before comprising all the $\frac{s\,.\,s-1}{1\,.\,2}$ combinations of the s indices taken in pairs.

Hence, the quantity on the right side of the equation (17), when we make $H=H_0$, becomes

$$\Sigma\frac{d\phi}{d\xi_r}\cdot\frac{a_r'^2}{a_r^3\sqrt{(\Sigma a_r'^2)}}+\Sigma\frac{d^2\phi}{d\xi_r^2}\cdot\frac{h^2}{a_r^2\Sigma a_r'^2}+\kappa\phi$$

$$+2\,(\Sigma\Sigma)\,\frac{d^2\phi}{d\xi_r d\xi_{r'}}\cdot\frac{h^2}{a_r a_{r'}\Sigma a_r'^2}+\left(n-\Sigma\frac{a_r'^2}{a_r^2}\right)\Sigma\frac{d\phi}{d\xi_r}\cdot\frac{1}{a_r\sqrt{(\Sigma a_r'^2)}}\ \dots\ (23).$$

But if we recollect that we have generally

$$\xi_r=\frac{a_r}{\sqrt{(\Sigma a_r'^2)}}\ \dots\dots\dots\dots\dots\dots\dots(24),$$

it is easy to perceive that in consequence of the equation (18) the quantity (23) will vanish, and therefore the foregoing value of H_0 will always satisfy the equation (17).

Having thus a particular value of H, we immediately get the general one by assuming

$$H=H_0\int zdh.$$

In fact, there thence results

$$H=KH_0\int\frac{h^{s-n}dh}{H_0^2\,a_1,\,a_2,\,a_3\,\dots\dots\,a_s},$$

the two arbitrary constants which the general integral ought to contain being K, and that which enters implicitly into the indefinite integral. But the condition $0=V''$ requires that H should vanish when h is infinite, and consequently the particular value adapted to the present investigation is

$$H_0=K\,.\,H_0\int_\infty\frac{h^{s-n}dh}{H_0^2\,a_1,\,a_2\,\dots\dots\,a_s}.$$

11. The value of ϕ and H being known, we may readily find the corresponding values of V and ρ'. For we have immediately

$$V = H\phi = K\phi H_0 \int_\infty \frac{h^{s-n}dh}{H_0^2 a_1, a_2, \ldots\ldots a_s} \quad \ldots\ldots\ldots\ldots\ldots (26),$$

and as the function ϕ is rational and entire, and the partial differential of V relative to h is finite, it follows that all the partial differentials of V are finite; and consequently, by what precedes (No. 7) the condition (9') is satisfied by the foregoing value of V, as well as the equation (2) and condition $0 = V''$. Hence the equations (b) and (c) No. 6 will give, since

$$\frac{dV}{du} = -v\left(1 - \Sigma_1^{s+1} \frac{a_r'^2 \xi_r^2}{a_r^2}\right)^{-1} \cdot \left(\Sigma_1^{s+1} \frac{h\xi_r}{a_r^2} \cdot \frac{dV}{d\xi_r} - \frac{dV}{dh}\right),$$

and h must be supposed equal to zero in these equations,

$$\rho' = \frac{-\Gamma\left(\frac{n-1}{2}\right)}{2\pi^{\frac{s}{2}}\Gamma\left(\frac{n-s+1}{2}\right)} v^{n-s-1} \cdot h^{n-s} \frac{dV}{dh} \ldots\ldots \text{(where } h = 0\text{)}:$$

since where $h = 0$, $a_r = a_r'$; and therefore

$$1 - \Sigma_1^{s+1} \frac{a_r'^2 \xi_r^2}{a^2 r} = 1 - \Sigma_1^{s+1} \xi_r^2 = v^2.$$

If now we substitute for V its value (26), and recollect that $n - s + 1$ is always positive, we get

$$\rho' = \frac{-\Gamma\left(\frac{n-1}{2}\right)}{2\pi^{\frac{s}{2}}\Gamma\left(\frac{n-s+1}{2}\right)} v^{n-s-1} \phi' \frac{K}{H_0' a_1', a_2', \ldots\ldots a_s'} \quad \ldots\ldots (27),$$

since it is clear from the form of H_0 that this quantity may always be expanded in a series of the entire powers of h^2. In the preceding expression, (27), H_0' indicates the value of H_0 when $h = 0$, and ϕ' the corresponding value of ϕ or that which would be obtained by simply changing the unaccented letter $\xi_1, \xi_2, \ldots\ldots \xi_s$ into the accented ones $\xi_1', \xi_2', \ldots\ldots \xi_s'$ deduced from

$$(\gamma) \qquad x' = a_1'\xi_1'; \qquad x_2' = a_2'\xi_2'; \qquad x_s' = a_s'\xi_s'.$$

It will now be easy to obtain the value of V corresponding to

$$\rho' = \left(1 - \frac{x_1'^2}{a_1'^2} - \frac{x_2'^2}{a_2'^2} - \ldots\ldots - \frac{x_s'^2}{a_s'^2}\right)^{\frac{n-s-1}{2}} F(x_1', x_2', \ldots\ldots x_s') \ldots(28)$$

without integrating the formula (1) No. 1, where F is the characteristic of any rational and entire function. In fact it is easy to see from what precedes (No. 9), that we may always expand F in a finite series of the form

$$F(x_1', x_2', \ldots\ldots x_s') = b_0\phi^{0'} + b_1\phi_1' + b_2\phi_2' + b_3\phi_3' + \&c.$$

after x_1', x_2', &c. have been replaced with their values (γ). Hence, we immediately get

$$\rho' = v^{n-s-1}. \{b_0\phi_0' + b_1\phi_1' + b_2\phi_2' + \&c.\} \ldots\ldots\ldots\ldots(29).$$

By comparing the formulæ (26) and (27) it is clear that any term, as $b_r\phi_r'$ for instance, of the series entering into ρ', will have for corresponding term in the required value of V, the quantity

$$-\frac{2\pi^{\frac{s}{2}}\Gamma\left(\frac{n-s+1}{2}\right)}{\Gamma\left(\frac{n-1}{2}\right)} H_0' a_1' a_2' \ldots a_s' . b_r\phi_r H_0 \int_\infty \frac{h^{s-n}dh}{H_0^2 a_1 a_2 \ldots a_s} \ldots.(30):$$

H_0 being a particular value of H satisfying the equation (17) and immediately deducible from ϕ by the method before explained.

12. All that now remains, is to demonstrate that

$$\nabla'\nabla\phi = \nabla\nabla'\phi \ldots\ldots\ldots\ldots\ldots\ldots\ldots\ldots(31),$$

whatever ϕ may be. For this purpose let us here resume the value of $\nabla\phi$, as immediately deduced from the equation (2″) No. 7, viz.

$$\nabla'\phi = (1 - \Sigma\xi^2)\,\Sigma\frac{d^2\phi}{a^2 d\xi^2} + (s - n - 1)\,\Sigma\frac{\xi d\phi}{a^2 d\xi}$$

$$+ h^2\Sigma\frac{\xi^2}{a^2} \times \Sigma\frac{d^2\phi}{a^2 d\xi^2} - h^2\Sigma\Sigma\frac{\xi\xi'}{a^2 a'^2}\frac{d^2\phi}{d\xi d\xi'}$$

$$+ h^2\Sigma\frac{\xi d\phi}{a^4 d\xi} - h^2\Sigma\frac{1}{a_2} \times \Sigma\frac{\xi d\phi}{a^2 d\xi} \ldots\ldots\ldots\ldots\ldots\ldots(32),$$

where for simplicity the indices at the foot of the letters ξ and a have been omitted, and their accents transferred to the letters themselves. Moreover all the finite integrals are supposed taken from 1 to $s+1$.

By making $h=0$ in the last expression we immediately get $\nabla'\phi$, and if for a moment, to prevent ambiguity, we write b_r in the place of the original a_r' and omit the lower indices as before, we obtain

$$\nabla'\phi = (1-\Sigma\xi^2)\,\Sigma\,\frac{d^2\phi}{a''^2 d\xi''^2} + (s-n-1)\,\Sigma\,\frac{\xi'' d\phi}{b''^2 d\xi'} \;\ldots\ldots(33);$$

where to avoid all risk of confusion r has been changed into r'', and the double accent of this index transferred to the letters ξ and b themselves.

We will now conceive the expression (32) to be written in the abridged form

$$\nabla\phi = \nabla_1\phi + h^2\nabla_2\phi - h^2\nabla_3\phi + h^2\nabla_4\phi - h^2\nabla_5\phi,$$

the order of the terms remaining unchanged.

If then we recollect that the accents have no other office to perform than to keep the various finite integrations quite distinct, and consequently that in the final results they may be permuted in any way at will, we shall readily get

$$\nabla'\nabla_1\phi - \nabla_1\nabla'\phi =$$

$$(1-\Sigma\xi^2)\left\{4\Sigma\Sigma\left(\frac{1}{a'^2b^2}-\frac{1}{a^2b'^2}\right)\frac{\xi' d^3\phi}{d\xi^2 d\xi'}_{(1)}\right.$$

$$\left. + 2\Sigma\,\frac{1}{a^2}\times\Sigma\,\frac{d^2\phi}{b^2d\xi^2}_{(2)} - 2\Sigma\,\frac{1}{b^2}\times\Sigma\,\frac{a^2\phi}{a^2d\xi^2}_{(3)}\right\}$$

$$+2\,(s-n-1)\left\{\Sigma\,\frac{\xi^2}{a^2}\times\Sigma\,\frac{d^2\phi}{b^2d\xi^2}_{(4)} - \Sigma\,\frac{\xi^2}{b^2}\times\Sigma\,\frac{d^2\phi}{a^2d\xi^2}_{(5)}\right\}$$

$$\nabla'\nabla_2\phi - \nabla_2\nabla'\phi =$$

$$(1-\Sigma\xi^2)\left\{4\Sigma\Sigma\,\frac{\xi'}{a^2a'^2b'^2}\cdot\frac{d^3\phi}{d\xi^2d\xi'}_{(6)} + 2\Sigma\,\frac{1}{a^2b^2}\times\Sigma\,\frac{d^2\phi}{a^2d\xi^2}_{(7)}\right\}$$

$$+4\Sigma\,\frac{\xi^2}{a^2}\times\Sigma\Sigma\,\frac{\xi}{a^2b'^2}\cdot\frac{d^3\phi}{d\xi d\xi'^2}_{(8)} + 2\Sigma\,\frac{1}{a_2}\times\Sigma\,\frac{\xi^2}{a_2}\times\Sigma\,\frac{d^2\phi}{b^2d\xi^2}_{(9)}$$

$$+2(s-n-1)\left\{\Sigma\frac{\xi^2}{a^2b^2}\times\Sigma\frac{a^2d\xi^2}{d^2\phi}_{(10)}-\Sigma\frac{\xi^2}{a^2}\times\Sigma\frac{1}{a^2b^2}\cdot\frac{d^2\phi}{d\xi^2}_{(11)}\right\}$$

$$\nabla_3\nabla'\phi-\nabla'\nabla_3\chi=$$

$$(1-\Sigma\xi)\left\{-4\Sigma\Sigma\frac{\xi d^3\phi}{a^2a'^2b'^2dyd\xi'^2}_{(12)}-2\Sigma\frac{1}{a^4b^2}\cdot\frac{d^2\phi}{d\xi^2}_{(13)}\right\}$$

$$-4\Sigma\frac{\xi^2}{a^2}\times\Sigma\Sigma\frac{\xi}{a^2b'^2}\cdot\frac{d^3\phi}{d\xi d'^2}_{(14)}-2\Sigma\frac{\xi^2}{a^2}\times\Sigma\frac{d^2\phi}{b^2d\xi^2}_{(15)}$$

$$\nabla'\nabla_4\phi-\nabla_4\nabla'\phi=$$

$$2(1-\Sigma^2)\Sigma\frac{1}{b^2a^4}\cdot\frac{d^2\phi}{d\xi^2}_{(16)}+2\Sigma\frac{\xi^2}{a^4}\times\Sigma\frac{d^2\phi}{b^2d\xi^2}_{(17)}$$

$$\nabla_5\nabla'\phi-\nabla'\nabla_5\phi=$$

$$-2.(1-\Sigma\xi^2)\Sigma\frac{1}{a^2}\times\Sigma\frac{1}{a^2b^2}\cdot\frac{d^2\phi}{d\xi^2}_{(18)}-2.\Sigma\frac{1}{a^2}\times\Sigma\frac{\xi^2}{a^2}+\Sigma\frac{d^2\phi}{b^2d\xi^2}_{(19)}$$

all the finite integrals being taken from $r=1$ to $r=s+1$, and from $r'=1$ to $r'=s+1$.

In order to obtain the required value

$$\nabla'\nabla\phi-\nabla\nabla'\phi,$$

it is clear that we shall only have to add the first of the five preceding quantities to the sum of the four following ones multiplied by h', and to render this more easy, we have appended to each of the terms in the preceding quantities a number inclosed in a small parenthesis.

Now since the accents may be permuted at will, and we have likewise $a^2=b^2+h^2$, it is easy to see that the terms marked (1), (6) and (12) mutually destroy each other. In like manner, (2), (3), (7) and (18) mutually destroy each other; the same may evidently be said of (13) and (16), of (15) and (17), of (9) and (19), and of (8) and (14). Moreover, the four quantities (4), (5), (10) and (11) will do so likewise, and consequently, we have

$$\nabla'\nabla\phi-\nabla\nabla'\phi=0.$$

Hence the truth of the equation (31) is manifest.

Application of the preceding General Theory to the Determination of the Attractions of Ellipsoids.

13. Suppose it is required to determine the attractions exerted by an ellipsoid whose semi-axes are a', b', c' whether the attracted point p is situated within the ellipsoid or not, the law of the attraction being inversely as the n'^{th} power of the distance. Then it is well known that the required attractions may always be deduced from the function

$$V = \int \frac{\rho' dx' dy' dz'}{\{(x-x')^2 + (y-y')^2 + (z-z')^2\}^{\frac{n'-1}{2}}};$$

ρ' being the density of the element $dx'dy'dz'$ of the ellipsoid, and x, y, z being the rectangular co-ordinates of p.

We may avoid the breach of the law of continuity which takes place in the value of V, when the point p passes from the interior of the ellipsoid into the exterior space, by adding the positive quantity u^2 to that inclosed in the braces, and may afterwards suppose u evanescent in the final result. Let us therefore now consider the function

$$V = \int \frac{\rho' dx' dy' dz'}{\{(x-x')^2 + (y-y')^2 + (z-z')^2 + u^2\}^{\frac{n'-1}{2}}};$$

this triple integral like the preceding including all the values of x', y', z', admitted by the condition

$$\frac{x'^2}{a'^2} + \frac{y'^2}{b'^2} + \frac{z'^2}{c'^2} < 1.$$

If now we suppose the density ρ' is of the form

$$\rho' = \left(1 - \frac{x'^2}{a'^2} - \frac{y'^2}{b'^2} - \frac{z'^2}{c'^2}\right)^{\frac{n'-2}{2}} f(x', y', z') \quad \ldots\ldots\ldots\ldots\ldots (34),$$

which will simplify $f(x', y', z')$ when ρ' is constant and $n' = 2$, and then compare this value with the one immediately deducible from the general expression (28) by supposing for a moment $n' = n$, viz.

$$\rho' = \left(1 - \frac{x'^2}{a'^2} - \frac{y'^2}{b'^2} - \frac{z'^2}{c'^2}\right)^{\frac{n'-4}{2}} F(x', y', z'),$$

we see that the function f will always be two degrees higher than F. But since our formulæ become more complicated in proportion as the degree of F is higher, it will be simpler to determine the differentials of V, because for these differentials the degree of F and f is the same. Let us therefore make

$$A = \frac{1}{1-n'}\frac{dV}{dx} = \int \frac{\rho'(x-x')\,dx'dy'dz'}{\{(x-x')^2 + (y-y')^2 + (z-z')^2 + u^2\}^{\frac{n'+1}{2}}},$$

then this quantity naturally divides itself into two parts, such that

$$A = xA' + A'',$$

where $$A' = +\int \frac{\rho' dx'dy'dz'}{\{(x-x')^2 + (y-y')^2 + (z-z')^2 + u^2\}^{\frac{n'+1}{2}}},$$

and $$A'' = -\int \frac{x'\rho' dx'dy'dz'}{\{(x-x')^2 + (y-y')^2 + (z-z')^2 + u^2\}^{\frac{n'+1}{2}}}.$$

By comparing these with the general formula (1), it is clear that $n-1 = n'+1$, and consequently $n = n'+2$. In this way the expression (28) gives

$$\rho' = \left(1 - \frac{x'^2}{a'^2} - \frac{y'^2}{b'^2} - \frac{z'^2}{c'^2}\right)^{\frac{n'-2}{2}} F(x', y', z'),$$

which coincides with (34) by supposing $F = f$.

The simplest case of the present theory is where $f(x',y',z') = 1$, and then by No. 11, we have $\phi_0' = 1$ and $b_0 = 1$, when A' is the quantity required, and as the general series (29), No. 11, then reduces itself to its first term, we immediately obtain from the formula (30), the value of A' following,

$$A' = -\frac{2\pi^{\frac{3}{2}}\Gamma\left(\frac{n'}{2}\right)}{\Gamma\left(\frac{n'+1}{2}\right)} a'b'c' \int_\infty \frac{h^{1-n}dh}{abc} \text{.....................} (35),$$

because in the present case $H_0 = 1$, $s = 3$, and $n = n'+2$.

Again, the same general theory being applied to the value of A'' given above, we get

$$F(x', y', z') = -x'f(x', y', z') = -x' \text{ (when } f=1),$$

and hence by No. 11, $F(x', y', z') = -a'\xi'$. In this way the series (29) again reduces itself to a single term, in which

$$\phi_0' = \xi', \text{ and } b_0 = -a',$$

and the particular value H_0 corresponding thereto, by omitting the superfluous constant $\frac{1}{\sqrt{(a'^2+b'^2+c'^2)}}$ will be (No. 10),

$$H_0 = a.$$

These substituted in the general formula (30) as before, immediately give

$$A'' = +\frac{2\pi^{\frac{3}{2}}\Gamma\left(\frac{n'}{2}\right)}{\Gamma\left(\frac{n'+1}{2}\right)} a'^3 b'c'\xi a \int_\infty \frac{h^{1-n'}dh}{a^3bc},$$

and consequently by reduction since $a\xi = x$,

$$A = xA' + A'' = -\frac{2\pi^{\frac{3}{2}}\Gamma\left(\frac{n'}{2}\right)}{\Gamma\left(\frac{n'+1}{2}\right)} a'b'c'x \int_\infty \frac{h^{3-n'}dh}{a^3bc} \text{......(36).}$$

The value of A just given belongs to the density

$$\rho' = \left(1 - \frac{x'^2}{a'^2} - \frac{y'^2}{b'^2} - \frac{z'^2}{c'^2}\right)^{\frac{n'-2}{2}}.$$

Hence we immediately obtain without calculation the corresponding values

$$B = \frac{1}{1-n'}\frac{dV}{dy} = -\frac{2\pi^{\frac{3}{2}}\Gamma\left(\frac{n'}{2}\right)}{\Gamma\left(\frac{n'+1}{2}\right)} a'b'c'y \int_\infty \frac{h^{3-n'}dh}{ab^3c},$$

$$C = \frac{1}{1-n'}\frac{dV}{dz} = -\frac{2\pi^{\frac{3}{2}}\Gamma\left(\frac{n'}{2}\right)}{\Gamma\left(\frac{n'+1}{2}\right)} a'b'c'z \int_\infty \frac{h^{3-n'}dh}{abc^3}.$$

If now we suppose moreover

$$D = \frac{1}{1-n'} \frac{dV}{du} = u \int \frac{\rho' dx' dy' dz'}{\{(x-x')^2 + (y-y')^2 + (z-z')^2 + u^2\}^{\frac{n'+1}{2}}},$$

the method before explained (No. 11) will immediately give

$$D = -\frac{2\pi^{\frac{3}{2}} \Gamma\left(\frac{n'}{2}\right)}{\Gamma\left(\frac{n'+1}{2}\right)} a'b'c'u \int_{\infty} \frac{h^{1-n'} dh}{abc},$$

and therefore if for abridgment we make

$$M = (n'-1) \frac{\pi^{\frac{3}{2}} \Gamma\left(\frac{n'}{2}\right)}{\Gamma\left(\frac{n'+1}{2}\right)} a'b'c',$$

the total differential of V may be written

$$dV = M \left\{ 2x\,dx \int_{\infty} \frac{h^{3-n'} dh}{a^3 bc} + 2y\,dy \int_{\infty} \frac{h^{3-n'} dh}{ab^3 c} \right.$$
$$\left. + 2z\,dz \int_{\infty} \frac{h^{3-n'} dh}{abc^3} + 2u\,du \int_{\infty} \frac{h^{1-n'} dh}{abc} \right\},$$

which being integrated in the usual way by first supposing h constant, and then completing the integral with a function of h, to be afterwards determined by making every thing in V variable, we get

$$V = M \left\{ x^2 \int_{\infty} \frac{h^{3-n} dh}{a^3 bc} + y^2 \int_{\infty} \frac{h^{3-n'} dh}{ab^3 c} + z^2 \int_{\infty} \frac{h^{3-n'} dh}{abc^3} + u^2 \int_{\infty} \frac{h^{1-n} dh}{abc} \right\} + k;$$

k being a quantity absolutely constant, which is equal to zero when $n' > 1$. What has just been advanced will be quite clear if we recollect that h may be regarded as a function of x, y, z and u, determined by the equation

$$1 = \frac{x^2}{a'^2 + h^2} + \frac{y^2}{b'^2 + h^2} + \frac{z^2}{c'^2 + h^2} + \frac{u^2}{h^2} = \xi^2 + \eta^2 + \zeta^2 + v^2 \ldots (37);$$

seeing that $a^2 = a'^2 + h^2$, $b^2 = b'^2 + h^2$, and $c^2 = c'^2 + h^2$.

After what precedes, it seems needless to enter into an examination of the values of V belonging to other values of the density ρ', since it must be clear that the general method is equally applicable when

$$\rho' = \left(1 - \frac{x'^2}{a'^2} - \frac{y'^2}{b'^2} - \frac{z'^2}{c'^2}\right)^{\frac{n'-2}{2}} f(x', y', z') ;$$

where f is the characteristic of any rational and entire function.

The quantity A before determined when we make $u = 0$, serves to express the attraction in the direction of the co-ordinate x of an ellipsoid on any point p, situated at will either within or without it. But by making $u = 0$ in (37) we have

$$1 = \frac{x^2}{a'^2 + h^2} + \frac{y^2}{b'^2 + h^2} + \frac{z^2}{c'^2 + h^2} + \frac{o^2}{h^2} \dots\dots\dots\dots (38),$$

and it is thence easy to perceive that when p is within the ellipsoid, h must constantly remain equal to zero, and the equation (38) will always be satisfied by the indeterminate positive quantity $\frac{o^2}{o^2}$. When on the contrary p is exterior to it, h can no longer remain equal to zero, but must be such a function of x, y, z, as will satisfy the equation (38), of which the last term now evidently vanishes in consequence of the numerator o^2. Thus the forms of the quantities A, B, C, D and V all remain unchanged, and the discontinuity in each of them falls upon the quantity h.

To compare the value of A here found with that obtained by the ordinary methods, we shall simply have to make $n' = 2$ in the expression (36), recollecting that $\Gamma(1) = 1$, and $\Gamma\left(\frac{3}{2}\right) = \frac{1}{2}\sqrt{\pi}$. In this way

$$A = -4\pi a'b'c'x \int_\infty \frac{hdh}{a^3bc} = -4\pi a'b'c'x \int_\infty \frac{da}{a^2bc}$$

$$= +4a'b'c'x \int_a^\infty \frac{da}{a^2bc} = 4\pi a'b'c' \int_a^\infty \frac{da}{a^2\sqrt{(a^2 - a'^2 + b'^2)(a^2 - a'^2 + c'^2)}}.$$

But the last quantity may easily be put under the form of a definite integral, by writing $\frac{a}{v}$ in the place of a under the sign of integration, and again inverting the limits. Thus there will result

$$A = \frac{4\pi a'b'c'}{a^3} \int_0^1 \frac{v^2 dv}{\sqrt{\left(1 + \frac{b'^2 - a'^2}{a^2} v^2\right)\left(1 + \frac{c'^2 - a'^2}{a^2} v^2\right)}},$$

which agrees with the ordinary formula, since the mass of the ellipsoid is $\frac{4\pi a'b'c'}{3}$ and $a^2 = a'^2 + h^2$.

Examination of a particular Case of the General Theory exposed in the former Part of this Paper.

14. There is a particular case of the general theory first considered, which merits notice, in consequence of the simplicity of the results to which it leads. The case in question is that where we have generally whatever r may be

$$a_r' = a'.$$

Then the equation (19) which serves to determine ϕ, becomes by supposing $k_0 = k \,.\, a'^2$

$$0 = (1 - \Sigma_1^{s+1} \xi_r^2)\, \Sigma_1^{s+1} \frac{d^2\phi}{d\xi_r^2} + (s - n - 1)\, \Sigma_1^{s+1} \xi_r \frac{d\phi}{d\xi_r} - k\phi \ldots (39).$$

If now we employ a transformation similar to that used in obtaining the formula (14), No. 6, by making

$$\xi_1 = \rho \cos\theta_1,\ \xi_2 = \rho \sin\theta_1 \cos\theta_2,\ \xi_3 = \rho \sin\theta_1 \sin\theta_2 \cos\theta_3,\ \&c.$$

and then conceive the equation (39) deduced from the condition that

$$\int d\xi_1 d\xi_2 \ldots d\xi_s (1 - \Sigma\xi_r^2)^{\frac{n-s+1}{2}} \left\{\Sigma_1^{s+1} \left(\frac{d\phi}{d\xi_r}\right)^2 + \frac{k\phi^2}{1 - \Sigma\xi_r^2}\right\}$$

must be a minimum (vide No. 8), we shall have

$$d\xi_1 d\xi_2 \dots d\xi_s = \rho^{s-1} \sin\theta_1^{s-2} \sin\theta_2^{s-3} \dots \sin\theta_{s-2}\, d\rho\, d\theta_1 d\theta_2 \dots d\theta_{s-1},$$

$$\Sigma_1^{s+1} \left(\frac{d\phi}{d\xi}\right)^2 = \left(\frac{d\phi}{d\rho}\right)^2 + \frac{1}{\rho^2} \Sigma_1^s \frac{\left(\frac{d\phi}{d\xi_r}\right)^2}{\sin\theta_1^2 \sin\theta_2^2 \dots \sin\theta_{r-1}^2},$$

$$\text{and } 1 - \Sigma\xi_r^2 = 1 - \rho^2.$$

Proceeding now in the manner before explained (No. 8), we obtain for the equivalent of (39) by reduction

$$0 = \frac{d^2\phi}{d\rho^2} + \frac{s-1-n\rho^2}{\rho(1-\rho^2)} \cdot \frac{d\phi}{d\rho}$$

$$+ \frac{1}{\rho^2} \Sigma_{1s} \frac{\frac{d^2\phi}{d\theta_r^2} + (s-r-1)\frac{\cos\theta_r}{\sin\theta_r}\frac{d\phi}{d\theta_r}}{\sin\theta_1^2 \sin\theta_2^2 \dots \sin\theta_{r-1}^2} - \frac{k}{1-\rho^2}\phi \dots\dots (40).$$

But this equation may be satisfied by a function of the form

$$\phi = P\Theta_1\Theta_2\Theta_3 \dots \Theta_{s-1};$$

P being a function of ρ only, and afterwards generally Θ_r a function of θ_r only. In fact, if we substitute this value of ϕ in (40), and then divide the result by ϕ, it is clear that it will be satisfied by the system

$$\frac{d^2\Theta_{s-1}}{\Theta_{s-1} d\theta_{s-1}^2} = \lambda_{s-1}$$

$$\frac{d^2\Theta_{s-2}}{\Theta_{s-2} d\theta_{s-2}^2} + 1 \cdot \frac{\cos\theta_{s-2}}{\sin\theta_{s-2}} \frac{d\Theta_{s-2}}{\Theta_{s-2} d\theta_{s-2}} + \frac{\lambda_{s-1}}{\sin\theta_{s-2}^2} = \lambda_{s-2} \dots\dots (41),$$

$$\frac{d^2\Theta_{s-3}}{\Theta_{s-3} d\theta_{s-3}^2} + 2 \cdot \frac{\cos\theta_{s-3}}{\sin\theta_{s-3}} \frac{d\Theta_{s-3}}{\Theta_{s-3} d\theta_{s-3}} + \frac{\lambda_{s-2}}{\sin\theta_{s-3}^2} = \lambda_{s-3}$$

&c. &c. &c. &c.

combined with the following equation,

$$\frac{d^2P}{P d\rho^2} + \frac{s-1-n\rho^2}{\rho(1-\rho^2)} \cdot \frac{dP}{P d\rho} + \frac{\lambda_1}{\rho^2} - \frac{k}{1-\rho^2} = 0 \dots\dots\dots (42),$$

where k, λ_1, λ_2, λ_3, &c. are constant quantities.

In order to resolve the system (41), let us here consider the general type of the equations therein contained, viz.

$$0 = \frac{d^2\Theta_{s-r}}{d\theta^2_{s-r}} + (r-1)\frac{\cos\theta_{s-r}}{\sin\theta_{s-r}} \cdot \frac{d\Theta_{s-r}}{d\theta_{s-r}} + \left(\frac{\lambda_{s-r+1}}{\sin\theta^2_{s-r}} - \lambda_{s-r}\right)\Theta_{s-r}.$$

Now if we reflect on the nature of the results obtained in a preceding part of this paper, it will not be difficult to see that Θ_{s-r} is of the form

$$\Theta_{s-r} = (\sin\theta_{s-r})^i p = (1-\mu^2)^{\frac{i}{2}} p;$$

where p is a rational and entire function of $\mu = \cos\theta_{s-r}$, and i a whole number.

By substituting this value in the general type and making

$$\lambda_{s-r+1} = -i(i+r-2) \dots\dots\dots\dots\dots\dots\dots (43)$$

we readily obtain

$$0 = (1-\mu^2)\frac{d^2p}{d\mu^2} - (2i+r)\mu\frac{dp}{d\mu} - \{\lambda_{s-r} + i(i+r-1)\}p.$$

To satisfy this equation, let us assume

$$p = \Sigma_0^\infty A_t \mu^{e-i-2t}.$$

Then by substituting in the above and equating separately the coefficients of the various powers of μ, we have in the first place from the highest

$$\lambda_{s-r} = -e(e+r-1) \dots\dots\dots\dots\dots\dots\dots (44),$$

and afterwards generally

$$A_{t+1} = -\frac{e-i-2t \,.\, e-i-2t-1}{2t+2 \times 2e+r-2t-3} A_t.$$

But the equation (43) may evidently be made to coincide with (44), by writing $i^{(r)}$ for i, and $i^{(r+1)}$ for e, since then both will be comprised in

$$\lambda_{s-r+1} = -i^{(r)}\{i^{(r)} + r - 2\} \dots\dots\dots\dots\dots\dots (45).$$

Hence we readily get for the general solution of the system (41),

$$\Theta_{s-r} = (1-\mu^2)^{\frac{i^{(r)}}{2}} \left[\mu^{i^{(r+1)} - i^{(r)}} - \frac{\{i^{(r+1)}\}\{i^{(r+1)} - i^{(r)} - 1\}}{2 \times 2i^{(r)} + r - 3} \mu^{i^{(r+1)} - i^{(r)} - 2} \right.$$

$$\left. + \frac{\{i^{(r+1)} - i^{(r)}\}\{i^{(r+1)} - i^{(r)} - 1\}\{i^{(r+1)} - i^{(r)} - 2\}\{i^{(r+1)} - i^{(r)} - 3\}}{2.4 \times \{2i^{(r)} + r - 3\}\{2i^{(r)} + r - 5\}} \mu^{i^{(r+1)} - i^{(r)} - 4} - \&c. \right];$$

where $\mu = \cos\theta_{s-r}$, and $i^{(r)}$ represents any positive integer whatever, provided $i^{(r)}$ is never greater than $i^{(r+1)}$.

Though we have thus the solution of every equation in the system (41), yet that of the first may be obtained under a simpler form by writing therein for λ_{s-1} its value $-i^{(2)^2}$ deduced from (45). We shall then immediately perceive that it is satisfied by

$$\Theta_{s-1} = \frac{\sin}{\cos}\left\{ i^{(2)}\theta_{s-1} \right\}.$$

In consequence of the formula (45), the equation (42) becomes

$$0 = \frac{d^2P}{d\rho^2} + \frac{s-1-n\rho^2}{\rho(1-\rho^2)} \cdot \frac{dP}{d\rho} - \left\{ \frac{i^{(s)}(i^{(s)}+s-2)}{\rho^2} + \frac{k}{1-\rho^2} \right\} P,$$

which is satisfied by making $k = -\lambda_1 - (i^{(s)} + 2\omega)(i^{(s)} + 2\omega + n - 1)$, and

$$P = \rho^{i^{(s)}} \left\{ \rho^{2\omega} - \frac{2\omega \times 2i^{(s)} + 2\omega + s - 2}{2 \times 2i^{(s)} + 4\omega + n - 3} \rho^{2\omega - 2} \right.$$

$$\left. + \frac{2\omega \,.\, 2\omega - 2 \times 2i^{(s)} + 2\omega + s - 2 \,.\, 2i^{(s)} + 2\omega + s - 4}{2 \,.\, 4 \times 2i^{(s)} + 4\omega + n + 3 \,.\, 2i + 4\omega + n - 5} \rho^{2\omega - 4} - \&c. \right\}$$

where ω represents any whole positive number.

Having thus determined all the factors of ϕ, it now only remains to deduce the corresponding value of H. But H_0 the particular value satisfying the differential equation in H, will be had from ϕ by simply making therein

$$\xi_r = \frac{a_r}{\sqrt{(\Sigma a_r'^2)}} = \frac{a}{a'\sqrt{s}},$$

since in the present case we have generally $a_r' = a'$.

Hence, it is clear that the proper values of θ_1, θ_2, θ_3, &c. to be here employed are all constant, and consequently the factor

$$\Theta_1 \Theta_2 \Theta_3 \ldots\ldots \Theta_{s-1}$$

entering into ϕ is likewise constant. Neglecting therefore this factor as superfluous, we get for the particular value of H,

$$H_0 = P_{\frac{a}{a'}};$$

$$\text{since } \rho^2 = \xi_1^2 + \xi_2^2 + \ldots\ldots + \xi_s^2 = \frac{sa^2}{sa'^2} = \frac{a^2}{a'^2},$$

and $P_{\frac{a}{a'}}$ represents what P becomes when ρ is changed into $\frac{a}{a'}$.

Substituting this value of H_0 in the equation (25), No. 10, there results since $a^2 = a'^2 + h^2$

$$H = K.P_{\frac{a}{a'}} \int_\infty \frac{h^{s-n}dh}{P_{\frac{a}{a'}}^2 (a'^2 + h^2)^{\frac{s}{2}}} \ldots\ldots\ldots\ldots\ldots\ldots (46),$$

K being an arbitrary constant quantity.

Thus the complete value of V for the particular case considered in the present number is

$$V = P\Theta_1\Theta_2 \ldots \Theta_{s-1} . KP_{\frac{a}{a'}} \int_\infty \frac{h^{s-n}dh}{P_{\frac{a}{a'}}^2 (a'^2 + h^2)^{\frac{s}{2}}} \ldots\ldots\ldots (47),$$

and the equation (27), No. 11, will give for the corresponding value of ρ',

$$\rho' = \frac{-\Gamma\left(\frac{n-1}{2}\right)}{2\pi^{\frac{s}{2}}\Gamma\left(\frac{n-s+1}{2}\right)} (1-\rho^2)^{\frac{n-s-1}{2}} \frac{K}{P_1 a'^s} P'\Theta_1'\Theta_2' \ldots \Theta'_{s-1} \ldots (48);$$

where P_1', Θ_1', Θ_2', &c. are the values which the functions P, Θ_1, Θ_2, &c. take when we change the unaccented variables $\xi_1, \xi_2, \ldots \xi_s$ into the corresponding accented ones $\xi_1', \xi_2', \ldots \xi_s'$, and

$$P_1 = \frac{n-s+1 \,.\, n-s+3 \ldots\ldots n-s+2\omega-1}{n+2i+2\omega-1 \,.\, n+2i+2\omega+1 \ldots\ldots n+2i+4\omega-3},$$

or the value of P when $\rho = 1$; where as well as in what follows i is written in the place of $i^{(s)}$.

The differential equation which serves to determine H when we introduce a instead of h as independent variable, may in the present case be written under the form

$$0 = a^2(a^2 - a'^2)\frac{d^2H}{da^2} + a^2\{na^2 - (s-1).a'^2\}\frac{dH}{ada}$$

$$+ \{i(i+s-2)a'^2 - (i+2\omega)(i+2\omega+n-1)a^2\}H,$$

and the particular integral here required is that which vanishes when h is infinite. Moreover it is easy to prove, by expanding in series, that this particular integral is

$$H = k'a^i\Delta^\omega . a^{2r}\int_\infty a^{1-2r-s-2i}da\,(a^2 - a'^2)^{\frac{s-1-n-2\omega}{2}};$$

provided we make the variable r to which Δ^ω refers vanish after all the operations have been effected.

But the constant k' may be determined by comparing the coefficient of the highest power of a in the expansion of the last formula with the like coefficient in that of the expression (46), and thus we have

$$k' = Ka'^{i+2\omega}(-1)^\omega \frac{n+2i+2\omega-1.n+2i+2\omega+1\ldots n+2i+4\omega-3}{2.4.6\ldots 2\omega}.$$

Hence we readily get for the equivalent of (47),

$$V = P\Theta_1\Theta_2\ldots\Theta_{s-1}\times\frac{n+2i+2\omega-1.n+2i+2\omega+1\ldots n+2i+4\omega-3}{2.4.6\ldots 2\omega}$$

$$\ldots\times Ka'^{i+2\omega}(-1)^\omega a^i\Delta^\omega a^{2r}\int_\infty da a^{1-2r-s-2i}(a^2-a'^2)^{\frac{s-1-n-2\omega}{2}}.$$

In certain cases the value of V just obtained will be found more convenient than the foregoing one (47). Suppose for instance we represent the value of V when $h=0$, or $a=a'$ by V_0. Then we shall hence get

$$V_0 = P\Theta_1\Theta_2\ldots\Theta_{s-1}\times\frac{n+2i+2\omega-1.n+2i+2\omega+1\ldots n+2i+4\omega-3}{2.4.6\ldots 2\omega}$$

$$\times Ka'^{2i+2\omega}(-1)^\omega.\Delta^\omega a'^{2r}\int_\infty^{a'} da a^{1-2r-s-2i}(a^2-a'^2)^{\frac{s-1-n-2\omega}{2}}\ldots.(\delta),$$

which in consequence of the well-known formula

$$\int_{\infty}^{a'} a^{-m} da\,(a^2 - a'^2)^{-p} = -\,a'^{1-m-2p} \times \frac{\Gamma(1-p)\,\Gamma\left(\frac{m+2p-1}{2}\right)}{2\Gamma\left(\frac{1+m}{2}\right)},$$

by reduction becomes

$$V_0 = -P\Theta_1\Theta_2\ldots\Theta_{s-1} \times \frac{\Gamma\left(\frac{1+s-n}{2}\right)\Gamma\left(\frac{n+2i+4\omega-1}{2}\right)}{2\Gamma(\omega+1)\,\Gamma\left(\frac{s+2i+2\omega}{2}\right)} Ka'^{1-n} \ldots (49);$$

since in the formula (δ), r ought to be made equal to zero at the end of the process.

By conceiving the auxiliary variable u to vanish, it will become clear from what has been advanced in the preceding number, that the values of the function V *within* circular planes and spheres are only particular cases of the more general one (49), which answer to $s = 2$ and $s = 3$ respectively. We have thus by combining the expressions (48) and (49), the means of determining V_0 when the density ρ' is given, and *vice versa;* and the present method of resolving these problems seems more simple if possible than that contained in the articles (4) and (5) of my former paper.

ON THE MOTION OF WAVES

IN A VARIABLE CANAL OF SMALL DEPTH AND WIDTH*.

* From the *Transactions of the Cambridge Philosophical Society*, 1838.
[Read *May* 15, 1837.]

ON THE MOTION OF WAVES IN A VARIABLE CANAL OF SMALL DEPTH AND WIDTH.

THE equations and conditions necessary for determining the motions of fluids in every case in which it is possible to subject them to Analysis, have been long known, and will be found in the First Edition of the *Mec. Anal.* of Lagrange. Yet the difficulty of integrating them is such, that many of the most important questions relative to this subject seem quite beyond the present powers of Analysis. There is, however, one particular case which admits of a very simple solution. The case in question is that of an indefinitely extended canal of small breadth and depth, both of which may vary very slowly, but in other respects quite arbitrarily. This has been treated of in the following paper, and as the results obtained possess considerable simplicity, perhaps they may not be altogether unworthy the Society's notice.

The general equations of motion of a non-elastic fluid acted on by gravity (g) in the direction of the axis z, are,

$$gz - \frac{p}{\rho} = \frac{d\phi}{dt} \text{ .... (1)},$$

$$0 = \frac{d^2\phi}{dx^2} + \frac{d^2\phi}{dy^2} + \frac{d^2\phi}{dz^2} \text{ .... (2)},$$

supposing the disturbance so small that the squares and higher powers of the velocities &c. may be neglected. In the above formulæ p = pressure, ρ = density, and ϕ is such a function of x, y, z and t, that the velocities of the fluid particles parallel to the three axes are

$$u = \left(\frac{d\phi}{dx}\right), \quad v = \left(\frac{d\phi}{dy}\right), \quad w = \left(\frac{d\phi}{dz}\right).$$

To the equations (1) and (2) it is requisite to add the conditions relative to the exterior surfaces of the fluid, and if $A = 0$ be the equation of one of these surfaces, the corresponding condition is [Lagrange, *Mec. Anal.* Tom. II. p. 303. (I.)],

$$0 = \frac{dA}{dt} + \frac{dA}{dx} u + \frac{dA}{dy} v + \frac{dA}{dz} w.$$

Hence

$$0 = \frac{dA}{dt} + \frac{dA}{dx} \cdot \frac{d\phi}{dx} + \frac{dA}{dy} \cdot \frac{d\phi}{dy} + \frac{dA}{dz} \cdot \frac{d\phi}{dz} \text{ (when } A = 0)\text{...(A).}$$

The equations (1) and (2) with the condition (A) applied to each of the exterior surfaces of the fluid will suffice to determine in every case the small oscillations of a non-elastic fluid, or at least in those where

$$udx + vdy + wdz$$

is an exact differential.

In what follows, however, we shall confine ourselves to the consideration of the motion of a non-elastic fluid, when two of the dimensions, viz. those parallel to y and z, are so small that ϕ may be expanded in a rapidly convergent series in powers of y and z, so that

$$\phi = \phi_0 + \phi' \frac{y}{1} + \phi_{,} \frac{z}{1} + \phi'' \frac{y^2}{1.2} + \phi_{,}' yz + \phi_{,,} \frac{z^2}{1.2} + \&c.$$

Then if we take the surface of the fluid in equilibrium as the plane of (x, y), and suppose the sides of the rectangular canal symmetrical with respect to the plane (x, z), ϕ will evidently contain none but even powers of y, and we shall have

$$\phi = \phi_0 + \phi_{,} z + \phi'' \frac{y^2}{1.2} + \phi_{,,} \frac{z^2}{1.2} + \&c. \text{ ....(3).}$$

Now if $$y = \pm \beta_x$$

represent the equation of the two sides of the canal, we need only satisfy one of them as

$$y - \beta_x = 0,$$

since the other will then be satisfied by the exclusion of the odd powers of y from ϕ.

The equation (A) gives, since here $A = y - \beta$,

$$0 = \frac{d\phi}{dy} - \frac{d\beta}{dx} \cdot \frac{d\phi}{dx} \text{ ...... (when } y = \beta) \text{ ....}(a).$$

Similarly, if $z - \gamma_x = 0$ is the equation of the bottom of the canal,

$$0 = \frac{d\phi}{dz} - \frac{d\gamma}{dx} \cdot \frac{d\phi}{dx} \text{ ...... (when } z = \gamma) \text{............ }(b).$$

If moreover $z - \zeta_{x.t} = 0$ be the equation of the upper surface,

$$\left.\begin{aligned} 0 &= \frac{d\phi}{dz} - \frac{d\zeta}{dx}\frac{d\phi}{dx} - \frac{d\zeta}{dt} \\ \text{But here } p = 0\text{; } \therefore \text{ also by (2) } g\zeta &= \frac{d\phi}{dt} \end{aligned}\right\} \text{ ... (when } z = \zeta) \text{...}(c).$$

Substituting from (3) in (b) we get

$$0 = \phi_{,} + \phi_{,,}\gamma + \text{\&c.} - \frac{d\gamma}{dx}\left\{\frac{d\phi_0}{dx} - \frac{d\phi_{,}}{dx}\frac{\gamma}{1} + \text{\&c.}\right\};$$

or neglecting quantities of the order γ^2,

$$0 = \phi_{,} + \phi_{,,}\gamma - \frac{d\gamma}{dx} \cdot \frac{d\phi_0}{dx} \text{ .... }(b').$$

Similarly (a) becomes

$$0 = \phi''\beta - \frac{d\beta}{dx} \cdot \frac{d\phi_0}{dx} \text{ .... }(a'),$$

and (c) becomes, since ζ is of the order of the disturbance,

$$\left.\begin{aligned} 0 &= \phi_{,,} - \frac{d\zeta}{dt} \\ g\zeta &= \frac{d\phi_0}{dt} \end{aligned}\right\} \left.\begin{matrix} \text{when } z = \zeta, \\ \text{or neglecting (disturbance)}^2\ z = 0 \end{matrix}\right\} \text{ ... }(c')$$

provided as above we neglect (disturbance)².

Again, the condition (2) gives by equating separately the coefficients of powers and products of y and z,

$$\left.\begin{aligned} 0 &= \frac{d^2\phi_0}{dx^2} + \phi'' + \phi_{,,} \\ 0 &= \text{\&c.} \end{aligned}\right\} \text{ ....(2').}$$

If now by means of (a'), (b'), (c') we eliminate $\phi''\phi_{,,}$ from (2'), there results

$$0=\frac{d^2\phi_0}{dx^2}+\left\{\frac{d\beta}{\beta dx}+\frac{d\gamma}{\gamma dx}\right\}\frac{d\phi_0}{dx}-\frac{1}{g\gamma}\left(\frac{d^2\phi_0}{dt^2}\right)\text{............}(4).$$

It now only remains to integrate this equation.

For this we shall suppose β and γ functions of x which vary very slowly, so that if written in their proper form we should have

$$\beta=\psi(\omega x),$$

where ω is a very small quantity. Then,

$$\frac{d\beta}{dx}=\omega\psi'(\omega x),\quad \frac{d^2\beta}{dx^2}=\omega^2\psi''(\omega x),\ \&c.$$

Hence if we allow ourselves to omit quantities of the order ω^2, and assume, to satisfy (4),

$$\phi_0=Af(t+X),$$

where A is a function of x of the same kind as β and γ, we have, omitting $\left(\frac{d^2A}{dx^2}\right)$,

$$\frac{d^2\phi_0}{dt^2}=Af'',$$

$$\frac{d\phi_0}{dx}=A\frac{dX}{dx}f'+\frac{dA}{dx}f,$$

$$\frac{d^2\phi_0}{dx^2}=A\left(\frac{dX}{dx}\right)^2f''+A\frac{d^2X}{dx^2}f'+2\frac{dA}{dx}\cdot\frac{dX}{dx}f'.$$

Substituting these in (4), and still neglecting quantities of the order ω^2, we get

$$0=\left\{A\left(\frac{dX}{dx}\right)^2-\frac{A}{g\gamma}\right\}f''$$

$$+\left\{A\frac{d^2X}{dx^2}+2\frac{dA}{dx}\frac{dX}{dx}+\left(\frac{d\beta}{\beta dx}+\frac{d\gamma}{\gamma dx}\right)A\frac{dX}{dx}\right\}f';$$

equating now separately the coefficients of f' and f'', we get

$$0=\left(\frac{dX}{dx}\right)^2-\frac{1}{g\gamma},$$

$$0=\frac{\frac{d^2X}{dx^2}}{\frac{dX}{dx}}+2\frac{dA}{Adx}+\frac{d\beta}{\beta dx}+\frac{d\gamma}{\gamma dx}.$$

The first, integrated, gives

$$X=\pm\int\frac{dx}{\sqrt{\gamma g}},$$

and the second

$$k=\frac{dX}{dx}\,A^2\beta\gamma=A^2\frac{\beta\gamma}{\sqrt{g\gamma}}=\frac{A^2\beta\gamma^{\frac{1}{2}}}{\sqrt{g}}.$$

Hence if we neglect the superfluous constant $k\sqrt{g}$, the general integral of (4) is, ($\because A=\beta^{-\frac{1}{2}}\gamma^{-\frac{1}{4}}$),

$$\phi_0=\beta^{-\frac{1}{2}}\gamma^{-\frac{1}{4}}\left\{f\left(t+\int\frac{dx}{\sqrt{g\gamma}}\right)+F\left(t-\int\frac{dx}{\sqrt{g\gamma}}\right)\right\};$$

therefore, by (c'),

$$\zeta=\frac{d\phi_0}{gdt}=\frac{\beta^{-\frac{1}{2}}\gamma^{-\frac{1}{4}}}{g}\left\{f'\left(t+\int\frac{dx}{\sqrt{g\gamma}}\right)+F'\left(t-\int\frac{dx}{\sqrt{g\gamma}}\right)\right\},$$

and the actual velocity of the fluid particles in the direction of the axis of x, is

$$u=\frac{d\phi}{dx}=\frac{d\phi_0}{dx}=\frac{\beta^{-\frac{1}{2}}\gamma^{-\frac{1}{4}}}{\sqrt{g\gamma}}\left\{f'\left(t+\int\frac{dx}{\sqrt{g\gamma}}\right)-F'\left(t-\int\frac{dx}{\sqrt{g\gamma}}\right)\right\},$$

neglecting quantities which are of the order (ω) compared with those retained.

If the initial values of ζ and u are given, we may then determine f' and F', and we thus see that a single wave, like a pulse of sound, divides into two, propagated in opposite directions. Considering, therefore, only that which proceeds in the direction of x positive, we have

$$\zeta=\frac{\beta^{-\frac{1}{2}}\gamma^{-\frac{1}{4}}}{g}F'\left(t-\int\frac{dx}{\sqrt{g\gamma}}\right)\ \ldots\ldots\ldots\ldots\ldots\ldots\ldots\ (5).$$

$$u=\frac{\beta^{-\frac{1}{2}}\gamma^{-\frac{3}{4}}}{g^{\frac{1}{2}}}F'\left(t-\int\frac{dx}{\sqrt{g\gamma}}\right)\ \ldots\ldots\ldots\ldots\ldots\ldots\ldots\ (6).$$

Suppose now the value of $F'(x) = 0$, except from $x = a$ to $x = a + \alpha$, and δx to be the corresponding length of the wave, we have

$$t - \int \frac{dx}{\sqrt{g\gamma}} = a + \alpha,$$

$$\text{and } t - \int \frac{dx}{\sqrt{g\gamma}} - \frac{\delta x}{\sqrt{g\gamma}} = a \text{ very nearly.}$$

Hence the variable length of the wave is

$$\delta x = \alpha \,.\, \sqrt{g\gamma} \quad \text{.........................} \quad (7).$$

Lastly, for any particular phase of the wave, we have

$$t - \int \frac{dx}{\sqrt{g\gamma}} = \text{const.}:$$

therefore

$$\frac{dx}{dt} = \sqrt{g\gamma} \quad \text{...............................} \quad (8)$$

is the velocity with which the wave, or more strictly speaking the particular phase in question, progresses.

From (5), (6), (7), and (8) we see that if β represent the variable breadth of the canal and γ its depth,

$$\zeta = \text{height of the wave} \propto \beta^{-\frac{1}{2}}\gamma^{-\frac{1}{4}},$$

$$u = \text{actual velocity of the fluid particles} \propto \beta^{-\frac{1}{2}}\gamma^{-\frac{3}{4}}.$$

$$dx = \text{length of the wave} \propto \gamma^{\frac{1}{2}},$$

$$\text{and } \frac{dx}{dt} = \text{velocity of the wave's motion} = \sqrt{g\gamma}.$$

ON THE REFLEXION AND REFRACTION OF SOUND.

From the *Transactions of the Cambridge Philosophical Society*, 1838.
[Read Dec. 11, 1837].

ON THE REFLEXION AND REFRACTION OF SOUND.

THE object of the communication which I have now the honour of laying before the Society, is to present, in as simple a form as possible, the laws of the reflexion and refraction of sound, and of similar phenomena which take place at the surface of separation of any two fluid media when a disturbance is propagated from one medium to the other. The subject has already been considered by Poisson (*Mém. de l'Acad.*, &c. Tome x. p. 317, &c.). The method employed by this celebrated analyst is one that he has used on many occasions with great success, and which he has explained very fully in several of his works, and recently in a digression on the Integrals of Partial Differential Equations (*Théorie de la Chaleur*, p. 129, &c.). In this way, the question is made to depend on sextuple definite integrals. Afterwards, by supposing the initial disturbance to be confined to a small sphere in one of the fluids, and to be everywhere the same at the same distance from its centre, the formulæ are made to depend on double definite integrals; from which are ultimately deduced the laws of the propagation of the motion at great distances from the centre of the sphere originally disturbed.

The chance of error in every very long analytical process, more particularly when it becomes necessary to use Definite Integrals affected with several signs of integration, induced me to think, that by employing a more simple method we should possibly be led to some useful result, which might easily be overlooked in a more complicated investigation. With this impression I endeavoured to ascertain how a plane wave of infinite extent, accompanied by its reflected and refracted waves, would be propagated in any two indefinitely extended media of which the surface of separation in a state of equilibrium should also be in a plane of infinite extent.

The suppositions just made simplify the question extremely. They may also be considered as rigorously satisfied when light is reflected. In which case the unit of space properly belonging to the problem is a quantity of the same order as $\lambda = \frac{1}{50,000}$ inch, and the unit of time that which would be employed by light itself in passing over this small space. Very often too, when sound is reflected, these suppositions will lead to sensibly correct results. On this last account, the problem has here been considered generally for all fluids whether *elastic* or *non-elastic* in the usual acceptation of these terms; more especially, as thus its solution is not rendered more complicated. One result of our analysis is so simple that I may perhaps be allowed to mention it here. It is this: If A be the ratio of the density of the reflecting medium to the density of the other, and B the ratio of the cotangent of the angle of refraction to the cotangent of the angle of incidence, then for all fluids

$$\frac{\text{the intensity of the reflected vibration}}{\text{the intensity of the incident vibration}} = \frac{A - B}{A + B}.$$

If now we apply this to the reflexion of sound at the surface of still water, we have $A > 800$, and the maximum value of $B < \frac{1}{4}$. Hence the intensity of the reflected wave will in every case be sensibly equal to that of the incident one. This is what we should naturally have anticipated. It is however noticed here because M. Poisson has inadvertently been led to a result entirely different.

When the velocity of transmission of a wave in the second medium, is greater than that in the first, we may, by sufficiently increasing the angle of incidence in the first medium, cause the refracted wave in the second to disappear. In this case the change in the intensity of the reflected wave is here shown to be such that, at the moment the refracted wave disappears, the intensity of the reflected becomes exactly equal to that of the incident one. If we moreover suppose the vibrations of the incident wave to follow a law similar to that of the cycloidal pendulum, as is usual in the Theory of Light, it is proved that on farther increasing the angle of incidence, the intensity of the reflected wave remains unaltered whilst the phase of the vibration gradually changes. The laws of the change of intensity, and of the subsequent alteration of phase, are given here for all media, *elastic* or *non-elastic*. When, however, both the media are *elastic*, it is remarkable that these laws are precisely the same as those for light polarized in a plane perpendicular to the plane of incidence. Moreover, the disturbance excited in the second medium, when, in the case of total reflexion, it ceases to transmit a wave in the regular way, is represented by a quantity of which one factor is a negative exponential. This factor, for light, decreases with very great rapidity, and thus the disturbance is not propagated to a sensible depth in the second medium.

Let the plane surface of separation of the two media be taken as that of (yz), and let the axis of z be parallel to the line of intersection of the plane *front* of the wave with (yz), the axis of x being supposed vertical for instance, and directed downwards; then, if Δ and Δ_1 are the densities of the two media under the constant pressure P and s, s_1 the condensations, we must have

$$\begin{cases} \Delta\,(1+s) = \text{density in the upper medium,} \\ \Delta_1(1+s_1) = \text{density in the lower medium.} \end{cases}$$

$$\begin{cases} P\,(1+As) \;\;= \text{pressure in the upper medium,} \\ P\,(1+A_1 s_1) = \text{pressure in the lower medium.} \end{cases}$$

Also, as usual, let ϕ be such a function of x, y, z, that the resolved parts of the velocity of any fluid particle parallel to the axes, may be represented by

$$\frac{d\phi}{dx},\quad \frac{d\phi}{dy},\quad \frac{d\phi}{dz}.$$

In the particular case, here considered, ϕ will be independent of z, and the general equations of motion in the upper fluid will be

$$0 = \frac{ds}{dt} + \frac{d^2\phi}{dx^2} + \frac{d^2\phi}{dy^2},$$

$$0 = \frac{d\phi}{dt} + \gamma^2 s;$$

where we have

$$\gamma^2 = \frac{PA}{\Delta},$$

or by eliminating s

$$\frac{d^2\phi}{dt^2} = \gamma^2 \left(\frac{d^2\phi}{dx^2} + \frac{d^2\phi}{dy^2}\right) \dots\dots\dots\dots\dots\dots (1).$$

Similarly, in the lower medium

$$\frac{d^2\phi_{,}}{dt^2} = \gamma_{,}^2 \left(\frac{d^2\phi_{,}}{dx^2} + \frac{d^2\phi_{,}}{dy^2}\right) \dots\dots\dots\dots\dots\dots (2),$$

where

$$s_{,} = \frac{-d\phi_{,}}{\gamma_{,}^2 dt}, \text{ and } \gamma_{,}^2 = \frac{PA_{,}}{\Delta_{,}}.$$

The above are the known general equations of fluid motion, which must be satisfied for all the internal points of both fluids; but at the surface of separation, the velocities of the particles perpendicular to this surface and the pressure there must be the same for both fluids. Hence we have the particular conditions

$$\left.\begin{aligned} \frac{d\phi}{dx} &= \frac{d\phi_{,}}{dx} \\ As &= A_{,}s_{,} \end{aligned}\right\} \text{ (where } x = 0\text{)},$$

neglecting such quantities as are very small compared with those retained, or by eliminating s and $s_{,}$, we get

$$\left.\begin{aligned} \frac{d\phi}{dx} &= \frac{d\phi_{,}}{dx} \\ \Delta \frac{d\phi}{dt} &= \Delta_{,} \frac{d\phi_{,}}{dt} \end{aligned}\right\} \text{ (when } x = 0\text{)} \dots\dots\dots\dots (A).$$

The general equations (1) and (2), joined to the particular conditions (*A*) which belong to the surface of separation (yz), only, are sufficient for completely determining the motion of our two fluids, when the velocities and condensations are independent of the co-ordinate z, whatever the initial disturbance may be. We shall not here attempt to give their complete solution, which would be complicated, but merely consider the propagation of a plane wave of indefinite extent, which is accompanied by its reflected and refracted wave.

Since the disturbance of all the particles, in any *front* of the incident plane wave, is the same at the same instant, we shall have for the incident wave

$$\phi = f(ax + by + ct),$$

retaining b and c unaltered, we may give to the *fronts* of the reflected and refracted waves, any position by making for them

$$\phi = F(a'x + by + ct),$$
$$\phi_{,} = f_{,}(a_{,}x + by + ct).$$

Hence, we have in the upper medium,

$$\phi = f(ax + by + ct) + F(a'x + by + ct) \text{ .... } (4),$$

and in the lower one

$$\phi_{,} = f_{,}(a_{,}x + by + ct) \text{ .... } (5).$$

These, substituted in the general equations (1) and (2), give

$$\left.\begin{aligned} c^2 &= \gamma^2 (a^2 + b^2) \\ c^2 &= \gamma^2 (a'^2 + b^2) \\ c^2 &= \gamma_{,}^2 (a_{,}^2 + b^2) \end{aligned}\right\} \text{ .... } (6).$$

Hence, $a' = \pm a$, where the lower signs must evidently be taken to represent the reflected wave. This value proves, that the angle of incidence is equal to that of reflexion. In like manner, the value of $a_{,}$, will give the known relation of sines for the incident and refracted wave, as will be seen afterwards.

Having satisfied the general equations (1) and (2), it only remains to satisfy the conditions (A), due to the surface of separation of the two media. But these by substitution give

$$af'(by + ct) - aF'(by + ct) = a_{,}f_{,}'(by + ct),$$
$$\Delta\{f'(by + ct) + F'(by + ct)\} = \Delta_{,}f_{,}'(by + ct),$$

because $a' = -a$, and $x = 0$.

Hence by writing, to abridge, the characteristics only of the functions

$$\left.\begin{aligned} f' &= \frac{1}{2}\left(\frac{\Delta_{,}}{\Delta} + \frac{a_{,}}{a}\right) f_{,}' \\ F' &= \frac{1}{2}\left(\frac{\Delta_{,}}{\Delta} - \frac{a_{,}}{a}\right) f_{,}' \end{aligned}\right\} \text{ .... } (7),$$

or if we introduce θ, $\theta_{,}$, the angle of incidence and refraction, since

$$\cot\theta = \frac{a}{b},$$

$$\cot\theta_{,} = \frac{a_{,}}{b},$$

$$f' = \frac{1}{2}\left(\frac{\Delta_{,}}{\Delta} + \frac{\cot\theta_{,}}{\cot\theta}\right) f_{,}',$$

$$F' = \frac{1}{2}\left(\frac{\Delta_{,}}{\Delta} - \frac{\cot\theta_{,}}{\cot\theta}\right) f_{,}',$$

and therefore $$\frac{F'}{f'} = \frac{\dfrac{\Delta_{,}}{\Delta} - \dfrac{\cot\theta_{,}}{\cot\theta}}{\dfrac{\Delta_{,}}{\Delta} + \dfrac{\cot\theta_{,}}{\cot\theta}},$$

which exhibits under a very simple form, the ratio between the intensities of the disturbances, in the incident and reflected wave.

But the equations (6) give

$$\gamma^2\left(\frac{a^2}{b^2}+1\right) = \gamma_{,}^2\left(\frac{a_{,}^2}{b^2}+1\right);$$

and hence

$$\frac{\gamma}{\sin\theta} = \frac{\gamma_{,}}{\sin\theta_{,}},$$

the ordinary law of sines.

The reflected wave will vanish when

$$0 = \frac{\Delta_{,}}{\Delta} - \frac{\cot\theta'}{\cot\theta};$$

which with the above gives

$$\cot\theta = \Delta\sqrt{\frac{\gamma^2 - \gamma_{,}^2}{(\gamma_{,}\Delta_{,})^2 - (\Delta\gamma)^2}}.$$

Hence the reflected wave may be made to vanish if $\gamma^2 - \gamma_{,}^2$ and $(\gamma\Delta)^2 - (\gamma_{,}\Delta_{,})^2$ have different signs.

For the ordinary elastic fluids, at least if we neglect the change of temperature due to the condensation, A is independent of the nature of the gas, and therefore

$$A = A_{,} \text{ or } \gamma^2\Delta = \gamma_{,}^2\Delta_{,}.$$

Hence

$$\tan\theta = \frac{\gamma}{\gamma_{,}},$$

which is the precise angle at which light polarized perpendicular to the plane of reflexion is wholly transmitted.

But it is not only at this particular angle that the reflexion of sound agrees in intensity with light polarized perpendicular to the plane of reflexion. For the same holds true for every angle of incidence. In fact, since

$$\gamma^2\Delta = \gamma_{,}^2\Delta_{,}; \quad \therefore \frac{\Delta_{,}}{\Delta} = \frac{\gamma^2}{\gamma_{,}^2} = \frac{\sin^2\theta}{\sin^2\theta_{,}},$$

and the formulæ (7) give

$$\frac{F'}{f'} = \frac{\dfrac{\sin^2\theta}{\sin^2\theta_{,}} - \dfrac{\tan\theta}{\tan\theta_{,}}}{\dfrac{\sin^2\theta}{\sin^2\theta_{,}} + \dfrac{\tan\theta}{\tan\theta_{,}}} = \frac{\tan(\theta - \theta_{,})}{\tan(\theta + \theta_{,})};$$

which is the same ratio as that given for light polarized perpendicular to the plane of incidence. (Vide Airy's *Tracts*, p. 356)*.

What precedes is applicable to all waves of which the *front* is plane. In what follows we shall consider more particularly the case in which the vibrations follow the law of the cycloidal pendulum, and therefore in the upper medium we shall have,

$$\phi = \alpha\sin(ax + by + ct) + \beta\sin(-ax + by + ct)\ \ldots\ldots\ldots(8).$$

Also, in the lower one,

$$\phi_{,} = \alpha_{,}\sin(a_{,}x + by + ct):$$

and as this is only a particular case of the more general one, before considered, the equation (7) will give

$$\alpha = \frac{1}{2}\left(\frac{\Delta_{,}}{\Delta} + \frac{a_{,}}{a}\right)\alpha_{,},$$

$$\beta = \frac{1}{2}\left(\frac{\Delta_{,}}{\Delta} - \frac{a_{,}}{a}\right)\alpha_{,}.$$

If $\gamma_{,} > \gamma$, or the velocity of transmission of a wave, be greater in the lower than in the upper medium, we may by decreasing a render $a_{,}$ imaginary. This last result merely indicates that the form of our integral must be changed, and that as far as

* [Airy on *The Undulatory Theory of Optics*, p. 111, Art. 129.]

regards the co-ordinate x an exponential must take the place of the circular function. In fact the equation,

$$\frac{d^2\phi_{,}}{dt^2} = \gamma_{,}^2 \left\{\frac{d^2\phi_{,}}{dx^2} + \frac{d^2\phi_{,}}{dy^2}\right\},$$

may be satisfied by

$$\phi_{,} = \epsilon^{-a_{,}'x} . B \sin \psi,$$

(where, to abridge, ψ is put for $by + ct$) provided

$$c^2 = \gamma_{,}^2 (- a_{,}'^2 + b^2);$$

when this is done it will not be possible to satisfy the conditions (A) due to the surface of separation, without adding constants to the quantities under the circular functions in ϕ. We must therefore take, instead of (8), the formula,

$$\phi = \alpha \sin (ax + by + ct + e) + \beta \sin (-ax + by + ct + e_{,}) \text{.... (9)}.$$

Hence when $x = 0$, we get

$$\frac{d\phi}{dx} = a\alpha \cos (\psi + e) - a\beta \cos (\psi + e_{,}),$$

$$\frac{d\phi}{dt} = c\alpha \cos (\psi + e) + c\beta \cos (\psi + e_{,}),$$

$$\frac{d\phi_{,}}{dx} = - a_{,}'B \sin \psi,$$

$$\frac{d\phi_{,}}{dt} = cB \cos \psi;$$

these substituted in the conditions (A), give

$$\alpha \cos (\psi + e) - \beta \cos (\psi + e_{,}) = - \frac{a_{,}'}{a} B \sin \psi,$$

$$\alpha \cos (\psi + e) + \beta \cos (\psi + e_{,}) = \frac{\Delta_{,}}{\Delta} B \cos \psi;$$

these expanded, give

$$\alpha \cos e - \beta \cos e_{,} = 0,$$

$$-\alpha \sin e + \beta \sin e_{,} = - \frac{a_{,}'}{a} B,$$

$$\alpha \cos e + \beta \cos e_{,} = \frac{\Delta_{,}}{\Delta} B,$$

$$\alpha \sin e + \beta \sin e_{,} = 0.$$

Hence, we get

$$2\alpha \sin e = \frac{a_{,}'}{a} B \dots\dots\dots\dots\dots\dots\dots (10),$$

$$2\alpha \cos e = \frac{\Delta_{,}}{\Delta} B,$$

$$2\beta \sin e_{,} = -\frac{a_{,}'}{a} B,$$

$$2\beta \cos e_{,} = \frac{\Delta_{,}}{\Delta} B,$$

and, consequently,

$$e = -e_{,}, \quad \beta = \alpha,$$

and

$$\tan e = +\frac{a_{,}'\Delta}{a\Delta_{,}}.$$

This result is general for all fluids, but if we would apply it to those only which are usually called *elastic*, we have, because in this case $\gamma^2\Delta = \gamma_{,}^2\Delta_{,}$,

$$\tan e = \frac{a_{,}'\Delta}{a\Delta_{,}} = \frac{a_{,}'\gamma_{,}^2}{a\gamma^2}.$$

But generally

$$c^2 = \gamma_{,}^2(-a_{,}'^2 + b^2) = \gamma^2(a^2 + b^2) \dots\dots\dots\dots (11);$$

and therefore, by substitution,

$$\tan e = \frac{a_{,}'\gamma_{,}^2}{a\gamma^2} = \frac{\gamma_{,}\sqrt{\gamma_{,}^2 b^2 - (a^2 + b^2)\gamma^2}}{a\gamma^2} = \mu\sqrt{\mu^2 \tan^2\theta - \sec^2\theta},$$

because $\mu = \frac{\gamma_{,}}{\gamma}$, and $\frac{b}{a} = \tan\theta$.

As $e = -e_{,}$, we see from equation (9), that $2e$ is the change of phase which takes place in the reflected wave; and this is precisely the same value as that which belongs to light polarized perpendicularly to the plane of incidence; (Vide Airy's *Tracts*, p. 362*.) We thus see, that not only the intensity of the reflected wave, but the change of phase also, when reflexion takes place at the surface of separation of two elastic media, is precisely the same as for light thus polarized.

* Airy, *ubi sup.* p. 114, Art. 133.

As $\alpha = \beta$, we see that when there is no transmitted wave the intensity of the reflected wave is precisely equal to that of the incident one. This is what might be expected: it is, however, noticed here because a most illustrious analyst has obtained a different result. (Poisson, *Mémoires de l'Academie des Sciences*, Tome x.) The result which this celebrated mathematician arrives at is, That at the moment the transmitted wave ceases to exist, the intensity of the reflected becomes precisely equal to that of the incident wave. On increasing the angle of incidence this intensity again diminishes, until it vanish at a certain angle. On still farther increasing this angle the intensity continues to increase, and again becomes equal to that of the incident wave, when the angle of incidence becomes a right angle.

It may not be altogether uninteresting to examine the nature of the disturbance excited in that medium which has ceased to transmit a wave in the regular way. For this purpose, we will resume the expression,

$$\phi_{,} = B\epsilon^{-a_{,}'x} \sin \psi = B\epsilon^{-a_{,}'x} \sin (by + ct);$$

or if we substitute for B, its value given by the last of the equations (10); and for $a_{,}'$, its value from (11); this expression, in the case of ordinary elastic fluids where $\gamma^2 \Delta = \gamma_{,}^2 \Delta_{,}$, will reduce to

$$\phi_{,} = 2\alpha\mu^2 \cos e \,.\, \epsilon^{-\frac{2\pi x}{\lambda}\sqrt{\frac{\mu^2 \sin^2\theta - 1}{\mu}}} \sin (by + ct),$$

λ being the length of the incident wave measured perpendicular to its own front, and θ the angle of incidence. We thus see with what rapidity in the case of light, the disturbance diminishes as the depth x below the surface of separation of the two media increases; and also that the rate of diminution becomes less as θ approaches the *critical* angle, and entirely ceases when θ is exactly equal to this angle, and the transmission of a wave in the ordinary way becomes possible.

ON THE LAWS

OF

REFLEXION AND REFRACTION OF LIGHT

AT THE COMMON SURFACE OF TWO NON-CRYSTALLIZED MEDIA.

From the *Transactions of the Cambridge Philosophical Society*, 1838.
[Read *December* 11, 1837.]

ON THE LAWS OF THE REFLEXION AND REFRACTION OF LIGHT AT THE COMMON SURFACE OF TWO NON-CRYSTALLIZED MEDIA.

M. CAUCHY seems to have been the first who saw fully the utility of applying to the Theory of Light those formulæ which represent the motions of a system of molecules acting on each other by mutually attractive and repulsive forces; supposing always that in the mutual action of any two particles, the particles may be regarded as points animated by forces directed along the right line which joins them. This last supposition, if applied to those compound particles, at least, which are separable by mechanical division, seems rather restrictive; as many phenomena, those of crystallization for instance, seem to indicate certain polarities in these particles. If, however, this were not the case, we are so perfectly ignorant of the mode of action of the elements of the luminiferous ether on each other, that it would seem a safer method to take some general physical principle as the basis of our reasoning, rather than assume certain modes of action, which, after all, may be widely different from the mechanism employed by nature; more especially if this principle include in itself, as a particular case, those before used by M. Cauchy and others, and also lead to a much more simple process of calculation. The principle selected as the basis of the reasoning contained in the following paper is this: In whatever way the elements of any material system may act upon each other, if all the internal forces exerted be multiplied by the elements of their respective directions, the total sum for any assigned portion of the mass will always be the exact differential of some function. But, this function being known, we can immediately apply the general method given in the *Mécanique Analytique*, and which appears to be more especially applicable to problems that relate to the motions of systems composed of an immense number of particles mutually acting upon each other. One of the advantages of this method, of great importance, is, that we are necessarily led by the mere process of the calculation, and with little care on our part, to all the equations and conditions which are *requisite* and *sufficient* for the complete solution of any problem to which it may be applied.

The present communication is confined almost entirely to the consideration of non-crystallized media; for which it is proved, that the function due to the molecular actions, in its most general form, contains only two arbitrary coefficients, A and B; the values of which depend of course on the unknown internal constitution of the medium under consideration, and it would be easy to shew, for the most general case, that any arbitrary disturbance, excited in a very small portion of the medium, would in general give rise to two spherical waves, one propagated entirely by normal, the other entirely by transverse, vibrations, and such that if the velocity of transmission of the former wave be represented by $\sqrt{A}$, that of the latter would be represented by $\sqrt{B}$. But in the transmission of light through a prism, though the wave which is propagated by normal vibrations were incapable itself of affecting the eye, yet it would be capable of giving rise to an ordinary wave of light propagated by transverse vibrations, except in the extreme cases where $\frac{A}{B} = 0$, or $\frac{A}{B}$ = a very large quantity; which, for the sake of simplicity, may be regarded as infinite; and it is not difficult to prove that the equilibrium of our medium would be unstable unless $\frac{A}{B} > \frac{4}{3}$. We are therefore compelled to adopt the latter value of $\frac{A}{B}$, and thus to admit that in the luminiferous ether, the velocity of transmission of waves propagated by normal vibrations is very great compared with that of ordinary light.

The principal results obtained in this paper relate to the intensity of the wave reflected at the common surface of two media, both for light polarized in and perpendicular to the plane of incidence; and likewise to the change of phase which takes place when the reflexion becomes total. In the former case, our values agree precisely with those given by Fresnel; supposing, as he has done, that the direction of the actual motion of the particles of the luminiferous ether is perpendicular to the plane of polarization. But it results from our formulæ, when the light is polarized perpendicular to the plane of incidence, that the expressions given by Fresnel are only very near approximations; and that the intensity of the reflected wave will never become absolutely null, but only attain a minimum value; which, in the case of reflexion from water at the proper angle, is $\frac{1}{151}$ part of that of the incident wave. This minimum value increases rapidly, as the index of refraction increases, and thus the quantity of light reflected at the polarizing angle, becomes considerable for highly refracting substances, a fact which has been long known to experimental philosophers.

It may be proper to observe, that M. Cauchy (*Bulletin des Sciences*, 1830) has given a method of determining the intensity of the waves reflected at the common surface of two media. He has since stated, (*Nouveaux Exercises des Mathématiques*,) that the hypothesis employed on that occasion is inadmissible, and has promised in a future memoir, to give a *new mechanical principle* applicable to this and other questions; but I have not been able to learn whether such a memoir has yet appeared. The first method consisted in satisfying a part, and only a part, of the conditions belonging to the surface of junction, and the consideration of the waves propagated by normal vibrations was wholly overlooked, though it is easy to perceive, that in general waves of this kind must necessarily be produced when the incident wave is polarized perpendicular to the plane of incidence, in consequence of the incident and refracted waves being in different planes. Indeed, without introducing the consideration of these last waves, it is impossible to satisfy the whole of the conditions due to the surface of junction of the two media. But when this consideration is introduced, the whole of the conditions may be satisfied, and the principles given in the *Mécanique Analytique* became abundantly sufficient for the solution of the problem.

In conclusion, it may be observed, that the radius of the sphere of sensible action of the molecular forces has been regarded as insensible with respect to the length λ of a wave of light, and thus, for the sake of simplicity, certain terms have been disregarded on which the different refrangibility of differently coloured rays might be supposed to depend. These terms, which are necessary to be considered when we are treating of the dispersion, serve only to render our formulæ uselessly complex in other investigations respecting the phenomena of light.

Let us conceive a mass composed of an immense number of molecules acting on each other by any kind of molecular forces, but which are sensible only at insensible distances, and let moreover the whole system be quite free from all extraneous action of every kind. Then x, y and z being the co-ordinates of any particle of the medium under consideration when in equilibrium, and

$$x + u, \quad y + v, \quad z + w,$$

the co-ordinates of the same particle in a state of motion (where u, v, and w are very small functions of the original co-ordinates (x, y, z), of any particle and of the time (t)), we get, by combining D'Alembert's principle with that of virtual velocities,

$$\Sigma Dm \left\{\frac{d^2u}{dt^2}\,\delta u + \frac{d^2v}{dt^2}\,\delta v + \frac{d^2w}{dt^2}\,\delta w\right\} = \Sigma Dv\,.\,\delta\phi \;\ldots\ldots\ldots\; (1);$$

Dm and Dv being exceedingly small corresponding elements of the mass and volume of the medium, but which nevertheless contain a very great number of molecules, and $\delta\phi$ the exact differential of some function and entirely due to the internal actions of the particles of the medium on each other. Indeed, if $\delta\phi$ were not an exact differential, a perpetual motion would be possible, and we have every reason to think, that the forces in nature are so disposed as to render this a natural impossibility.

Let us now take any element of the medium, rectangular in a state of repose, and of which the sides are dx, dy, dz; the length of the sides composed of the same particles will in a state of motion become

$$dx' = dx\,(1 + s_1), \quad dy' = dy\,(1 + s_2), \quad dz' = dz\,(1 + s_3);$$

where s_1, s_2, s_3 are exceedingly small quantities of the first order. If, moreover, we make,

$$\alpha = \cos < \frac{dy'}{dz'}, \quad \beta = \cos < \frac{dx'}{dz'}, \quad \gamma = \cos < \frac{dx'}{dy'};$$

α, β, and γ will be very small quantities of the same order. But, whatever may be the nature of the internal actions, if we represent by

$$\delta\phi\, dx\, dy\, dz,$$

the part of the second member of the equation (1), due to the molecules in the element under consideration, it is evident, that ϕ will remain the same when all the sides and all the angles of the parallelopiped, whose sides are $dx'\, dy'\, dz'$, remain unaltered, and therefore its most general value must be of the form

$$\phi = \text{function } \{s_1, s_2, s_3, \alpha, \beta, \gamma\}.$$

But s_1, s_2, s_3, α, β, γ being very small quantities of the first order, we may expand ϕ in a very convergent series of the form

$$\phi = \phi_0 + \phi_1 + \phi_2 + \phi_3 + \&c.:$$

ϕ_0, ϕ_1, ϕ_2, &c. being homogeneous functions of the six quantities α, β, γ, s_1, s_2, s_3 of the degrees 0, 1, 2, &c. each of which is very great compared with the next following one. If now, ρ represent the primitive density of the element $dx\, dy\, dz$, we may write $\rho\, dx\, dy\, dz$ in the place of Dm in the formula (1), which will thus become, since ϕ_0 is constant,

$$\iiint \rho \, dx \, dy \, dz \left\{ \frac{d^2u}{dt^2} \delta u + \frac{d^2v}{dt^2} dv + \frac{d^2w}{dt^2} \delta w \right\}$$

$$= \iiint dx \, dy \, dz \, (\delta\phi_1 + \delta\phi_2 + \&c.) \, ;$$

the triple integrals extending over the whole volume of the medium under consideration.

But by the supposition, when $u = 0$, $v = 0$ and $w = 0$, the system is in equilibrium, and hence

$$0 = \iiint dx \, dy \, dz \, \delta\phi_1 :$$

seeing that ϕ_1 is a homogeneous function of s_1, s_2, s_3, α, β, γ of the *first* degree only. If therefore we neglect ϕ_3, ϕ_4, &c. which are exceedingly small compared with ϕ_2, our equation becomes

$$\iiint \rho \, dx \, dy \, dz \left\{ \frac{d^2u}{dt^2} \delta u + \frac{d^2v}{dt^2} \delta v + \frac{d^2w}{dt^2} \delta w \right\} = \iiint dx dy \, dz \, \delta\phi_2 \ldots (2) \, ;$$

the integrals extending over the whole volume under consideration. The formula just found is true for any number of media comprised in this volume, provided the whole system be perfectly free from all extraneous forces, and subject only to its own molecular actions.

If now we can obtain the value of ϕ_2, we shall only have to apply the general methods given in the *Mécanique Analytique.* But ϕ_2 being a homogeneous function of six quantities of the second degree, will in its most general form contain 21 arbitrary coefficients. The proper value to be assigned to each will of course depend on the internal constitution of the medium. If, however, the medium be a non-crystallized one, the form of ϕ_2 will remain the same, whatever be the directions of the co-ordinate axes in space. Applying this last consideration, we shall find that the most general form of ϕ_2 for non-crystallized bodies contains only two arbitrary coefficients. In fact, by neglecting quantities of the higher orders, it is easy to perceive that

$$s_1 = \frac{du}{dx}, \quad s_2 = \frac{dv}{dy}, \quad s_3 = \frac{dw}{dz},$$

$$\alpha = \frac{dw}{dy} + \frac{dv}{dz}, \quad \beta = \frac{dw}{dx} + \frac{du}{dz}, \quad \gamma = \frac{du}{dy} + \frac{dv}{dx},$$

and if the medium is symmetrical with regard to the plane (xy) only, ϕ_2 will remain unchanged when $-z$ and $-w$ are written for z and w. But this alteration evidently changes α and β to $-\alpha$ and $-\beta$. Similar observations apply to the planes (xz) (yz). If therefore the medium is merely symmetrical with respect to each of the three co-ordinate planes, we see that ϕ_2 must remain unaltered when

$$\left.\begin{array}{l}\text{or } -z, \ -w, \ -\alpha, \ -\beta \\ \text{or } -y, \ -v, \ -\alpha, \ -\gamma \\ \text{or } -x, \ -u, \ -\beta, \ -\gamma\end{array}\right\} \text{are written for} \left\{\begin{array}{l}z, \ w, \ \alpha, \ \beta \\ y, \ v, \ \alpha, \ \gamma \\ x, \ u, \ \beta, \ \gamma.\end{array}\right.$$

In this way the 21 coefficients are reduced to 9, and the resulting function is of the form

$$G\left(\frac{du}{dx}\right)^2 + H\left(\frac{dv}{dy}\right)^2 + I\left(\frac{dw}{dz}\right)^2 + L\alpha^2 + M\beta^2 + N\gamma^2$$

$$+ 2P\frac{dv}{dy}\cdot\frac{dw}{dz} + 2Q\frac{du}{dx}\cdot\frac{dw}{dz} + 2R\frac{du}{dx}\cdot\frac{dv}{dy} = \phi_2 \ldots (A).$$

Probably the function just obtained may belong to those crystals which have three axes of elasticity at right angles to each other.

Suppose now we further restrict the generality of our function by making it symmetrical all round one axis, as that of z for instance. By shifting the axis of x through the infinitely small angle $\delta\theta$,

$$\left.\begin{array}{c}x \\ y \\ z\end{array}\right\} \text{becomes} \left\{\begin{array}{l}x + y\delta\theta \\ y - x\delta\theta, \\ z\end{array}\right.$$

$$\left.\begin{array}{c}\dfrac{d}{dx}\\[2ex]\dfrac{d}{dy}\\[2ex]\dfrac{d}{dz}\end{array}\right\}\text{ becomes }\left\{\begin{array}{l}\dfrac{d}{dx}+\delta\theta\dfrac{d}{dy}\\[2ex]\dfrac{d}{dy}-d\theta\dfrac{d}{dx},\\[2ex]\dfrac{d}{dz}\end{array}\right.$$

and

$$\left.\begin{array}{r}u\\v\\w\end{array}\right\}\text{ becomes }\left\{\begin{array}{l}u+v\delta\theta\\v-u\delta\theta.\\w\end{array}\right.$$

Making these substitutions in (A), we see that the form of ϕ_2 will not remain the same for the new axes, unless

$$G=H=2N+R,$$
$$L=M,$$
$$P=Q;$$

and thus we get

$$\phi_2=G\left\{\left(\frac{du}{dx}\right)^2+\left(\frac{dv}{dy}\right)^2\right\}+I\left(\frac{dw}{dz}\right)^2+L(\alpha^2+\beta^2)$$
$$+N\gamma^2+2P\left(\frac{dv}{dy}+\frac{du}{dx}\right)\frac{dw}{dz}+(2G-4N)\frac{du}{dx}\cdot\frac{dv}{dy}\ \ldots\ (B);$$

under which form it may possibly be applied to uniaxal crystals.

Lastly, if we suppose the function ϕ_2 symmetrical with respect to all three axes, there results

$$G=H=I=2N+R,$$
$$L=M=N,$$
$$P=Q=R;$$

and consequently,

$$\phi^2=G\left\{\left(\frac{du}{dx}\right)^2+\left(\frac{dv}{dy}\right)^2+\left(\frac{dw}{dz}\right)^2\right\}+L(\alpha^2+\beta^2+\gamma^2)$$
$$+(2G-4L)\left\{\frac{dv}{dy}\cdot\frac{dw}{dz}+\frac{du}{dx}\cdot\frac{dw}{dz}+\frac{du}{dx}\cdot\frac{dv}{dy}\right\};$$

or, by merely changing the two constants and restoring the values of α, β, and γ,

$$2\phi_2 = -A\left(\frac{du}{dx}+\frac{dv}{dy}+\frac{dw}{dz}\right)^2$$

$$-B\left\{\left(\frac{du}{dy}+\frac{dv}{dx}\right)^2+\left(\frac{du}{dz}+\frac{dw}{dx}\right)^2+\left(\frac{dv}{dz}+\frac{dw}{dy}\right)^2\right.$$

$$\left.-4\left(\frac{dv}{dy}\cdot\frac{dw}{dz}+\frac{du}{dx}\cdot\frac{dw}{dz}+\frac{du}{dx}\cdot\frac{dv}{dy}\right)\right\} \text{......} (C).$$

This is the most general form that ϕ_2 can take for non-crystallized bodies, in which it is perfectly indifferent in what directions the rectangular axes are placed. The same result might be obtained from the most general value of ϕ_2, by the method before used to make ϕ_2 symmetrical all round the axis of z, applied also to the other two axes. It was, indeed, thus I first obtained it. The method given in the text, however, and which is very similar to one used by M. Cauchy, is not only more simple, but has the advantage of furnishing two intermediate results, which may possibly be of use on some future occasion.

Let us now consider the particular case of two indefinitely extended media, the surface of junction when in equilibrium being a plane of infinite extent, horizontal (suppose), and which we shall take as that of (yz), and conceive the axis of x positive directed downwards. Then if ρ be the constant density of the upper, and $\rho_{\prime}$ that of the lower medium, ϕ_2 and $\phi_2^{(1)}$ the corresponding functions due to the molecular actions; the equation (2) adapted to the present case will become

$$\iiint \rho\, dx\, dy\, dz \left\{\frac{d^2u}{dt^2}\,\delta u+\frac{d^2v}{dt^2}\,\delta v+\frac{d^2w}{dt^2}\,\delta w\right\}$$

$$+\iiint \rho_{\prime}\, dx\, dy\, dz \left\{\frac{d^2u_{\prime}}{dt^2}\,\delta u_{\prime}+\frac{d^2v_{\prime}}{dt^2}\,\delta v_{\prime}+\frac{d^2w_{\prime}}{dt^2}\,dw_{\prime}\right\},$$

$$=\iiint dx\, dy\, dz\, \phi_2+\iiint dx\, dy\, dz\, \phi_2^{(1)} \text{.........} (3);$$

$u_{\prime}$, $v_{\prime}$, $w_{\prime}$ belonging to the lower fluid, and the triple integrals being extended over the whole volume of the fluids to which they respectively belong.

It now only remains to substitute for ϕ_2 and $\phi_2^{(1)}$ their values, to effect the integrations by parts, and to equate separately to zero the coefficients of the independent variations. Substituting therefore for ϕ_2 its value (C), we get

$$\iiint dx\,dy\,dz\,\delta\phi_2$$

$$= -A\iiint dx\,dy\,dz\left\{\left(\frac{du}{dx}+\frac{dv}{dy}+\frac{dw}{dz}\right)\left(\frac{d\delta u}{dx}+\frac{d\delta v}{dy}+\frac{d\delta w}{dz}\right)\right\}$$

$$-B\iiint dx\,dy\,dz\left\{\left(\frac{du}{dy}+\frac{dv}{dx}\right)\left(\frac{d\delta u}{dy}+\frac{d\delta v}{dx}\right)+\left(\frac{du}{dz}+\frac{dw}{dx}\right)\left(\frac{d\delta u}{dz}+\frac{d\delta w}{dx}\right)+\left(\frac{dv}{dz}+\frac{dw}{dy}\right)\left(\frac{d\delta v}{dz}+\frac{d\delta w}{dy}\right)\right.$$

$$\left.-2\left[\left(\frac{dv}{dy}\cdot\frac{d\delta w}{dz}+\frac{dw}{dz}\cdot\frac{d\delta v}{dy}\right)+\left(\frac{du}{dx}\cdot\frac{d\delta w}{dz}+\frac{dw}{dz}\cdot\frac{d\delta u}{dx}\right)+\left(\frac{du}{dx}\cdot\frac{d\delta v}{dy}+\frac{dv}{dy}\cdot\frac{d\delta u}{dx}\right)\right]\right\}$$

$$= -\iint dy\,dz\left\{A\,.\left(\frac{du}{dx}+\frac{dv}{dy}+\frac{dw}{dz}\right)-2B\left(\frac{dv}{dy}+\frac{dw}{dz}\right)\right\}.\,\delta u$$

$$-\iint dy\,dz\left\{B\left(\frac{du}{dy}+\frac{dv}{dx}\right)\delta v+B\left(\frac{du}{dz}+\frac{dw}{dx}\right)\delta w\right\}$$

$$+\iiint dx\,dy\,dz\left\{A\,\frac{d}{dx}\cdot\left(\frac{du}{dx}+\frac{dv}{dy}+\frac{dw}{dz}\right)+B\left[\frac{d^2u}{dy^2}+\frac{d^2u}{dz^2}-\frac{d}{dx}\left(\frac{dv}{dy}+\frac{dw}{dz}\right)\right]\right\}.\,\delta u$$

$$+\left\{A\,\frac{d}{dy}\cdot\left(\frac{du}{dx}+\frac{dv}{dy}+\frac{dw}{dz}\right)+B\left[\frac{d^2v}{dx^2}+\frac{d^2v}{dz^2}-\frac{d}{dy}\left(\frac{du}{dx}+\frac{dw}{dz}\right)\right]\right\}\delta v$$

$$+\left\{A\,\frac{d}{dz}\cdot\left(\frac{du}{dx}+\frac{dv}{dy}+\frac{dw}{dz}\right)+B\left[\frac{d^2w}{dx^2}+\frac{d^2w}{dy^2}-\frac{d}{dz}\cdot\left(\frac{du}{dx}+\frac{dv}{dy}\right)\right]\right\}\delta w\,;$$

seeing that we may neglect the double integrals at the limits $x=-\infty$, $y=\pm\infty$, $z=\pm\infty$; as the conditions imposed at these limits cannot affect the motion of the system at any *finite* distance from the origin; and thus the double integrals belong only to the surface of junction, of which the equation, in a state of equilibrium, is

$$0 = x.$$

In like manner we get

$$\iiint dx\,dy\,dz\,\delta\phi_2^{(1)}$$

$$= + \iint dy\,dz \left\{A_{,}\left(\frac{du_{,}}{dx} + \frac{dv_{,}}{dy} + \frac{dw_{,}}{dz}\right) - 2B_{,}\left(\frac{dv_{,}}{dy} + \frac{dw_{,}}{dz}\right)\right\}\delta u_{,}$$

$$+ \iint dy\,dz \left\{B_{,}\left(\frac{du_{,}}{dy} + \frac{dv_{,}}{dx}\right)\delta v_{,} + B_{,}\left(\frac{du_{,}}{dz} + \frac{dw_{,}}{dx}\right)\delta w_{,}\right\}$$

+ the triple integral;

since it is the *least* value of x which belongs to the surface of junction in the *lower* medium, and therefore the double integrals belonging to the limiting surface must have their signs changed.

If, now, we substitute the preceding expression in (3), equate separately to zero the coefficients of the independent variation δu, δv, δw, under the triple sign of integration, there results for the upper medium

$$\rho\frac{d^2u}{dt^2} = A\frac{d}{dx}\cdot\left(\frac{du}{dx} + \frac{dv}{dy} + \frac{dw}{dz}\right) + B\left\{\frac{d^2u}{dy^2} + \frac{d^2u}{dz^2} - \frac{d}{dx}\cdot\left(\frac{dv}{dy} + \frac{dw}{dz}\right)\right\};$$

$$\rho\frac{d^2v}{dt^2} = A\frac{d}{dy}\cdot\left(\frac{du}{dx} + \frac{dv}{dy} + \frac{dw}{dz}\right) + B\left\{\frac{d^2v}{dx^2} + \frac{d^2v}{dz^2} - \frac{d}{dy}\cdot\left(\frac{du}{dx} + \frac{dw}{dz}\right)\right\} (4);$$

$$\rho\frac{d^2w}{dt^2} = A\frac{d}{dz}\cdot\left(\frac{du}{dx} + \frac{dv}{dy} + \frac{dw}{dz}\right) + B\left\{\frac{d^2w}{dx^2} + \frac{d^2w}{dy^2} - \frac{d}{dz}\cdot\left(\frac{du}{dx} + \frac{dv}{dy}\right)\right\};$$

and by equating the coefficients of $\delta u_{,}$, $\delta v_{,}$, $\delta w_{,}$, we get three similar equations for the lower medium.

To the six general equations just obtained, we must add the conditions due to the surface of junction of the two media; and at this surface we have first,

$$u = u_{,},\quad v = v_{,},\quad w = w_{,},\quad \text{(when } x = 0) \text{ .... } (5);$$

and consequently,

$$\delta u = \delta u_{,};\quad \delta v = \delta v_{,};\quad \delta w = \delta w_{,}.$$

But the part of the equation (3) belonging to this surface, and which yet remains to be satisfied, is

$$0 = -\iint dy\,dz \left\{ A\left(\frac{du}{dx} + \frac{dv}{dy} + \frac{dw}{dz}\right) - 2B\left(\frac{dv}{dy} + \frac{dw}{dz}\right)\right\} \delta u$$

$$+ \iint dy\,dz \left\{ A_{,}\left(\frac{du_{,}}{dx} + \frac{dv_{,}}{dy} + \frac{dw_{,}}{dz}\right) - 2B_{,}\left(\frac{dv_{,}}{dy} + \frac{dw_{,}}{dz}\right)\right\} \delta u_{,}$$

$$+ \iint dy\,dz \left\{ B\left(\frac{du}{dy} + \frac{dv}{dx}\right)\delta v + B\left(\frac{du}{dz} + \frac{dw}{dx}\right)\delta w\right\}$$

$$+ \iint dy\,dz \left\{ B_{,}\left(\frac{du_{,}}{dy} + \frac{dv_{,}}{dx}\right)\delta v_{,} + B_{,}\left(\frac{du_{,}}{dz} + \frac{dw_{,}}{dx}\right)\delta w_{,}\right\};$$

and as $\delta u = \delta u_{,}$, &c., we obtain, as before,

$$A\left(\frac{du}{dx} + \frac{dv}{dy} + \frac{dw}{dz}\right) - 2B\left(\frac{dv}{dy} + \frac{dw}{dz}\right)$$

$$= A_{,}\left(\frac{du_{,}}{dx} + \frac{dv_{,}}{dy} + \frac{dw_{,}}{dz}\right) - 2B_{,}\left(\frac{dv_{,}}{dy} + \frac{dw_{,}}{dz}\right)$$

$$B\left(\frac{du}{dy} + \frac{dv}{dx}\right) = B_{,}\left(\frac{du_{,}}{dy} + \frac{dv_{,}}{dx}\right) \dots\dots\dots\dots\dots (6),$$

$$B\left(\frac{du}{dz} + \frac{dw}{dx}\right) = B_{,}\left(\frac{du_{,}}{dz} + \frac{dw_{,}}{dx}\right);$$

and these belong to the particular value $x = 0$.

The six particular conditions (5) and (6), belonging to the surface of junction of the two media, combined with the six general equations before obtained, are *necessary* and *sufficient* for the complete determination of the motion of the two media, supposing the initial state of each given. We shall not here attempt their general solution, but merely consider the propagation of a plane wave of infinite extent, accompanied by its reflected and refracted waves, as in the preceding paper on Sound.

Let the direction of the axis of z, which yet remains arbitrary, be taken parallel to the intersection of the plane of the incident wave with the surface of junction, and suppose the disturbance of the particles to be wholly in the direction of the axis of z, which is the case with light polarized in the plane of incidence, according to Fresnel. Then we have

$$0 = u, \quad 0 = v, \quad 0 = u_{\prime}, \quad 0 = v_{\prime};$$

and supposing the disturbance the same for every point of the same front of a wave, w and $w_{\prime}$ will be independent of z, and thus the three general equations (4) will all be satisfied if

$$\rho \frac{d^2 w}{dt^2} = B \left\{ \frac{d^2 w}{dx^2} + \frac{d^2 w}{dy^2} \right\},$$

or by making $\frac{B}{\rho} = \gamma^2$,

$$\frac{d^2 w}{dt^2} = \gamma^2 \left\{ \frac{d^2 w}{dx^2} + \frac{d^2 w}{dy^2} \right\} \text{.....................(7)}.$$

Similarly in the lower medium we have

$$\frac{d^2 w_{\prime}}{dt^2} = \gamma_{\prime}^2 \left\{ \frac{d^2 w_{\prime}}{dx^2} + \frac{d^2 w_{\prime}}{dy^2} \right\} \text{.....................(8)},$$

$w_{\prime}$ and $\gamma_{\prime}$ belonging to this medium.

It now remains to satisfy the conditions (5) and (6). But these are all satisfied by the preceding values provided

$$w = w_{\prime},$$

$$B \frac{dw}{dx} = B_{\prime} \frac{dw_{\prime}}{dx}.$$

The formulæ which we have obtained are quite general, and will apply to the ordinary elastic fluids by making $B = 0$. But for all the known gases, A is independent of the nature of the gas, and consequently $A = A_{\prime}$. If, therefore, we suppose $B = B_{\prime}$, at least when we consider those phenomena only which depend merely on different states of the same medium, as is the case with light, our conditions become*

$$\left. \begin{aligned} w &= w_{\prime} \\ \frac{dw}{dx} &= \frac{dw_{\prime}}{dx} \end{aligned} \right\} \text{ (when } x = 0\text{)} \text{..................(9)}.$$

* Though for all known gases A is independent of the nature of the gas, perhaps it is extending the analogy rather too far, to assume that in the lumini-

The disturbance in the upper medium which contains the incident and reflected wave, will be represented, as in the case of Sound, by

$$w = f(ax + by + ct) + F(-ax + by + ct);$$

f belonging to the incident, F to the reflected plane wave, and c being a negative quantity. Also in the lower medium,

$$w_{,} = f_{,}(a_{,}x + by + ct).$$

These values evidently satisfy the general equation (7) and (8), provided $c^2 = \gamma^2 (a^2 + b^2)$, and $c^2 = \gamma_{,}^2 (a_{,}^2 + b^2)$; we have therefore only to satisfy the conditions (9), which give

$$f(by + ct) + F(by + ct) = f_{,}(by + ct),$$

$$af'(by + ct) - aF'(by + ct) = a_{,}f'(by + ct).$$

Taking now the differential coefficient of the first equation, and writing to abridge the characteristics of the functions only, we get

$$2f' = \left(1 + \frac{a_{,}}{a}\right) f'_{,}, \text{ and } 2F' = \left(1 - \frac{a_{,}}{a}\right) f'_{,},$$

and therefore

$$\frac{F'}{f'} = \frac{1 - \frac{a_{,}}{a}}{1 + \frac{a_{,}}{a}} = \frac{a - a_{,}}{a + a_{,}} = \frac{\cot\theta - \cot\theta_{,}}{\cot\theta + \cot\theta_{,}} = \frac{\sin(\theta_{,} - \theta)}{\sin(\theta_{,} + \theta)};$$

θ and $\theta_{,}$ being the angles of incidence and refraction.

This ratio between the intensity of the incident and reflected

ferous ether the constants A and B must always be independent of the state of the ether, as found in different refracting substances. However, since this hypothesis greatly simplifies the equations due to the surface of junction of the two media, and is itself the most simple that could be selected, it seemed natural first to deduce the consequences which follow from it before trying a more complicated one, and, as far as I have yet found, these consequences are in accordance with observed facts.

waves is exactly the same as that for light polarized in the plane of incidence (vide Airy's *Tracts*, p. 356*), and which Fresnel supposes to be propagated by vibrations perpendicular to the plane of incidence, agreeably to what has been assumed in the foregoing process.

We will now limit the generality of the functions f, F and $f_{,}$, by supposing the law of the motion to be similar to that of a cycloidal pendulum; and if we farther suppose the angle of incidence to be increased until the refracted wave ceases to be transmitted in the regular way, as in our former paper on Sound, the proper integral of the equation

$$\frac{d^2 w_{,}}{dt^2} = \gamma_{,}^2 \left\{ \frac{d^2 w_{,}}{dx^2} + \frac{d^2 w_{,}}{dy^2} \right\}$$

will be

$$w_{,} = \epsilon^{-a_{,}' x} B \sin \psi \ldots\ldots\ldots\ldots\ldots\ldots\ldots (10);$$

where $\psi = by + ct$, and $a_{,}'$ is determined by

$$\gamma_{,}^2 (b^2 - a_{,}'^2) = c^2 = \gamma^2 (b^2 + a^2) \ldots\ldots\ldots\ldots\ldots (11).$$

But one of the conditions (9) will introduce *sines* and the other *cosines*, in such a way that it will be impossible to satisfy them unless we introduce both *sines* and *cosines* into the value of w, or, which amounts to the same, unless we make

$$w = \alpha \sin (ax + by + ct + e) + \beta \sin (-ax + by + ct + e_{,}) \ldots (12),$$

in the first medium, instead of

$$w = \alpha \sin (ax + by + ct) + \beta \sin (-ax + by + ct),$$

which would have been done had the refracted wave been transmitted in the usual way, and consequently no exponential been introduced into the value of $w_{,}$. We thus see the analytical reason for what is called the change of phase which takes place when the reflexion of light becomes total.

* [Airy on the Undulatory Theory of Optics, p. 109, Art. 128.]

Substituting now (10) and (12), in the equations (9), and proceeding precisely as for Sound, we get

$$0 = \alpha \cos e - \beta \cos e_{,,}$$

$$0 = \alpha \sin e + \beta \sin e_{,,}$$

$$\frac{a_{,}'}{a} B = \alpha \sin e - \beta \sin e_{,}$$

$$B = \alpha \cos e + \beta \cos e_{,}.$$

Hence there results $\alpha = \beta$, and $e_{,} = -e$, and

$$\tan e = \frac{a_{,}'}{a} = \frac{a_{,}'}{b} \div \frac{a}{b} = \frac{a_{,}'}{b} \tan \theta.$$

But by (11),

$$\frac{a_{,}'}{b} = \sqrt{\left\{1 - \frac{\gamma^2}{\gamma_{,}^2} \cdot \left(1 + \frac{a^2}{b^2}\right)\right\}} = \sqrt{\left(1 - \frac{1}{\mu^2 \sin^2 \theta}\right)};$$

by introducing μ the index of refraction, and θ the angle of incidence. Thus,

$$\tan e = \frac{\sqrt{(\mu^2 \sin^2 \theta - 1)}}{\mu \cos \theta};$$

and as e represents half the alteration of phase in passing from the incident to the reflected wave, we see that here also our result agrees precisely with Fresnel's for light polarized in the plane of incidence. (Vide Airy's *Tracts*, p. 362*.)

Let us now conceive the direction of the transverse vibrations in the incident wave to be perpendicular to the direction in the case just considered; and therefore that the actual motions of the particles are all parallel to the intersection of the plane of incidence (xy) with the front of the wave. Then, as the planes of the incident and refracted waves do not coincide, it is easy to perceive that at the surface of junction there will, in this case, be a resolved part of the disturbance in the direction of the

* [Airy, *ubi sup.* p. 114, Art. 133.]

normal; and therefore, besides the incident wave, there will, in general, be an accompanying reflected and refracted wave, in which the vibrations are transverse, and another pair of accompanying reflected and refracted waves, in which the directions of the vibrations are normal to the fronts of the waves. In fact, unless the consideration of the two latter waves is also introduced, it is impossible to satisfy all the conditions at the surface of junction; and these are as essential to the complete solution of the problem, as the general equations of motion.

The direction of the disturbance being in plane (xy) $w = 0$, and as the disturbance of every particle in the same front of a wave is the same, u and v are independent of z. Hence, the general equations (4) for the first medium become

$$\frac{d^2u}{dt^2} = g^2 \frac{d}{dx}\left(\frac{du}{dx} + \frac{dv}{dy}\right) + \gamma^2 \frac{d}{dy}\left(\frac{du}{dy} - \frac{dv}{dx}\right),$$

$$\frac{d^2v}{dt^2} = g^2 \frac{d}{dy}\left(\frac{du}{dx} + \frac{dv}{dy}\right) + \gamma^2 \frac{d}{dx}\left(\frac{dv}{dx} - \frac{du}{dy}\right),$$

where $g^2 = \frac{A}{\rho}$, and $\gamma^2 = \frac{B}{\rho}$.

These equations might be immediately employed in their present form; but they will take a rather more simple form, by making

$$\left.\begin{aligned} u &= \frac{d\phi}{dx} + \frac{d\psi}{dy} \\ v &= \frac{d\phi}{dy} - \frac{d\psi}{dx} \end{aligned}\right\} \quad \ldots\ldots\ldots\ldots\ldots\ldots(13);$$

ϕ and ψ being two functions of x, y, and t, to be determined.

By substitution, we readily see that the two preceding equations are equivalent to the system

$$\left.\begin{aligned} \frac{d^2\phi}{dt^2} &= g^2\left(\frac{d^2\phi}{dx^2} + \frac{d^2\phi}{dy^2}\right) \\ \frac{d^2\psi}{dt^2} &= \gamma^2\left(\frac{d^2\psi}{dx^2} + \frac{d^2\psi}{dy^2}\right) \end{aligned}\right\} \quad \ldots\ldots\ldots\ldots\ldots\ldots(14).$$

In like manner, if in the second medium we make

$$\left.\begin{aligned} u_{,} &= \frac{d\phi_{,}}{dx} + \frac{d\psi_{,}}{dy} \\ v_{,} &= \frac{d\phi_{,}}{dy} - \frac{d\psi_{,}}{dx} \end{aligned}\right\} \quad \text{.......................} \ (15),$$

we get to determine $\phi_{,}$ and $\psi_{,}$ the equations

$$\left.\begin{aligned} \frac{d^2\phi_{,}}{dt^2} &= g_{,}^{2}\left(\frac{d^2\phi_{,}}{dx^2} + \frac{d^2\phi_{,}}{dy^2}\right) \\ \frac{d^2\psi_{,}}{dt^2} &= \gamma_{,}^{2}\left(\frac{d^2\psi_{,}}{dx^2} + \frac{d^2\psi_{,}}{dy^2}\right) \end{aligned}\right\} \quad \text{.................} (16),$$

and as we suppose the constants A and B the same for both media, we have

$$\frac{\gamma}{\gamma_{,}} = \frac{g}{g_{,}}.$$

For the complete determination of the motion in question, it will be necessary to satisfy all the conditions due to the surface of junction of the two media. But, since $w = 0$ and $w_{,} = 0$, also, since $u, v, u_{,}, v_{,}$ are independent of z, the equations (5) and (6) become

$$u = u_{,}, \quad v = v_{,};$$

$$A\left(\frac{du}{dx} + \frac{dv}{dy}\right) - 2B\frac{dv}{dy} = A\left(\frac{du_{,}}{dx} + \frac{dv_{,}}{dy}\right) - 2B\frac{dv_{,}}{dy},$$

$$\frac{du}{dy} + \frac{dv}{dx} = \frac{du_{,}}{dy} + \frac{dv_{,}}{dx},$$

provided $x = 0$. But since $x = 0$ in the last equations, we may differentiate them with regard to any of the independent variables except x, and thus the two latter, in consequence of the two former, will become

$$\frac{du}{dx} = \frac{du_{,}}{dx}, \quad \frac{dv}{dx} = \frac{dv_{,}}{dx}.$$

Substituting now for u, v, &c., their values (13) and (15), in ϕ and ψ, the four resulting conditions relative to the surface of junction of the two media may be written,

$$\left.\begin{aligned}
\frac{d\phi}{dx}+\frac{d\psi}{dy}&=\frac{d\phi_{,}}{dx}+\frac{d\psi_{,}}{dy}\\
\frac{d\phi}{dy}-\frac{d\psi}{dx}&=\frac{d\phi_{,}}{dy}-\frac{d\psi_{,}}{dx}\\
\frac{d^2\phi}{dx^2}+\frac{d^2\psi}{dx\,dy}&=\frac{d^2\phi_{,}}{dx^2}+\frac{d^2\psi_{,}}{dx\,dy}\\
\frac{d^2\phi}{dx\,dy}-\frac{d^2\psi}{dx^2}&=\frac{d^2\phi_{,}}{dx\,dy}-\frac{d^2\psi_{,}}{dx^2}
\end{aligned}\right\}\text{(when } x=0\text{);}$$

or since we may differentiate with respect to y, the first and fourth equations give

$$\frac{d^2\psi}{dx^2}+\frac{d^2\psi}{dy^2}=\frac{d^2\psi_{,}}{dx^2}+\frac{d^2\psi_{,}}{dy^2};$$

in like manner, the second and third give

$$\frac{d^2\phi}{dx^2}+\frac{d^2\phi}{dy^2}=\frac{d^2\phi_{,}}{dx^2}+\frac{d^2\phi_{,}}{dy^2},$$

which, in consequence of the general equations (14) and (16), become

$$\frac{d^2\psi}{\gamma^2dt^2}=\frac{d^2\psi_{,}}{\gamma_{,}^2dt^2},\quad\text{and}\quad\frac{d^2\phi}{g^2dt^2}=\frac{d^2\phi_{,}}{g_{,}^2dt^2}.$$

Hence, the equivalent of the four conditions relative to the surface of junction may be written

$$\left.\begin{aligned}
\frac{d\phi}{dx}+\frac{d\psi}{dy}&=\frac{d\phi_{,}}{dx}+\frac{d\psi_{,}}{dy}\\
\frac{d\phi}{dy}-\frac{d\psi}{dx}&=\frac{d\phi_{,}}{dy}-\frac{d\psi_{,}}{dx}\\
\frac{d^2\phi}{g^2dt^2}&=\frac{d^2\phi_{,}}{g_{,}^2dt^2}\\
\frac{d^2\psi}{\gamma^2dt^2}&=\frac{d^2\psi_{,}}{\gamma_{,}^2dt^2}
\end{aligned}\right\}\text{(when } x=0\text{) .... (17).}$$

If we examine the expressions (13) and (15), we shall see that the disturbances due to ϕ and $\phi_{,}$ are normal to the front of the wave to which they belong, whilst those which are due to ψ,

$\psi_{,}$ are transverse or wholly in the front of the wave. If the coefficients A and B did not differ greatly in magnitude, waves propagated by both kinds of vibrations must in general exist, as was before observed. In this case, we should have in the upper medium

$$\left.\begin{aligned} &\psi = f(ax+by+ct) + F(-ax+by+ct) \\ \text{and} \quad &\phi = \chi_{,}(-a'x+by+ct) \end{aligned}\right\}\text{......(18)};$$

and for the lower one

$$\left.\begin{aligned} \psi_{,} &= f_{,}(a_{,}x+by+ct) \\ \phi_{,} &= \chi_{,}(a_{,}'x+by+ct) \end{aligned}\right\}\text{..................(19)}.$$

The coefficients b and c being the same for all the functions to simplify the results, since the indeterminate coefficients $a_{,}'a_{,}a'$ will allow the fronts of the waves to which they respectively belong, to take any position that the nature of the problem may require. The coefficient of x in F belonging to that reflected wave, which, like the incident one, is propagated by transverse vibrations would have been determined exactly like $a_{,}'a_{,}a'$, as, however, it evidently $=-a$, it was for the sake of simplicity introduced immediately into our formulæ.

By substituting the values just given in the general equations (14) and (16), there results

$$c^2 = (a^2+b^2)\,\gamma^2 = (a_{,}^2+b^2)\,\gamma_{,}^2 = (a'^2+b^2)\,g^2 = (a_{,}'^2+b^2)\,g'^2,$$

we have thus the position of the fronts of the reflected and refracted waves.

It now remains to satisfy the conditions due to the surface of junction of the two media. Substituting, therefore, the values (18) and (19) in the equations (17), we get

$$f''+F''=\frac{\gamma^2}{\gamma_{,}^2}f_{,}'',$$

$$\chi''=\frac{g^2}{g_{,}^2}\chi_{,}'';$$

$$-a'\chi'+b\,(f'+F')=a_{,}'\chi_{,}'+bf_{,}',$$

$$b\chi'-a\,(f'-F')=b\chi_{,}'-a_{,}f_{,}';$$

where to abridge, the characteristics only of the functions are written.

By means of the last four equations, we shall readily get the values of $F''\chi'' f_{\prime}''\chi_{\prime}''$ in terms of f'', and thus obtain the intensities of the two reflected and two refracted waves, when the coefficients A and B do not differ greatly in magnitude, and the angle which the incident wave makes with the plane surface of junction is contained within certain limits. But in the introductory remarks, it was shewn that $\frac{A}{B}=$ a very great quantity, which may be regarded as infinite, and therefore g and $g_{\prime}$ may be regarded as infinite compared with γ and $\gamma_{\prime}$. Hence, for all angles of incidence except such as are infinitely small, the waves dependent on ϕ and $\phi_{\prime}$ cease to be transmitted in the regular way. We shall therefore, as before, restrain the generality of our functions by supposing the law of the motion to be similar to that of a cycloidal pendulum, and as two of the waves cease to be transmitted in the regular way, we must suppose in the upper medium

$$\left.\begin{aligned} \psi &= \alpha \sin (ax + by + ct + e) + \beta \sin (-ax + by + ct + e_{\prime}) \\ \text{and} \quad \phi &= \epsilon^{a'x} (A \sin \psi_0 + B \cos \psi_0) \end{aligned}\right\} \quad (20);$$

and in the lower one

$$\left.\begin{aligned} \psi_{\prime} &= \alpha_{\prime} \sin (a_{\prime}x + by + ct) \\ \phi_{\prime} &= \epsilon^{-a_{\prime}'x} (A_{\prime} \sin \psi_0 + B_{\prime} \cos \psi_0) \end{aligned}\right\} \ldots\ldots\ldots\ldots (21),$$

where to abridge $\psi_0 = by + ct$.

These substituted in the general equations (14) and (15), give

$$c^2 = \gamma^2 (a^2 + b^2) = \gamma_{\prime}^2 (a_{\prime}^2 + b^2) = g^2 (-a'^2 + b^2) = g_{\prime}^2 (-a_{\prime}'^2 + b^2),$$

or, since g and $g_{\prime}$ are both infinite,

$$b = a' = a_{\prime}'.$$

It only remains to substitute the values (20), (21) in the equations (17), which belong to the surface of junction, and thus we get

$$bA \sin\psi_0 + bB \cos\psi_0 + b\alpha \cos(\psi_0 + e) + b\beta \cos(\psi_0 + e_{\prime})$$
$$= -bA_{\prime} \sin\psi_0 - bB_{\prime} \cos\psi_0 + b\alpha_{\prime} \cos\psi_0,$$
$$bA \cos\psi_0 - bB \sin\psi_0 - a\alpha \cos(\psi_0 + e) + a\beta \cos(\psi_0 + e_{\prime})$$
$$= bA_{\prime} \cos\psi_0 - bB_{\prime} \sin\psi_0 - a_{\prime}\alpha_{\prime} \cos\psi_0 \;\ldots\ldots\ldots(22).$$
$$\frac{1}{g^2}(A \sin\psi_0 + B\cos\psi_0) = \frac{1}{g_{\prime}^2}(A_{\prime} \sin\psi_0 + B_{\prime}\cos\psi_0),$$
$$\frac{1}{\gamma^2}\{\alpha \sin(\psi_0 + e) + \beta \sin(\psi_0 + e_{\prime})\} = \frac{1}{\gamma_{\prime}^2}\alpha_{\prime} \sin\psi_0.$$

Expanding the two last equations, comparing separately the coefficients of $\cos\psi_0$ and $\sin\psi_0$, and observing that

$$\frac{g}{g_{\prime}} = \frac{\gamma}{\gamma_{\prime}} = \mu \text{ suppose,}$$

we get

$$\left.\begin{aligned} A &= \mu^2 A_{\prime} \\ B &= \mu^2 B_{\prime} \\ \alpha \cos e + \beta \cos e_{\prime} &= \mu^2 \alpha_{\prime} \\ \alpha \sin e + \beta \sin e_{\prime} &= 0 \end{aligned}\right\} \;\ldots\ldots\ldots\ldots\ldots(23).$$

In like manner the two first equations of (22) will give

$$0 = A + A_{\prime} - \alpha \sin e - \beta \sin e_{\prime},$$
$$0 = A - A_{\prime} + \frac{a_{\prime}\alpha_{\prime}}{b} + \frac{a}{b}(\beta \cos e_{\prime} - \alpha \cos e),$$
$$0 = B + B_{\prime} + \alpha \cos e + \beta \cos e_{\prime} - \alpha_{\prime},$$
$$0 = B - B_{\prime} + \frac{a}{b}(\beta \sin e_{\prime} - \alpha \sin e);$$

combining these with the system (23), there results

$$\left.\begin{aligned} 0 &= A + A_{\prime} \\ 0 &= B + B_{\prime} + (\mu^2 - 1)\alpha_{\prime} \\ 0 &= A - A_{\prime} + \frac{a_{\prime}\alpha_{\prime}}{b} + \frac{a}{b}(\beta \cos e_{\prime} - \alpha \cos e) \\ 0 &= B - B_{\prime} + \frac{a}{b}(\beta \sin e_{\prime} - \alpha \sin e) \end{aligned}\right\} \;\ldots(24).$$

Again, the systems (23) and (24) readily give

$$\left.\begin{aligned} \alpha \sin e &= -\tfrac{1}{2}\cdot\frac{(\mu^2-1)^2}{\mu^2+1}\frac{b}{a}\alpha_{,} \\ \alpha \cos e &= \tfrac{1}{2}\cdot\left(\mu^2+\frac{a_{,}}{a}\right)\alpha_{,} \\ \beta \sin e_{,} &= \tfrac{1}{2}\cdot\frac{(\mu^2-1)^2}{\mu^2+1}\frac{b}{a}\alpha_{,} \\ \beta \cos e_{,} &= \tfrac{1}{2}\cdot\left(\mu^2-\frac{a_{,}}{a}\right)\alpha_{,} \end{aligned}\right\} \quad \ldots\ldots\ldots\ldots (25);$$

and therefore

$$\frac{\beta^2}{\alpha^2} = \frac{(\mu^2+1)^2\cdot\left(\mu^2-\dfrac{a_{,}}{a}\right)^2 + (\mu^2-1)^4\dfrac{b^2}{a^2}}{(\mu^2+1)^2\cdot\left(\mu^2+\dfrac{a_{,}}{a}\right)^2 + (\mu^2-1)^4\dfrac{b^2}{a^2}} \quad \ldots\ldots\ldots (26).$$

When the refractive power in passing from the upper to the lower medium is not very great, μ does not differ much from 1. Hence, $\sin e$ and $\sin e_{,}$ are small, and $\cos e$, $\cos e_{,}$ do not differ sensibly from unity; we have, therefore, as a first approximation,

$$\frac{\beta}{\alpha} = \frac{\mu^2-\dfrac{a_{,}}{a}}{\mu^2+\dfrac{a_{,}}{a}} = \frac{\dfrac{\sin^2\theta}{\sin^2\theta_{,}}-\dfrac{\cot\theta_{,}}{\cot\theta}}{\dfrac{\sin^2\theta}{\sin^2\theta_{,}}+\dfrac{\cot\theta_{,}}{\cot\theta}} = \frac{\sin 2\theta - \sin 2\theta_{,}}{\sin 2\theta + \sin 2\theta_{,}} = \frac{\tan(\theta-\theta_{,})}{\tan(\theta+\theta_{,})},$$

which agrees with the formula in Airy's *Tracts*, p. 358*, for light polarized perpendicular to the plane of reflexion. This result is only a near approximation: but the formula (26) gives the correct value of $\dfrac{\beta^2}{\alpha^2}$, or the ratio of the intensity of the reflected to the incident light; supposing, with all optical writers, that the intensity of light is properly measured by the square of the actual velocity of the molecules of the luminiferous ether.

From the rigorous value (26), we see that the intensity of the reflected light never becomes absolutely null, but attains a minimum value nearly when

* [Airy, *ubi sup.* p. 110.]

$$0 = \mu^2 - \frac{a_{\prime}}{a}, \text{ i.e., when } \tan(\theta + \theta_{\prime}) = \infty,$$

which agrees with experiment, and this minimum value is, since (27) gives $\frac{b}{a} = \mu$,

$$\frac{\beta^2}{\alpha^2} = \frac{(\mu^2 - 1)^4 \frac{b^2}{a^2}}{4(\mu^2+1)^2 \mu^4 + (\mu^2 - 1)^4 \frac{b^2}{a^2}} = \frac{(\mu^2 - 1)^4}{4\mu^2(\mu^2+1)^2 + (\mu^2-1)^4} \quad \dots(28).$$

If $\mu = \frac{4}{3}$, as when the two media are air and water, we get

$$\frac{\beta^2}{\alpha^2} = \frac{1}{151} \text{ nearly.}$$

It is evident from the formula (28), that the magnitude of this minimum value increases very rapidly as the index of refraction increases, so that for highly refracting substances, the intensity of the light reflected at the polarizing angle becomes very sensible, agreeably to what has been long since observed by experimental philosophers. Moreover, an inspection of the equations (25) will shew, that when we gradually increase the angle of incidence so as to pass through the polarizing angle, the change which takes place in the reflected wave is not due to an alteration of the sign of the coefficient β, but to a change of phase in the wave, which for ordinary refracting substances is very nearly equal to 180°; the minimum value of β being so small as to cause the reflected wave sensibly to disappear. But in strongly refracting substances like diamond, the coefficient β remains so large that the reflected wave does not seem to vanish, and the change of phase is considerably less than 180°. These results of our theory appear to agree with the observations of Professor Airy. (*Camb. Phil. Trans.* Vol. IV. p. 418, &c.)

Lastly, if the velocity $\gamma_{\prime}$ of transmission of a wave in the lower exceed γ that in the upper medium, we may, by sufficiently augmenting the angle of incidence, cause the refracted wave to disappear, and the change of phase thus produced in the reflected wave may readily be found. As the calculation is extremely easy after what precedes, it seems sufficient to give the result. Let therefore, here, $\mu = \frac{\gamma_{,}}{\gamma}$, also e, $e_{,}$ and θ as before, then $e_{,} = -e$, and the accurate value of e is given by

$$\tan e = \mu \sqrt{\mu^2 \tan^2 \theta - \sec^2 \theta} - \frac{(\mu^2 - 1)^2 \tan \theta}{\mu^2 + 1}.$$

The first term of this expression agrees with the formula of page 362, Airy's *Tracts**, and the second will be scarcely sensible except for highly refracting substances.

* [Airy, *ubi sup.* p. 114, Art. 133.]

NOTE

ON THE

MOTION OF WAVES IN CANALS*.

* [From the *Transactions of the Cambridge Philosophical Society*, 1839.]
[Read *February* 18, 1839.]

NOTE ON THE MOTION OF WAVES IN CANALS.

In a former communication* I have endeavoured to apply the ordinary Theory of Fluid Motion to determine the law of the propagation of waves in a rectangular canal, supposing ζ the depression of the actual surface of the fluid below that of equilibrium very small compared with its depth; the depth γ as well as the breadth β of the canal being small compared with the length of a wave. For greater generality, β and γ are supposed to vary very slowly as the horizontal co-ordinate x increases, compared with the rate of the variation of ζ, due to the same cause. These suppositions are not always satisfied in the propagation of the tidal wave, but in many other cases of propagation of what Mr Russel denominates the "Great Primary Wave," they are so, and his results will be found to agree very closely with our theoretical deductions. In fact, in my paper on the Motion of Waves, it has been shown that the height of a wave varies as

$$\beta^{-\frac{1}{2}}\gamma^{-\frac{1}{4}}.$$

With regard to the effect of the breadth β, this is expressly stated by Mr Russel (vide Seventh Report of the British Association, p. 425), and the results given in the tables, p. 494, of the same work, seem to agree with our formula as well as could be expected, considering the object of the experiments there detailed.

In order to examine more particularly the way in which the Primary Wave is propagated, let us resume the formulæ,

$$\phi = \beta^{-\frac{1}{2}}\gamma^{-\frac{1}{4}} F\left(t - \int \frac{dx}{\sqrt{g\gamma}}\right),$$

$$\zeta = \frac{d\phi}{gdt} = \frac{\beta^{-\frac{1}{2}}\gamma^{-\frac{1}{4}}}{g} F'\left(t - \int \frac{\delta x}{\sqrt{g\gamma}}\right);$$

* *Supra*, p. 223.

where we have neglected the function f, which relates to the wave propagated in the direction of x negative.

Suppose, for greater simplicity, that β and γ are constant, the origin of x being taken at the point where the wave commences when $t = 0$. Then we may, without altering in the slightest degree the nature of our formulæ, take the values,

$$\phi = F(x - t\sqrt{g\gamma}) \ldots\ldots\ldots\ldots\ldots\ldots\ldots (1),$$

$$\zeta = \frac{d\phi}{gdt} = -\sqrt{\frac{\gamma}{g}} \,.\, F'(x - t\sqrt{g\gamma}).$$

But for all small oscillations of a fluid, if (a, b, c) are the co-ordinates of any particle P in its primitive state, that of equilibrium suppose; (x, y, z) the co-ordinates of P at the end of the time t, and $\Phi = \int \phi dt$ when (x, y, z) are changed into (a, b, c), we have (vide *Mécanique Analytique*, Tome II. p. 313),

$$x = a + \frac{d\Phi}{da}, \quad y = b + \frac{d\Phi}{db}, \quad z = c + \frac{d\Phi}{dc} \ldots\ldots (2).$$

Applying these general expressions to the formulæ (1) we get

$$\Phi = -\frac{1}{\sqrt{g\gamma}} {}'F(a - t\sqrt{g\gamma}),$$

and $$x = a - \frac{1}{\sqrt{g\gamma}} F(a - t\sqrt{g\gamma}).$$

Neglecting (disturbance)², we have

$$\zeta = -\sqrt{\frac{\gamma}{g}} F'(a - t\sqrt{g\gamma}),$$

and consequently,

$$\int_0^a \zeta(a - t\sqrt{g\gamma}) \,.\, da = -\sqrt{\frac{\gamma}{g}} F(a - t\sqrt{g\gamma}),$$

supposing for greater simplicity that the origin of the integral is at $a = 0$.

Hence the value of x becomes

$$x = a + \frac{1}{\gamma} \int_0^a da\, \zeta(a - t\sqrt{g\gamma}).$$

Suppose α = length of the wave when $t=0$; then $\zeta(a)=0$, except when a is between the limits 0 and α. If therefore we consider a point P before the wave has reached it,

$$\int_0^\alpha da\,\zeta(a-t\sqrt{g\gamma})=\int_0^\alpha da\,\zeta(a)=V;$$

the whole volume of the fluid which would be required to fill the hollow caused by the depression ζ below the surface of equilibrium when $t=0$. Hence we get

$$x'=a+\frac{V}{\gamma};$$

x' being the horizontal co-ordinate of P, before the wave reaches P.

Also, let x'' be the value of this co-ordinate after the wave has passed completely over P, then

$$\int_0^\alpha da\,\zeta(a-t\sqrt{g\gamma})=0, \quad \text{and } x''=a.$$

If ζ were wholly negative, or the wave were elevated above the surface of equilibrium, we should only have to write $-V$ for V, and thus

$$x'=a-\frac{V}{\gamma}, \quad \text{and } x''=a.$$

We see therefore, in this case, that the particles of the fluid by the transit of the wave are transferred forwards in the direction of the wave's motion, and permanently deposited at rest in a new place at some distance from their original position, and that the extent of the transference is sensibly equal throughout the whole depth. These waves are called by Mr Russel, positive ones, and this result agrees with his experiments, vide p. 423. If however ζ were positive, or the wave wholly depressed, it follows from our formula, that the transit of the fluid particles would be in the opposite direction. The experimental investigation of those waves, called by Mr Russel, negative ones, has not yet been completed, p. 445, and the last result cannot therefore be compared with experiment.

The value $\frac{V}{\gamma}$ which we have obtained analytically for the extent over which the fluid particles are transferred, suggests a simple physical reason for the fact. For previous to the transit of a positive wave over any particle P, a volume of fluid behind P, and equal to V, is elevated above the surface of equilibrium. During the transit, this descends within the surface of equilibrium, and must therefore force the fluid about P forward through the space

$$\left(\frac{V}{\gamma}\right);$$

admitting as an experimental fact, that after the transit of the wave the fluid particles always remain absolutely at rest.

Mr Russel, p. 425, is inclined to infer from his experiments, that the velocity of the Great Primary Wave is that due to gravity acting through a height equal to the depth of the centre of gravity of the transverse section of the channel below the surface of the fluid. When this section is a triangle of which one side is vertical, as in channel (H), p. 443, the ordinary Theory of Fluid Motion may be applied with extreme facility. For if we take the lowest edge of the horizontal channel as the axis of x, and the axis of z vertical and directed upwards, the general equations for small oscillations in this case become

$$0 = gz + \frac{p}{\rho} + \frac{d\phi}{dt} \dots\dots\dots\dots\dots\dots\dots (A),$$

$$0 = \frac{d^2\phi}{dx^2} + \frac{d^2\phi}{dy^2} + \frac{d^2\phi}{dz^2} \dots\dots\dots\dots\dots\dots (B),$$

we have, also, the conditions

$$v = \frac{d\phi}{dy} = 0 \text{ (when } y = 0) \dots\dots\dots\dots\dots\dots (a),$$

$$\frac{w}{v} = \frac{\frac{d\phi}{dz}}{\frac{d\phi}{dy}} = \frac{z}{y} \text{ (when } \frac{z}{y} = \cot\alpha) \dots\dots\dots\dots (b),$$

α being the angle which the inclined side of the channel makes with the vertical.

The first of these conditions is due to the vertical side, and the second to the inclined one, since at these extreme limits the fluid particles must move along the sides.

Now from what has been shown in our memoir, it is clear that we may satisfy the equation (B) and the two conditions just given, by

$$\phi = \phi_0 + \phi_, (y^2 + z^2) \text{ .... } (c),$$

ϕ and $\phi_,$ being two such functions of x and t only that

$$0 = \frac{d^2\phi_0}{dx^2} + 4\phi_, \text{ .... } (C).$$

It now only remains to satisfy the condition due to the upper surface. Let therefore

$$0 = z - \zeta_{x.t}$$

be the equation of this surface. Then the formula (A) of our paper before cited gives

$$0 = \frac{d\phi}{dz} - \frac{d\zeta}{dt} - \frac{d\zeta}{dx}\frac{d\phi}{dx} \text{ (when } z = c + \zeta),$$

or neglecting (disturbance)^2

$$0 = \frac{d\phi}{dz} - \frac{d\zeta}{dt} \text{ (when } z = c);$$

c being the vertical depth of the fluid in equilibrium.

Also at the upper surface $p = 0$, therefore by continuing to neglect (disturbance)^2 (A) gives

$$0 = g\zeta + \frac{d\phi}{dt} \text{ (when } z = c).$$

Hence, by eliminating ζ, we get

$$0 = g\frac{d\phi}{dz} + \frac{d^2\phi}{dt^2} \text{ (when } z = c),$$

which by (c) becomes, when we neglect terms of the order y^2 and z^2 compared with those retained,

$$0 = 2gc\phi_, + \frac{d^2\phi_0}{dt^2}.$$

Or eliminating $\phi_,$ by means of (C),

$$0 = \frac{d^2\phi_0}{dt^2} - \frac{gc}{2} \cdot \frac{d^2\phi_0}{dx^2}.$$

The particular integral of which belonging to the wave that proceeds in the direction of x positive is

$$\phi_0 = f\left(x - t\sqrt{\frac{gc}{2}}\right),$$

and hence the velocity of propagation of the wave is

$$v' = \sqrt{\frac{gc}{2}} \quad \text{.............................} (D).$$

Mr Russel gives $\sqrt{\frac{2gc}{3}}$ as the velocity, but at the same time remarks, that in consequence of the attraction of the sides of the canal fixing a portion of the fluid in its lower angle, we ought in employing any formula to calculate for an *effective* depth in place of the real one, p. 442. Instead of adopting this method, let us compare the formula (D) given by the common Theory of Fluid Motion, with Mr Russel's experiments. And as in our theory we have considered those waves only in which the elevation above the surface of equilibrium is very small compared with the depth c, it will be necessary to select those waves in which this condition is nearly satisfied. I have therefore taken from the Table, p. 443, all the waves in which

$$\zeta < \frac{c}{20},$$

and have supposed $g = 32\frac{1}{6}$ feet: the results are given below.

Observation.	Value of c.	Observed Vel. viz. feet per second.	Velocity by formula (D).
lviii........	4, in.	2,19	2,313
lxii	5,11	2,58	2,617
lxvi........	6,04	2,85	2,845
lxvii........	6,05	2,88	2,847
lxxv........	7,04	3,03	3,072
lxxii........	7,04	3,05	3,072
lxxiii......	7,04	3,04	3,072
lxxi........	7,04	3,02	3,072
lxxiii......	7,04	3,02	3,072

A more perfect agreement with theory than this could scarcely be expected. Had the formula $\sqrt{\frac{2gc}{3}} = v$ been used, the errors would have been much greater.

The theory of the motion of waves in a deep sea, taking the most simple case, in which the oscillations follow the law of the cycloidal pendulum, and considering the depth as infinite, is extremely easy, and may be thus exhibited.

Take the plane (xz) perpendicular to the ridge of one of the waves supposed to extend indefinitely in the direction of the axis y, and let the velocities of the fluid particles be independent of the co-ordinate y. Then if we conceive the axis z to be directed vertically downwards, and the plane (xy) to coincide with the surface of the sea in equilibrium, we have generally,

$$gz - \frac{p}{\rho} = \frac{d\phi}{dt},$$

$$0 = \frac{d^2\phi}{dx^2} + \frac{d^2\phi}{dz^2}.$$

The condition due to the upper surface, found as before, is

$$0 = g\frac{d\phi}{dz} - \frac{d^2\phi}{dt^2}.$$

From what precedes, it will be clear that we have now only to satisfy the second of the general equations in conjunction with the condition just given. This may be effected most conveniently by taking

$$\phi = H\epsilon^{-\frac{2\pi}{\lambda}z} \sin\frac{2\pi}{\lambda}(v't - x),$$

by which the general equation is immediately satisfied, and the condition due to the surface gives

$$g = \frac{2\pi}{\lambda}v'^2, \quad \text{or } v' = \sqrt{\frac{g\lambda}{2\pi}},$$

where λ is evidently the length of a wave. Hence, the velocity of these waves varies as $\sqrt{\lambda}$, agreeably to what Newton asserts. But the velocity assigned by the correct theory exceeds Newton's value in the ratio $\sqrt{\pi}$ to $\sqrt{2}$, or of 5 to 4 nearly.

What immediately precedes is not given as new, but merely on account of the extreme simplicity of the analysis employed. We shall, moreover, be able thence to deduce a singular consequence which has not before been noticed, that I am aware of.

Let (a, b, c) be the co-ordinates of any particle P of the fluid when in equilibrium. Then, since

$$\phi = H\epsilon^{-\frac{2\pi}{\lambda}z} \sin \frac{2\pi}{\lambda}(v't - x);$$

$$\therefore \ \Phi = \frac{-H\lambda}{2\pi v'} \epsilon^{-\frac{2\pi c}{\lambda}} \cos \frac{2\pi}{\lambda}(v't - a),$$

and the general formulæ (2) give

$$x = a + \frac{d\Phi}{da} = a - \frac{H}{v'} \epsilon^{-\frac{2\pi c}{\lambda}} \sin \frac{2\pi}{\lambda}(v't - a),$$

$$z = c + \frac{d\Phi}{dc} = c + \frac{H}{v'} \epsilon^{-\frac{2\pi c}{\lambda}} \cos \frac{2\pi}{\lambda}(v't - a).$$

Hence,

$$(x - a)^2 + (z - c)^2 = \left(\frac{H}{v'} \epsilon^{-\frac{2\pi c}{\lambda}}\right)^2,$$

and therefore any particle P revolves continually in a circular orbit, of which the radius is

$$\frac{H}{v'} \epsilon^{-\frac{2\pi}{\lambda}c},$$

round the point which it would occupy in a state of equilibrium. The radius of this circle, and consequently the agitation of the fluid particles, decreases very rapidly as the depth c increases, and much more rapidly for short than long waves, agreeably to observation.

Moreover, the direction of the rotation is such, that in the upper part of the circle the point P moves in the direction of the motion of the wave. Hence, as in the propagation of the Great Primary Wave, the actual motion of the fluid particles is direct where the surface of the fluid rises above that of equilibrium, and retrograde in the contrary case.

SUPPLEMENT TO A MEMOIR

ON THE REFLEXION AND REFRACTION OF LIGHT.

From the *Transactions of the Cambridge Philosophical Society*, 1839.
[Read May 6, 1839].

SUPPLEMENT TO A MEMOIR ON THE REFLEXION AND REFRACTION OF LIGHT.

In a paper which the Society did me the honour to publish some time ago*, I endeavoured to determine the laws of Reflexion and Refraction of a plane wave at the surface of separation of two elastic media, supposing this surface perfectly plane, and both media to terminate there abruptly: neglecting also all extraneous forces, whether due to the action of the solid particles of transparent bodies on the elastic medium, which is supposed to pervade their interstices, or to extraneous pressures. I am inclined to think that in the case of non-crystallized bodies the latter cause would not alter the *form* of our results in the slightest degree; and possibly there would be some difficulty in submitting the effects of the former to calculation. Moreover, should the radius of the sphere of sensible action of the molecular forces bear any finite ratio to λ, the length of a wave of light, as some philosophers have supposed, in order to explain the phenomena of dispersion, instead of an abrupt termination of our two media we should have a continuous though rapid change of state of the ethereal medium in the immediate vicinity of their surface of separation. And I have here endeavoured to shew, by probable reasoning, that the effect of such a change would be to diminish greatly the quantity of light reflected at the polarizing angle, even for highly refracting substances: supposing the light polarized perpendicular to the plane of incidence. The same reasoning would go to prove that in this case the quantity of the reflected light would depend greatly on minute changes in the state of the reflecting surface. I have on the present occasion merely noticed, but not insisted upon, these inferences, feeling persuaded that in researches like the present, little confidence is due to such consequences as are not supported by a rigorous analysis.

* *Supra*, p. 243.

The principal object of this supplement has been to put the equations due to the surface of junction of two media, and belonging to light polarized perpendicular to the plane of incidence, under a more simple form. The resulting expressions have here been made to depend on those before given in our paper on Sound, and thus the determination of the intensities of the reflected and refracted waves becomes in every case a matter of extreme facility. As an example of the use of the new formulæ, the intensities of the refracted waves have been determined for both kinds of light: the consideration of which waves had inadvertently been omitted in a former communication.

Perhaps I may be permitted on the present occasion to state, that though I feel great confidence in the truth of the fundamental principle on which our reasonings concerning the vibrations of elastic media have been based, the same degree of confidence is by no means extended to those adventitious suppositions which have been introduced for the sake of simplifying the analysis.

Let us here resume the equations of the paper before mentioned, namely,

$$\left.\begin{aligned} \frac{d\phi}{dx}+\frac{d\psi}{dy}&=\frac{d\phi_{,}}{dx}+\frac{d\psi_{,}}{dy} \\ \frac{d\phi}{dy}-\frac{d\psi}{dx}&=\frac{d\phi_{,}}{dy}-\frac{d\psi_{,}}{dx} \\ \frac{d^2\phi}{g^2dt^2}&=\frac{d^2\phi_{,}}{g_{,}^2dt^2} \\ \frac{d^2\psi}{\gamma^2dt^2}&=\frac{d^2\psi_{,}}{\gamma_{,}^2dt^2} \end{aligned}\right\}\text{(when } x=0)\ldots\ldots\ldots\ldots(17),$$

where u and v, the disturbances in the upper medium parallel to the axes x and y, are given by

$$u=\frac{d\phi}{dx}+\frac{d\psi}{dy},$$

$$v=\frac{d\phi}{dy}-\frac{d\psi}{dx};$$

$u_{,}$ and $v_{,}$ the disturbances in the lower medium being expressed by similar formulæ in $\phi_{,}$ and $\psi_{,}$.

The two last equations of (17) give, since

$$\mu = \frac{g}{g_{,}} = \frac{\gamma}{\gamma_{,}},$$

$$\phi' = \mu^2 \phi_{,}', \qquad \psi' = \mu^2 \psi_{,}';$$

ϕ and $\phi_{,}$ being accented for a moment to distinguish between the particular values belonging to the plane (yz) and their more general values

$$\phi = \epsilon^{bx} \phi' \text{ and } \phi_{,} = \epsilon^{-bx} \phi_{,}'.$$

The correctness of these values will be evident on referring to the Memoir, formulæ (20), (21), and recollecting that

$$b = a' = a_{,}'.$$

Hence the first equation gives, since $x = 0$,

$$b(\mu^2 + 1)\, \phi_{,}' = \frac{d\psi_{,}}{dy} - \frac{d\psi}{dy} = -(\mu^2 - 1) \frac{d\psi_{,}}{dy};$$

$$\therefore \phi_{,}' = -\frac{\mu^2 - 1}{b(\mu^2 + 1)} \frac{d\psi_{,}}{dy}, \text{ and } \phi' = \frac{-\mu^2(\mu^2 - 1)}{b(\mu^2 + 1)} \frac{d\psi_{,}}{dy}.$$

Also the second equation may be written,

$$\frac{d\psi}{dx} - \frac{d\psi_{,}}{dx} = \frac{d\phi'}{dy} - \frac{d\phi_{,}'}{dy} = -\frac{(\mu^2 - 1)^2}{b(\mu^2 + 1)} \frac{d^2\psi_{,}}{dy^2}.$$

And since we may differentiate or integrate the equations (17) relative to any variable except x, we get for the conditions requisite to complete the determination of ψ and $\psi_{,}$,

$$\left.\begin{aligned} &\psi = \mu^2 \psi_{,} \\ &\frac{d\psi}{dx} = \frac{d\psi_{,}}{dx} - \frac{(\mu^2 - 1)^2}{(\mu^2 + 1)\, b} \frac{d^2\psi_{,}}{dx^2} \end{aligned}\right\} \text{(when } x = 0\text{)} \ldots\ldots (29).$$

Or neglecting the term which is insensible except for highly refracting substances,

$$\left.\begin{aligned} \psi &= \mu^2 \psi_, \\ \frac{d\psi}{dx} &= \frac{d\psi_,}{dx} \end{aligned}\right\} \text{ (when } x = 0) \ldots\ldots\ldots\ldots (30),$$

*where $\mu = \frac{\gamma}{\gamma_,}$ is the index of refraction.

These equations belong to light polarized in a plane perpendicular to that of incidence, and as ϕ and $\phi_,$ are insensible at sensible distances from the surface of junction of the two media, we have, except in the immediate vicinity of this surface,

$$\left.\begin{aligned} u &= \frac{d\psi}{dy} \\ v &= -\frac{d\psi}{dx} \end{aligned}\right\} \ldots\ldots\ldots\ldots\ldots\ldots\ldots\ldots (31).$$

When light is polarized in the plane of incidence, the conditions at the surface of junction have been shewn to be

$$\left.\begin{aligned} w &= w_, \\ \frac{dw}{dx} &= \frac{dw_,}{dx} \end{aligned}\right\} \text{ (when } x = 0) \ldots\ldots\ldots\ldots (32).$$

Since in these conditions we may differentiate or integrate relative to any of the independent variables except x, we see that the expressions (30) and (32) are reduced to a form equiva-

* Though these equations have been obtained on the supposition that the vibrations of the media follow the law of the cycloidal pendulum, yet as (b) has disappeared, they are equally applicable for all plane waves whatever.

In fact, instead of using the value

$$\psi_, = \alpha_, \sin (a_, x + by + ct),$$

and corresponding values of the other quantities, we might have taken the infinite series

$$\psi_, = \Sigma \alpha_, \sin n (a_, x + by + ct),$$

where α and n may have any series of values at will. But the last expression is the equivalent of an arbitrary function of

$$a_, x + by + ct.$$

Or the same equations might have been immediately obtained from (17), without introducing this consideration. The method in the text has been employed for the sake of the intermediate result (29), of which we shall afterwards make use.

lent to that marked (A) in our paper on Sound; and the general equations in ψ and w being the same, we may immediately obtain the intensity of the reflected or refracted waves, by merely writing in the simple formulæ contained in that paper,

$\Delta = 1$ and $\Delta_, = 1$ for light polarized in the plane of incidence;

or $\Delta = \frac{1}{\gamma^2}$ and $\Delta_, = \frac{1}{\gamma_,^2}$ for light polarized perpendicular to the plane of incidence.

As an example, we will here deduce the intensity of the refracted wave for both kinds of light.

Representing, therefore, the parts of w and $w_,$ due to the disturbances in the Incident Reflected and Refracted waves by

$$f(ax + by + ct), \quad F(-ax + by + ct), \quad \text{and } f_,(a_,x + by + ct)$$

respectively, and resuming the first of our expressions (7) in the paper on Sound, viz.—

$$f' = \frac{1}{2}\left(\frac{\Delta_,}{\Delta} + \frac{a_,}{a}\right) f_,',$$

we get for light polarized in the plane of incidence, where

$$\Delta = \Delta_, = 1,$$

$$\frac{f_,}{f} = \frac{2}{1 + \frac{a_,}{a}} = \frac{2}{1 + \frac{\cot\theta_,}{\cot\theta}} = \frac{2\cos\theta\sin\theta_,}{\sin(\theta_, + \theta)},$$

which agrees with the value given in Airy's *Tracts*, p. 356*.

For light polarized perpendicular to the plane of incidence we have $\Delta = \frac{1}{\gamma^2}$ and $\Delta_, = \frac{1}{\gamma_,^2}$. If, therefore, we here represent the parts of ψ and $\psi_,$ due to the same disturbances by f, F and $f_,$, we get

$$\frac{f_,'}{f'} = \frac{2}{\frac{\gamma^2}{\gamma_,^2} + \frac{\cot\theta_,}{\cot\theta}} = \frac{\sin\theta_,\cos\theta}{\sin\theta\cos\theta_,} \cdot \frac{2}{\frac{\cos\theta\sin\theta}{\cos\theta_,\sin\theta_,} + 1}.$$

[* Airy, *ubi sup.* p. 109.]

Also, if D be the disturbance of the incident wave in its own plane, and $D_{,}$ the like disturbance in the refracted wave, we have by first equation of (31),

$$D \sin\theta = u = \frac{d\psi}{dy} = bf'(ax + by + ct),$$

$$\text{and } D_{,} \sin\theta_{,} = u_{,} = \frac{d\psi_{,}}{dy} = bf_{,}'(ax + by + ct),$$

retaining in ψ the part due to the incident wave only.

Thus by writing the characteristics merely,

$$\frac{D_{,}}{D} = \frac{\sin\theta}{\sin\theta_{,}}\frac{f_{,}'}{f'} = \frac{\cos\theta}{\cos\theta_{,}} \cdot \frac{2}{\dfrac{\cos\theta\sin\theta}{\cos\theta_{,}\sin\theta_{,}} + 1}$$

$$= \frac{\cos\theta}{\cos\theta_{,}}\left\{1 + \frac{-\dfrac{\cos\theta\sin\theta}{\cos\theta_{,}\sin\theta_{,}} + 1}{\dfrac{\cos\theta\sin\theta}{\cos\theta_{,}\sin\theta_{,}} + 1}\right\} = \frac{\cos\theta}{\cos\theta_{,}}\left\{1 + \frac{\tan(\theta_{,} - \theta)}{\tan(\theta + \theta_{,})}\right\},$$

which agrees with the formula in use. (Vide Airy's *Tracts*, p. 358*.)

In our preceding paper, the two media have been supposed to terminate abruptly at their surface of junction, which would not be true of the luminiferous ether, unless the radius of the sphere of sensible action of the molecular forces was exceedingly small compared with λ, the length of a wave of light.

In order, therefore, to form an estimate of the effect which would be produced by a continuous though rapid change of state of the ethereal medium in the immediate vicinity of the surface of junction, we will resume the conditions (29), which belong to light polarized in a plane perpendicular to that of Reflexion, viz.

$$\psi = \mu^2\psi_{,}, \text{ and } \frac{d\psi}{dx} = \frac{d\psi_{,}}{dx} - \frac{(\mu^2-1)^2}{(\mu^2+1)b}\frac{d^2\psi_{,}}{dx^2}\,(x=0)\;\ldots(29);$$

and instead of supposing the index of refraction to change suddenly from 0 to μ, we will conceive it to pass through the

[* Airy, *ubi sup.* p. 110.]

regular series of gradations,

$$\mu_0,\ \mu_1,\ \mu_2,\ \mu_3 \ldots\ldots\ldots\ldots\ldots\ldots \mu_n;$$

τ being the common thickness of each of these successive media.

Then it is clear we should have to replace the last system by

$$\left.\begin{array}{l}
\mu_0\psi_0 = \mu_1^2\psi_1, \text{ and } \dfrac{d\psi_0}{dx} = \dfrac{d\psi_1}{dx} - \dfrac{(\mu_1^2 - \mu_0^2)^2}{\mu_0^2(\mu_1^2 + \mu_0^2)\,b}\dfrac{d^2\psi_1}{dx^2}\ (x = 0) \\
\mu_1^2\psi_1 = \mu_2^2\psi_2, \text{ and } \dfrac{d\psi_1}{dx} = \dfrac{d\psi_2}{dx} - \dfrac{(\mu_2^2 - \mu_1^2)^2}{\mu_1^2(\mu_2^2 + \mu_1^2)\,b}\dfrac{d^2\psi_2}{dx^2}\ (x = \tau) \\
\mu_2^2\psi_2 = \mu_3^2\psi_3, \text{ and } \dfrac{d\psi_2}{dx} = \dfrac{d\psi_3}{dx} - \dfrac{(\mu_3^2 - \mu_2^2)^2}{\mu_2^2(\mu_3^2 + \mu_2^2)\,b}\cdot\dfrac{d^2\psi_3}{dx^2}\ (x = 2\tau)
\end{array}\right\}\ \ldots (33),$$

....

$$\mu^2_{n-1}\psi_{n-1} = \mu_n^2\psi_n,$$

$$\text{and } \frac{d\psi_{n-1}}{dx} = \frac{d\psi_n}{dx} - \frac{(\mu_n^2 - \mu^2_{n-1})^2}{\mu^2_{n-1}(\mu_n^2 + \mu^2_{n-1})\,b}\frac{d^2\psi_n}{dx^2}\ \{x = (n - 1 \,.\, \tau)\}.$$

But it is evident from the form of the equations on the right side of system (33), that the total effect due to the last terms of their second members will be far less when n is great, than that due to the corresponding term in the second equation of system (29)*. If, therefore, we reject these second terms, and conceive the common interval τ so small that the result due to the first terms may not differ very sensibly from that which would be produced by a single refraction, we should have to replace the system (29) by (30), and the intensity of the reflected wave would then agree with the law assigned by Fresnel. In virtue of this law, however highly refracting any substance may be, homogeneous light will always be completely polarized at a certain angle of incidence; and Sir David Brewster states

* In fact, in the system (33) each of the last terms will, in consequence of the factors $(\mu_1^2 - \mu_0^2)$, &c. be quantities of the order $\frac{1}{n^2}$ compared with the last term of (29′), and as their number is only n, their joint effect will be a quantity of the order $\frac{1}{n}$ compared with that of the term just mentioned.

that this is the case with diamond at the proper angle. But the phenomena observed by Professor Airy appears to him entirely inconsistent with this result (Vide *Camb. Phil. Trans.*, Vol. IV. p. 423); what immediately precedes seems to render it probable that considerable differences in this respect may be due to slight changes in the reflecting surface.

ON THE

PROPAGATION OF LIGHT

IN CRYSTALLIZED MEDIA*.

* From the *Transactions of the Cambridge Philosophical Society*, 1839. [Read *May* 20, 1839.]

ON THE PROPAGATION OF LIGHT IN CRYSTALLIZED MEDIA.

In a former paper * I endeavoured to determine in what way a plane wave would be modified when transmitted from one non-crystallized medium to another; founding the investigation on this principle: In whatever manner the elements of any material system may act upon each other, if all the internal forces be multiplied by the elements of their respective directions, the total sums for any assigned portion of the mass will always be the exact differential of some function. This principle requires a slight limitation, and when the necessary limitation is introduced, appears to possess very great generality. I shall here endeavour to apply the same principle to crystallized bodies, and shall likewise introduce the consideration of the effects of extraneous pressures, which had been omitted in the former communication. Our problem thus becomes very complicated, as the function due to the internal forces, even when there are no extraneous pressures, contains twenty-one coefficients. But with these pressures we are obliged to introduce six additional coefficients; so that without some limitation, it appears quite hopeless thence to deduce any consequences which could have the least chance of a physical application. The absolute necessity of introducing some arbitrary restrictions, and the desire that their number should be as small as possible, induced me to examine how far our function would be limited by confining ourselves to the consideration of those media only in which the directions of the transverse vibrations shall always be *accurately* in the front of the wave. This fundamental principle of Fresnel's Theory gives fourteen relations between the twenty-one constants originally entering into our function; and it seems worthy of remark, that when there are no extraneous pressures, the directions of polarization and the wave-velocities given by our theory, when thus limited, are identical with those assigned by Fresnel's general construction for biaxal crystals; provided we suppose the actual direction of disturbance in the particles

* Supra, p. 243.

of the medium is *parallel* to the plane of polarization, agreeably to the supposition first advanced by M. Cauchy.

If we admit the existence of extraneous pressures, it will be necessary in addition to the single restriction before noticed, to suppose that for three plane waves parallel to three orthogonal sections of our medium, and which may be denominated principal sections, the wave-velocities shall be the same for any two of the three waves whose fronts are parallel to these sections, provided the direction of the corresponding disturbances are parallel to the line of their intersection. With this additional supposition, the directions of the actual disturbances by which any plane wave will propagate itself without subdivision, and the wave-velocities, agree exactly with those given by Fresnel, supposing, with him, that these directions are *perpendicular* to the plane of polarization. The last, or Fresnel's hypothesis, was adopted in our former paper. But as that paper relates merely to the intensities of the waves reflected and refracted at the surface of separation of two media, and as these intensities may depend upon physical circumstances, the consideration of which was not introduced into our former investigations, it seems right, in the present paper, considering the actual situation of the theory of light, when the partial differential equations on which the determination of the motion of the luminiferous ether depends are yet to discover, to state fairly the results of both hypotheses.

It is hoped the analysis employed on the present occasion will be found sufficiently simple, as a method has here been given of passing immediately and without calculation from the function due to the internal forces of our medium to the equation of an ellipsoidal surface, of which the semi-axes represent in magnitude the reciprocals of the three wave-velocities, and in direction the directions of the three corresponding disturbances by which a wave can propagate itself in one medium without subdivision. This surface, which may be properly styled the ellipsoid of elasticity, must not be confounded with the one whose section by a plane parallel to the wave's front gives the reciprocals of the wave-velocities, and the corresponding directions of polarization. The two surfaces have only this section in common *, and a very simple application of our theory would shew that no force perpendicular to the wave's front is rejected, as in the ordinary one, but that the force in question is absolutely null †.

Let us conceive a system composed of an immense number of particles mutually acting on each other, and moreover subjected to the influence of extraneous pressures. Then if x, y, z are the co-ordinates of any particle of this system in its primitive state, (that of equilibrium under pressure for example), the co-ordinates of the same particle at the end of the time t will become x', y', z', where x', y', z' are functions of x, y, z and t. If now we consider an element of this medium, of which the primitive form is that of a rectangular parallelopiped, whose sides are dx, dy, dz, this element in its new state will assume the form of an oblique-angled parallelopiped, the lengths of the three edges being (dx'), (dy'), (dz'), these edges being composed of the same particles which formed the three edges dx, dy, dz in the primitive state of the element. Then will

$$\left.\begin{aligned}(dx')^2&=\left\{\left(\frac{dx'}{dx}\right)^2+\left(\frac{dy'}{dx}\right)^2+\left(\frac{dz'}{dx}\right)^2\right\}dx^2=a^2dx^2\\(dy')^2&=\left\{\left(\frac{dx'}{dy}\right)^2+\left(\frac{dy'}{dy}\right)^2+\left(\frac{dz'}{dy}\right)^2\right\}dy^2=b^2dy^2\\(dz')^2&=\left\{\left(\frac{dx'}{dz}\right)^2+\left(\frac{dy'}{dz}\right)^2+\left(\frac{dz'}{dz}\right)^2\right\}dz^2=c^2dz^2\end{aligned}\right\}\text{ suppose.}$$

Again, let

$$\alpha=\cos\angle\binom{dy'}{dz'}=\frac{\dfrac{dx'}{dy}\dfrac{dx'}{dz}+\dfrac{dy'}{dy}\dfrac{dy'}{dz}+\dfrac{dz'}{dy}\dfrac{dz'}{dz}}{\sqrt{\left\{\left(\dfrac{dx'}{dy}\right)^2+\left(\dfrac{dy'}{dy}\right)^2+\left(\dfrac{dz'}{dy}\right)^2\right\}\left\{\left(\dfrac{dx'}{dz}\right)^2+\left(\dfrac{dy'}{dz}\right)^2+\left(\dfrac{dz'}{dz}\right)^2\right\}}},$$

* [It will be seen that this remark is not strictly correct, as the surface must necessarily have another common plane section.]

† [Referring to the values of u, v, w given in p. 301, we see that, since the direction of vibration is supposed to be in the front of the wave, we have

$$au+bv+cw=0.$$

But the force perpendicular to the wave's front is $a\dfrac{d^2u}{dt^2}+b\dfrac{d^2v}{dt^2}+w\dfrac{d^2w}{dt^2}$, which is equal to $e^2(au+bv+cw)$, and is therefore null.]

$$\beta=\cos\angle\binom{dx'}{dz'}=\frac{\frac{dx'}{dx}\frac{dx'}{dz}+\frac{dy'}{dx}\frac{dy'}{dz}+\frac{dz'}{dx}\frac{dz'}{dz}}{\sqrt{\left\{\left(\frac{dx'}{dx}\right)^2+\left(\frac{dy'}{dx}\right)^2+\left(\frac{dz'}{dx}\right)^2\right\}\left\{\left(\frac{dx'}{dz}\right)^2+\left(\frac{dy'}{dz}\right)^2+\left(\frac{dz'}{dz}\right)^2\right\}}},$$

$$\gamma=\cos\angle\binom{dx'}{dy'}=\frac{\frac{dx'}{dx}\frac{dx'}{dy}+\frac{dy'}{dx}\frac{dy'}{dy}+\frac{dz'}{dx}\frac{dz'}{dy}}{\sqrt{\left\{\left(\frac{dx'}{dx}\right)^2+\left(\frac{dy'}{dx}\right)^2+\left(\frac{dz'}{dx}\right)^2\right\}\left\{\left(\frac{dx'}{dy}\right)^2+\left(\frac{dy'}{dy}\right)^2+\left(\frac{dz'}{dy}\right)^2\right\}}};$$

or we may write

$$\alpha'=bc\alpha=\frac{dx'}{dy}\frac{dx'}{dz}+\frac{dy'}{dy}\frac{dy'}{dz}+\frac{dz'}{dy}\frac{dz'}{dz},$$

$$\beta'=ac\beta=\frac{dx'}{dx}\frac{dx'}{dz}+\frac{dy'}{dx}\frac{dy'}{dz}+\frac{dz'}{dx}\frac{dz'}{dz},$$

$$\gamma'=ab\gamma=\frac{dx'}{dx}\frac{dx'}{dy}+\frac{dy'}{dx}\frac{dy'}{dy}+\frac{dz'}{dx}\frac{dz'}{dy}.$$

Suppose now, as in a former paper, that $\phi\,dx\,dy\,dz$ is the function due to the mutual actions of the particles which compose the element whose primitive volume $= dx\,dy\,dz$. Since ϕ must remain the same, when the sides (dx'), (dy'), (dz') and the cosines α, β, γ of the angles of the elementary oblique-angled parallelopiped remain unchanged, its most general form must be

$$\phi=\text{function }(a,\ b,\ c,\ \alpha,\ \beta,\ \gamma),$$

or since a, b, and c are necessarily positive, also

$$\alpha'=bc\alpha,\quad \beta'=ac\beta,\quad\text{and}\quad \gamma'=ab\gamma,$$

we may write $\qquad \phi=f(a^2,\ b^2,\ c^2,\ \alpha',\ \beta',\ \gamma')$.................(1).

This expression is the equivalent of the one immediately preceding, and is here adopted for the sake of introducing greater symmetry into our formulæ.

We will in the first place suppose that ϕ is symmetrical with regard to three planes at right angles to each other, which we shall take as the co-ordinate planes. The condition of symmetry with respect to the plane (yz), will require ϕ to remain unchanged, when we change

$$\left.\begin{matrix} x \\ x' \end{matrix}\right\} \text{ into } \left\{\begin{matrix} -x \\ -x'. \end{matrix}\right.$$

But thus a^2, b^2, c^2 and α' evidently remain unaltered; moreover

$$\left.\begin{matrix} \beta' \\ \gamma' \end{matrix}\right\} \text{ become } \left\{\begin{matrix} -\beta' \\ -\gamma'. \end{matrix}\right.$$

Hence we get

$$\phi = f(a^2, b^2, c^2, \alpha'^2, \beta'^2, \gamma'^2).$$

Applying the like reasoning to the other co-ordinate planes, we see that the ultimate result will be

$$\phi = f(a^2, b^2, c^2, \alpha'^2, \beta'^2, \gamma'^2) \text{ ....(2)}.$$

The foregoing values are perfectly general, whatever the disturbance may be; but if we consider this disturbance as very small, we may make

$$x' = x + u,$$

$$y' = y + v,$$

$$z' = z + w,$$

u, v, and w being very small functions of x, y, z, and t of the first order. Then by substitution we get

$$\left.\begin{aligned} a^2 &= 1 + 2\frac{du}{dx} + \left(\frac{du}{dx}\right)^2 + \left(\frac{dv}{dx}\right)^2 + \left(\frac{dw}{dx}\right)^2 = 1 + s_1 \\ b^2 &= 1 + 2\frac{dv}{dy} + \left(\frac{du}{dy}\right)^2 + \left(\frac{dv}{dy}\right)^2 + \left(\frac{dw}{dy}\right)^2 = 1 + s_2 \\ c^2 &= 1 + 2\frac{dw}{dz} + \left(\frac{du}{dz}\right)^2 + \left(\frac{dv}{dz}\right)^2 + \left(\frac{dw}{dz}\right)^2 = 1 + s_3 \end{aligned}\right\} \text{ suppose ...(3)},$$

$$\alpha' = \frac{dv}{dz} + \frac{dw}{dy} + \frac{du}{dy}\frac{du}{dz} + \frac{dv}{dy}\frac{dv}{dz} + \frac{dw}{dy}\frac{dw}{dz},$$

$$\beta' = \frac{du}{dz} + \frac{dw}{dx} + \frac{du}{dx}\frac{du}{dz} + \frac{dv}{dx}\frac{dv}{dz} + \frac{dw}{dx}\frac{dw}{dz},$$

$$\gamma' = \frac{du}{dy} + \frac{dv}{dx} + \frac{du}{dx}\frac{du}{dy} + \frac{dv}{dx}\frac{dv}{dy} + \frac{dw}{dx}\frac{dw}{dy};$$

we thus see that s_1, s_2, s_3, α', β', γ', are very small quantities of the first order, and that the general formula (1) by substituting the preceding values would take the form

$$\phi = \text{function } (s_1, s_2, s_3, \alpha', \beta', \gamma'),$$

which may be expanded in a very convergent series of the form

$$\phi = \phi_0 + \phi_1 + \phi_2 + \phi_3 + \&\text{c.}:$$

ϕ_0, ϕ_1, ϕ_2, &c. being homogeneous functions of s_1, s_2, s_3 α', β', γ', of the degrees 0, 1, 2, 3, &c. each of which is very great compared with the next following one.

But ϕ_0 being constant, if $\rho =$ the primitive density of the element, the general formula of Dynamics will give

$$\iiint \rho\, dxdydz \left\{\frac{d^2u}{dt^2}\delta u + \frac{d^2v}{dt^2}\delta v + \frac{d^2w}{dt^2}\delta w\right\} = \iiint dxdydz (\delta\phi_1 + \delta\phi_2 + \&\text{c.}).$$

If there were no extraneous pressures, the supposition that the primitive state was one of equilibrium would require $\phi_1 = 0$, as was observed in a former paper; but this is not the case if we introduce the consideration of extraneous pressures. However, as in the first case, the terms ϕ_3, ϕ_4, &c. will be insensible, and the preceding formula may be written

$$\iiint \rho\, dx\, dy\, dz \left\{\frac{d^2u}{dt^2}\delta u + \frac{d^2v}{dt^2}\delta v + \frac{d^2w}{dt^2}\delta w\right\} = \iiint dx\, dy\, dz\, (\delta\phi_1 + \delta\phi_2).$$

Supposing ρ the primitive density constant, the most general form of ϕ_1 will be

$$\phi_1 = -\frac{1}{2}(As_1 + Bs_2 + Cs_3 + 2D\alpha' + 2E\beta' + 2F\gamma'),$$

A, B, C, D, E, and F being constant quantities.

In like manner the most general form of ϕ_2 will contain twenty-one coefficients. But if we first employ the more particular value (2), we shall get

$$-2\phi_1 = As_1 + Bs_2 + Cs_3$$

$$-2\phi_2 = Gs_1^2 + Hs_2^2 + Is_3^2 + 2Ps_2s_3 + 2Qs_1s_3 + 2Rs_1s_2$$
$$+ L\alpha'^2 + M\beta'^2 + N\gamma'^2.$$

Or by substituting for s_1, s_2, s_3, α', β', γ' their values, given by system (3), continuing to neglect quantities of the third order, we get

$$-2\phi = -2\phi_1 - 2\phi_2$$

$$= 2A\frac{du}{dx} + 2B\frac{dv}{dy} + 2C\frac{dw}{dz}$$

$$+ A\left\{\left(\frac{du}{dx}\right)^2 + \left(\frac{dv}{dx}\right)^2 + \left(\frac{dw}{dx}\right)^2\right\}$$

$$+ B\left\{\left(\frac{du}{dy}\right)^2 + \left(\frac{dv}{dy}\right)^2 + \left(\frac{dw}{dy}\right)^2\right\}$$

$$+ C\left\{\left(\frac{du}{dz}\right)^2 + \left(\frac{dv}{dz}\right)^2 + \left(\frac{dw}{dz}\right)^2\right\}$$

$$+ G\left(\frac{du}{dx}\right)^2 + H\left(\frac{dv}{dy}\right)^2 + I\left(\frac{dw}{dz}\right)^2 + 2P\frac{dv}{dy}\frac{dw}{dz} + 2Q\frac{du}{dx}\frac{dw}{dz} + 2R\frac{du}{dx}\frac{dv}{dy}$$

$$+ L\left(\frac{dv}{dz} + \frac{dw}{dy}\right)^2 + M\left(\frac{du}{dz} + \frac{dw}{dx}\right)^2 + N\left(\frac{du}{dy} + \frac{dv}{dx}\right)^2 \dots\dots(4).$$

Having thus the form of the function due to the internal actions of the particles, we have merely to substitute it in the general formula of Dynamics, and to effect the integrations by parts, agreeably to the method of Lagrange. Thus,

$$\iiint dx\,dy\,dz\,\delta\phi =$$

$$-\iint dy\,dz \left\{A\delta u + A\left(\frac{du}{dx}\delta u + \frac{dv}{dx}\delta v + \frac{dw}{dx}\delta w\right) + \left(G\frac{du}{dx} + R\frac{dv}{dy} + Q\frac{dw}{dz}\right)\delta u + M\left(\frac{du}{dz} + \frac{dw}{dx}\right)\delta w + N\left(\frac{du}{dy} + \frac{dv}{dx}\right)\delta v\right\}$$

$$-\iint dx\,dz \left\{B\delta v + B\left(\frac{du}{dy}\delta u + \frac{dv}{dy}\delta v + \frac{dw}{dy}\delta w\right) + \left(R\frac{du}{dx} + H\frac{dv}{dy} + P\frac{dw}{dz}\right)\delta v + L\left(\frac{dv}{dz} + \frac{dw}{dy}\right)\delta w + N\left(\frac{du}{dy} + \frac{dv}{dx}\right)\delta u\right\}$$

$$-\iint dx\,dy \left\{C\delta w + C\left(\frac{du}{dz}\delta u + \frac{dv}{dz}\delta v + \frac{dw}{dz}\delta w\right) + \left(Q\frac{du}{dx} + P\frac{dv}{dy} + I\frac{dw}{dz}\right)\delta w + L\left(\frac{dv}{dz} + \frac{dw}{dy}\right)\delta v + M\left(\frac{du}{dz} + \frac{dw}{dx}\right)\delta u\right\}$$

$$+\iiint dx\,dy\,dz\,\delta u \left\{(G+A)\frac{d^2u}{dx^2} + (N+B)\frac{d^2u}{dy^2} + (M+C)\frac{d^2u}{dz^2} + (R+N)\frac{d^2v}{dx\,dy} + (Q+M)\frac{d^2w}{dx\,dz}\right\}$$

$$+\iiint dx\,dy\,dz\,\delta v \left\{(N+A)\frac{d^2v}{dx^2} + (H+B)\frac{d^2v}{dy^2} + (L+C)\frac{d^2v}{dz^2} + (N+R)\frac{d^2u}{dx\,dy} + (P+L)\frac{d^2w}{dy\,dz}\right\}$$

$$+\iiint dx\,dy\,dz\,\delta w \left\{(M+A)\frac{d^2w}{dx^2} + (L+B)\frac{d^2w}{dy^2} + (I+C)\frac{d^2w}{dz^2} + (M+Q)\frac{d^2u}{dx\,dz} + (L+P)\frac{d^2v}{dy\,dz}\right\}.$$

Neglecting the double integrals which relate to the extreme boundaries of the medium, and which we will suppose situated at an infinite distance, we get for the general equations of motion,

$$\left.\begin{aligned}
\rho\frac{d^2u}{dt^2} &= (G+A)\frac{d^2u}{dx^2} + (N+B)\frac{d^2u}{dy^2} + (M+C)\frac{d^2u}{dz^2} + (R+N)\frac{d^2v}{dx\,dy} + (Q+M)\frac{d^2w}{dx\,dz}\\
\rho\frac{d^2v}{dt^2} &= (N+A)\frac{d^2v}{dx^2} + (H+B)\frac{d^2v}{dy^2} + (L+C)\frac{d^2v}{dz^2} + (N+R)\frac{d^2u}{dx\,dy} + (P+L)\frac{d^2w}{dy\,dz}\\
\rho\frac{d^2w}{dt^2} &= (M+A)\frac{d^2w}{dx^2} + (L+B)\frac{d^2w}{dy^2} + (I+C)\frac{d^2w}{dz^2} + (M+Q)\frac{d^2u}{dx\,dz} + (L+P)\frac{d^2v}{dy\,dz}
\end{aligned}\right\} \quad \text{..........} (5).$$

If now in our indefinitely extended medium we wish to determine the laws of propagation of plane waves, we must take, to satisfy the last equations,

$$u = \alpha f\,(ax + by + cz + et),$$
$$v = \beta f\,(ax + by + cz + et),$$
$$w = \gamma f\,(ax + by + cz + et)\,;$$

a, b, and c being the cosines of the angles which a normal to the wave's front makes with the co-ordinate axes, α, β, γ constant coefficients, and e the velocity of transmission of a wave perpendicular to its own front, and taken with a contrary sign.

Substituting these values in the equations (5), and making to abridge

$$A' = (G + A)\,a^2 + (N + B)\,b^2 + (M + C)\,c^2,$$
$$B' = (N + A)\,a^2 + (H + B)\,b^2 + (L + C)\,c^2,$$
$$C' = (M + A)\,a^2 + (L + B)\,b^2 + (I + C)\,c^2\,;$$
$$D' = (L + P)\,bc,$$
$$E' = (M + Q)\,ac,$$
$$F' = (N + R)\,ab\,;$$

we get

$$\left.\begin{array}{l} 0 = (A' - \rho e^2)\,\alpha + F'\beta + E'\gamma \\ 0 = F'\alpha + (B' - \rho e^2)\,\beta + D'\gamma \\ 0 = E'\alpha + D'\beta + (C' - \rho e^2)\,\gamma \end{array}\right\}\ldots(6).$$

These last equations will serve to determine three values of ρ^2, and three corresponding ratios of the quantities α, β, γ; and hence we know the directions of the disturbance by which a plane wave will propagate itself without subdivision, and also the corresponding velocities of propagation. From the form of the equations (6), it is well known, that if we conceive an ellipsoid whose equation is

$$1 = A'x^2 + B'y^2 + C'z^2 + 2D'yz + 2E'xz + 2F'xy^{*}\ldots\ldots(7),$$

* If we reflect on the connexion of the operations by which we pass from the function (4) to the equation (7), it will be easy to perceive that the right side of the equation (7) may always be immediately deduced from that portion of the function

and represent its three semi-axes by r', r'', and r''', the directions of these axes will be the required directions of the disturbance, and the corresponding velocities of propagation will be given by

$$\rho e^2 = \frac{1}{r^2}.$$

Fresnel supposes those vibrations of the particles of the luminiferous ether which affect the eye, to be *accurately* in the front of the wave.

Let us therefore investigate the relation which must exist between our coefficients, in order to satisfy this condition for two out of our three waves, the remaining one in consequence being necessarily propagated by normal vibrations.

For this we may remark, that the equation of a plane parallel to the wave's front is

$$0 = ax' + by' + cz' \ldots\ldots (a)$$

If therefore we make

$$x = x' + a\lambda,$$

$$y = y' + b\lambda,$$

$$z = z' + c\lambda,$$

and substitute these values in the equation (7) of the ellipsoid; restoring the values of

$$A', B', C', D', E', F',$$

the odd powers of λ ought to disappear in consequence of the equation (a), whatever may be the position of the wave's front. We thus get

$$G = H = I = \mu \text{ suppose,}$$

and

$$P = \mu - 2L,$$

$$Q = \mu - 2M,$$

$$R = \mu - 2N.$$

which is of the second degree, by changing u, v, and w into x, y, and z. Also $\frac{d}{dx}$, $\frac{d}{dy}$ and $\frac{d}{dz}$ into a, b, c.

This remark will be of use to us afterwards, when we come to consider the most general form of the function due to the internal actions.

In fact, if we substitute these values in the function (4), there will result

$$\begin{aligned}
-2\phi &= -2\phi_1 - 2\phi_2 \\
&= 2A\frac{du}{dx} + 2B\frac{dv}{dy} + 2C\frac{dw}{dz} \\
&+ A\left\{\left(\frac{du}{dx}\right)^2 + \left(\frac{dv}{dx}\right)^2 + \left(\frac{dw}{dx}\right)^2\right\} \\
&+ B\left\{\left(\frac{du}{dy}\right)^2 + \left(\frac{dv}{dy}\right)^2 + \left(\frac{dw}{dy}\right)^2\right\} \\
&+ C\left\{\left(\frac{du}{dz}\right)^2 + \left(\frac{dv}{dz}\right)^2 + \left(\frac{dw}{dz}\right)^2\right\} \quad \ldots\ldots\ldots\ldots (A), \\
&+ \mu\left(\frac{du}{dx} + \frac{dv}{dy} + \frac{dw}{dz}\right)^2 \\
&+ L\left\{\left(\frac{dv}{dz} + \frac{dw}{dy}\right)^2 - 4\frac{dv}{dy}\frac{dw}{dz}\right\} \\
&+ M\left\{\left(\frac{du}{dz} + \frac{dw}{dx}\right)^2 - 4\frac{du}{dx}\frac{dw}{dz}\right\} \\
&+ N\left\{\left(\frac{du}{dy} + \frac{dv}{dx}\right)^2 - 4\frac{du}{dy}\frac{dv}{dx}\right\},
\end{aligned}$$

which, when $0 = A$, $0 = B$, $0 = C$, reduces to the last four lines.

Making the same substitution in the equation (7), we get

$$\left.\begin{aligned}
1 &= \mu\,(ax + by + cz)^2 \\
&+ (Aa^2 + Bb^2 + Cc^2)\,(x^2 + y^2 + z^2) \\
&+ L\,(cy - bz)^2 + M\,(az - ca)^2 + N\,(bx - ay)^2
\end{aligned}\right\} \ldots\ldots (8).$$

Let us in the first place suppose the system free from all extraneous pressure.

Then $\qquad A = 0, \quad B = 0, \quad C = 0,$

and the above equation, combined with that of a plane parallel to the wave's front, will give

$$0 = ax + by + cz \quad \ldots\ldots\ldots\ldots\ldots\ldots\ldots (9),$$

$$1 = L\,(cy - bz)^2 + M\,(az - cx)^2 + N\,(bx - ay)^2,$$

the equations of an infinite number of ellipses which, in general, do not belong to the same curve surface. If, however, we cause each ellipsis to turn 90^0 in its own plane, the whole system will belong to an ellipsoid, as may be thus shewn: Let (xyz) be the co-ordinates of any point p in its original position, and $(x'y'z')$ the co-ordinates of the point p' which would coincide with p when the ellipsis is turned 90^0 in its own plane. Then

$$x^2+y^2+z^2=x'^2+y'^2+z'^2,$$

since the distance from the origin O is unaltered,

$$0=ax'+by'+cz', \text{ since the plane is the same,}$$

$$0=xx'+yy'+zz', \text{ since } pOp'=90^0.$$

The two last equations give

$$\frac{x'}{cy-bz}=\frac{y'}{az-cx}=\frac{z'}{bx-ay}=w \text{ suppose.}$$

Hence the last of the equations (9) becomes

$$\omega^2=Lx'^2+My'^2+Nz'^2.$$

But

$$x'^2+y'^2+z'^2=\omega^2\{(cy-bz)^2+(az-cx)^2+(bx-ay)^2\}$$
$$=\omega^2\{(b^2+a^2)z^2+(c^2+a^2)y^2+(b^2+c^2)x^2-2(bcyz+abxy+acxz\}$$
$$=\omega^2\{(a^2+b^2+c^2)(x^2+y^2+z^2)-(ax+by+cz)^2\}$$
$$=\omega^2(x^2+y^2+z^2)=x^2+y^2+z^2.$$

Therefore $\omega^2=1,$

and our equation finally becomes

$$1=Lx'^2+My'^2+Nz'^2 \text{..................(10).}$$

We thus see that if we conceive a section made in the ellipsoid to which the equation (10) belongs, by a plane passing through its centre and parallel to the wave's front, this section, when turned 90 degrees in its own plane, will coincide with a similar section of the ellipsoid to which the equation (8) belongs, and which gives the directions of the disturbance that will cause a plane wave to propagate itself without subdivision, and the velocity of propagation parallel to its own front. The change of position here made in the elliptical section, is evidently equivalent to supposing the actual disturbances of the ethereal particles to be parallel to the plane usually denominated the *plane of polarization.*

This hypothesis, at first advanced by M. Cauchy, has since been adopted by several philosophers; and it seems worthy of remark, that if we suppose an elastic medium free from all extraneous pressure, we have merely to suppose it so constituted that two of the wave-disturbances shall be *accurately* in the wave's front, agreeably to Fresnel's fundamental hypothesis, thence to deduce his general construction for the propagation of waves in biaxal crystals. In fact, we shall afterwards prove that the function ϕ_2, which in its most general form contains twenty-one coefficients, is, in consequence of this hypothesis, reduced to one containing only seven coefficients; and that, from this last form of our function, we obtain for the directions of the disturbance and velocities of propagation precisely the same values as given by Fresnel's construction.

The above supposes, that in a state of equilibrium every part of the medium is quite free from pressure. When this is not the case, A, B, and C will no longer vanish in the equation (8). In the first place, conceive the plane of the wave's front parallel to the plane (yz); then $a = 1$, $b = 0$, $c = 0$, and the equation (8) of our ellipsoid becomes

$$1 = \mu x^2 + A(x^2 + y^2 + z^2) + Mz^2 + Ny^2;$$

and that of a section by a plane through its centre parallel to the wave's front, will be

$$1 = (A + N)\, y^2 + (A + M)\, z^2:$$

and hence, by what precedes, the velocities of propagation of our two polarized waves will be

$\sqrt{A+N}$. The disturbance being parallel to the axis of y,

$\sqrt{A+M}$. .... to the axis of z.

Similarly, if the plane of the wave's front is parallel to the plane (xz), the wave-velocities are

$\sqrt{B+N}$. The disturbance being parallel to the axis x,

$\sqrt{B+L}$. .... to the axis z.

Or if the plane of the wave's front is parallel to (xy), the velocities are

$\sqrt{C+M}$. The disturbance being parallel to x,

$\sqrt{C+L}$. .... y.

Fresnel supposes that the wave-velocity depends on the direction of the disturbance only, and is independent of the position of the wave's front. Instead of assuming this to be generally true, let us merely suppose it holds good for these three principal waves. Then we shall have

$$N+A=C+L,\quad M+A=B+L,\ \text{and}\ B+N=C+M;$$

or we may write

$$A-L=B-M=C-N=\nu\ \text{(suppose)}.$$

Thus our equation (8) becomes, since $a^2+b^2+c^2=1$,

$$\begin{aligned}1=\mu\,(ax+by+cz)^2&+\nu\,(x^2+y^2+z^2)\\&+(La^2+Mb^2+Nc^2)\,(x^2+y^2+z^2)\\&+L\,(cy-bz)^2+M\,(az-cx)^2+N\,(bx-ay)^2.\end{aligned}$$

But the two last lines of this formula easily reduce to

$$\begin{aligned}&(M+N)\,x^2+(N+L)\,y^2+(L+M)\,z^2\\&+L\,\{a^2x^2-(by+cz)^2\}+M\,\{b^2y^2-(ax+cz)^2\}\\&+N\,\{c^2z^2-(ax+by)^2\}.\end{aligned}$$

And hence our last equation becomes

$$\begin{aligned}1=(\nu+M+N)\,x^2+(\nu+N+L)\,y^2+(\nu+L+M)\,z^2+\mu\,(ax+by+cz)^2\\+L\,\{a^2x^2-(by+cz)^2\}+M\,\{b^2y^2-(ax+cz)^2\}+N\,\{c^2z^2-(ax+by)^2\}\\ \text{..........................(11).}\end{aligned}$$

In consequence of the condition which was satisfied in forming the equation (8), it is evident that two of its semi-axes are in a plane parallel to the wave's front, and of which the equation is

$$0 = ax + by + cz \text{ ....(12)};$$

the same therefore will be true for the ellipsoid whose equation is (11), as this is only a particular case of the former. But the section of the last ellipsoid by the plane (12) is evidently given by

$$\left.\begin{aligned} 1 &= (\nu + M + N)\, x^2 + (\nu + L + N)\, y^2 + (\nu + L + M)\, z^2 \\ 0 &= ax + by + cz \end{aligned}\right\} \text{...(12, 1)}.$$

By what precedes, the two axes of this elliptical section will give the two directions of disturbance which will cause a wave to be propagated without subdivision, and the velocity of propagation of each wave will be inversely as the corresponding semi-axes of the section: which agrees with Fresnel's construction, supposing, as he has done, the actual direction of the disturbance of the particles of the ether is perpendicular to the plane of polarization.

Let us again consider the system as quite free from extraneous pressure, and take the most general value of ϕ_2 containing twenty-one coefficients. Then, if to abridge, we make

$$\frac{du}{dx} = \xi, \qquad \frac{du}{dy} = \eta, \qquad \frac{du}{dz} = \zeta;$$

$$\frac{dv}{dz} + \frac{dw}{dy} = \alpha, \quad \frac{du}{dz} + \frac{dw}{dx} = \beta, \quad \frac{du}{dy} + \frac{dv}{dx} = \gamma,$$

we shall have

$$\begin{aligned} -\phi_2 = {} & (\xi^2)\,\xi^2 + (\eta^2)\,\eta^2 + (\zeta^2)\,\zeta^2 + 2\,(\eta\zeta)\,\eta\zeta + 2\,(\xi\zeta)\,\xi\zeta + 2\,(\xi\eta)\,\xi\eta \\ & + (\alpha^2)\,\alpha^2 + (\beta^2)\,\beta^2 + (\gamma^2)\,\gamma^2 + 2\,(\beta\gamma)\,\beta\gamma + 2\,(\alpha\gamma)\,\alpha\gamma + 2\,(\alpha\beta)\,\alpha\beta \\ & + 2\,(\alpha\xi)\,\alpha\xi + 2\,(\beta\xi)\,\beta\xi + 2\,(\gamma\xi)\,\gamma\xi \\ & + 2\,(\alpha\eta)\,\alpha\eta + 2\,(\beta\eta)\,\beta\eta + 2\,(\gamma\eta)\,\gamma\eta \\ & + 2\,(\alpha\zeta)\,\alpha\zeta + 2\,(\beta\zeta)\,\beta\zeta + 2\,(\gamma\zeta)\,\gamma\zeta, \end{aligned}$$

where (ξ^2), (α^2), &c. are the twenty-one coefficients which enter into ϕ_2. Suppose now the equation to the front of a wave is

$$0 = ax + by + cz.$$

Then, by what was before observed, the right side of the equation of the ellipsoid, which gives the directions of disturbance of the three polarized waves and their respective velocities, will be had from ϕ_2 by changing u, v, and w into x, y, and z; also

$$\frac{d}{dx}, \ \frac{d}{dy}, \text{ and } \frac{d}{dz} \text{ into } a, b, \text{ and } c.$$

We shall thus get

$$1 = Ax^2 + By^2 + Cz^2 + 2Dyz + 2Exz + 2Fxy.$$

Provided

$$A = (\xi^2)\, a^2 + (\beta^2)\, c^2 + (\gamma^2)\, b^2 + 2\,(\beta\gamma)\, bc + 2\,(\xi\beta)\, ac + 2\,(\xi\gamma)\, ab,$$

$$B = (\eta^2)\, b^2 + (\alpha^2)\, c^2 + (\gamma^2)\, a^2 + 2\,(\alpha\gamma)\, ac + 2\,(\eta\alpha)\, bc + 2\,(\eta\gamma)\, ab,$$

$$C = (\zeta^2)\, c^2 + (\alpha^2)\, b^2 + (\beta^2)\, a^2 + 2\,(\alpha\beta)\, ab + 2\,(\zeta\alpha)\, bc + 2\,(\zeta\beta)\, ac,$$

$$D = (\eta\zeta)\, bc + (\alpha^2)\, bc + (\beta\gamma)\, a^2 + (\alpha\beta)\, ac + (\alpha\gamma)\, ab$$
$$+ (\alpha\eta)\, b^2 + (\alpha\zeta)\, c^2 + (\beta\eta)\, ab + (\gamma\zeta)\, ac,$$

$$E = (\xi\zeta)\, ac + (\beta^2)\, ac + (\alpha\gamma)\, b^2 + (\alpha\beta)\, bc + (\beta\gamma)\, ab$$
$$+ (\beta\xi)\, a^2 + (\beta\zeta)\, c^2 + (\alpha\xi)\, ab + (\gamma\zeta)\, bc,$$

$$F = (\xi\eta)\, ab + (\gamma^2)\, ab + (\alpha\beta)\, c^2 + (\alpha\gamma)\, bc + (\beta\gamma)\, ac$$
$$+ (\gamma\xi)\, a^2 + (\gamma\eta)\, b^2 + (\alpha\xi)\, ac + (\beta\eta)\, bc.$$

But if the directions of two of the disturbances are rigorously in the front of a wave, a plane parallel to this front passing through the centre of the ellipsoid, and whose equation is

$$0 = ax + by + cz,$$

must contain two of the semi-axes of this ellipsoid; and therefore a system of chords perpendicular to the plane will be bisected by it; and hence we get

$$0 = (A - C)\, ac + E\,(c^2 - a^2) + Fbc - Dab,$$

$$0 = (B - C)\, bc + D\,(c^2 - b^2) + Fac - Eab.$$

Substituting in these the values of A, B, &c., before given, we shall obtain the fourteen relations following between the coefficients of ϕ_2, viz.

$$0=(\alpha\eta),\ 0=(\beta\xi),\ 0=(\gamma\xi),\ 0=(\alpha\zeta),\ 0=(\beta\zeta),\ 0=(\gamma\eta),$$

$$(\alpha\xi)=-2(\beta\gamma),\ (\beta\eta)=-2(\alpha\gamma),\ (\gamma\zeta)=-2(\alpha\beta),$$

$$(\xi^2)=(\eta^2)=(\zeta^2)=2(\alpha^2)+(\eta\zeta)=2(\beta^2)+(\xi\zeta)=2(\gamma^2)+(\xi\eta).$$

Hence, we may readily put the function ϕ_2 under the following form,

$$(\xi^2)(\xi+\eta+\zeta)^2+(\alpha^2)(\alpha^2-4\eta\zeta)+(\beta^2)(\beta^2-4\xi\zeta)+(\gamma^2)(\gamma^2-4\xi\eta)$$
$$+2(\beta\gamma)(\beta\gamma-2\alpha\xi)+2(\alpha\gamma)(\alpha\gamma-2\beta\eta)+2(\alpha\beta)(\alpha\beta-2\gamma\zeta),$$

or by restoring the values of ξ, η, &c., and making $G=(\xi^2)$, $L=(\alpha^2)$, &c., our function will become

$$G\left(\frac{du}{dx}+\frac{dv}{dy}+\frac{dw}{dz}\right)^2+$$

$$L\left\{\left(\frac{dv}{dz}+\frac{dw}{dy}\right)^2-4\frac{dv}{dy}\frac{dw}{dz}\right\}+M\left\{\left(\frac{du}{dz}+\frac{dw}{dx}\right)^2-4\frac{du}{dx}\frac{dw}{dz}\right\}+N\left\{\left(\frac{du}{dy}+\frac{dv}{dz}\right)^2-4\frac{du}{dx}\frac{dv}{dy}\right\}$$

$$+2P\left\{\left(\frac{du}{dz}+\frac{dw}{dx}\right)\left(\frac{du}{dy}+\frac{dv}{dx}\right)-2\frac{du}{dx}\left(\frac{dv}{dz}+\frac{dw}{dy}\right)\right\}$$

$$+2Q\left\{\left(\frac{dv}{dz}+\frac{dw}{dy}\right)\left(\frac{du}{dy}+\frac{dv}{dx}\right)-2\frac{dv}{dy}\left(\frac{du}{dz}+\frac{dw}{dx}\right)\right\}$$

$$+2R\left\{\left(\frac{dv}{dz}+\frac{dw}{dy}\right)\left(\frac{du}{dz}+\frac{dw}{dx}\right)-2\frac{dw}{dz}\left(\frac{du}{dy}+\frac{dv}{dx}\right)\right\}\ \ldots\ldots.(12),$$

and hence we get for the equation of the corresponding ellipsoid,

$$1=G(ax+by+cz)^2+L(bz-cy)^2+M(az-cx)^2+N(ay-bx)^2$$
$$+2P(cx-az)(ay-bx)+2Q(bz-cy)(ay-bx)+2R(bz-cy)(cx-az)\ldots(13).$$

But if in equation (8) and corresponding function (A), we suppose $A=0$, $B=0$, and $C=0$, and then refer the equation to axes taken arbitrarily in space, we shall thus introduce three new coefficients, and evidently obtain a result equivalent to equation (13) and function (12). We therefore see that the single supposition of the wave-disturbance, being always *accurately* in the wave's front, leads to a result equivalent to that given by the former process; and we are thus assured that by employing the simpler method we do not, in the case in question, eventually lessen the generality of our result, but merely, in effect, select the three rectangular axes, which may be called the axes of elasticity of the medium, for our co-ordinate axes. From the general form of ϕ, it is clear that the same observation applies to it, and therefore the consequences before deduced possess all the requisite generality.

The same conclusions may be obtained, whether we introduce the consideration of extraneous pressures or not, by direct calculation. In fact, when these pressures vanish, and we conceive a section of the ellipsoid whose equation is (13), made by a plane parallel to the wave's front, to turn 90 degrees in its own plane, the same reasoning by which equation (10) was before found, immediately gives, in the present case,

$$1 = Lx'^2 + My'^2 + Nz'^2 + 2Py'z' + 2Qx'z' + 2Rx'y' \dots (14),$$

for the equation of the surface in which all the elliptical sections in their new situations, and corresponding to every position of the wave's front, will be found.

Lastly, when we introduce the consideration of extraneous pressures, it is clear, from what precedes, that we shall merely have to add to the function on the right side of the equation (13), the quantity

$$(Aa^2 + Bb^2 + Cc^2 + 2Dbc + 2Eac + 2Fab)\,(x^2 + y^2 + z^2),$$

which would arise from changing u, v, and w into x, y, and z. Also $\frac{d}{dx}$, $\frac{d}{dy}$, $\frac{d}{dz}$ into a, b, c, in that part of ϕ which is of the second degree in u, v, w, agreeably to the remark in a foregoing note. Afterwards, when we determine the values of A, B, &c., by the same condition which enabled us to deduce the system (12, 1), we shall have, in the place of this system, the following:

$$\left.\begin{array}{l} 1=K(x^2+y^2+z^2)-\{Lx^2+My^2+Nz^2+2Pyz+2Qxz+2Rxy\} \\ 0=ax+by+cz \end{array}\right\}\ldots(15),$$

which is applicable to the more general case just considered*.

* Vide Professor Stokes' Report on Double Refraction (British Association, 1862, p. 265).

RESEARCHES ON THE VIBRATION OF PENDULUMS IN FLUID MEDIA*.

* From the *Transactions of the Royal Society of Edinburgh.*

[Read Dec. 16, 1833.]

RESEARCHES ON THE VIBRATION OF PENDULUMS IN FLUID MEDIA.

PROBABLY no department of Analytical Mechanics presents greater difficulties than that which treats of the motion of fluids; and hitherto the success of mathematicians therein has been comparatively limited. In the theory of waves, as presented by MM. Poisson and Cauchy, and in that of sound, their success appears to have been more complete than elsewhere; and if to these investigations we join the researches of Laplace concerning the tides, we shall have the principal important applications hitherto made of the general equations upon which the determination of this kind of motion depends. The same equations will serve to resolve completely a particular case of the motion of fluids, which is capable of a useful practical application; and as I am not aware that it has yet been noticed, I shall endeavour, in the following paper, to consider it as briefly as possible.

In the case just alluded to, it is required to determine the circumstances of the motion of an indefinitely extended non-elastic fluid, when agitated by a solid ellipsoidal body, moving parallel to itself, according to any given law, always supposing the body's excursions very small, compared with its dimensions. From what will be shown in the sequel, the general solution of this problem may very easily be obtained. But as the principal object of our paper is to determine the alteration produced in the motion of a pendulum by the action of the surrounding medium, we have insisted more particularly on the case where the ellipsoid moves in a right line parallel to one of its axes, and have thence proved, that, in order to obtain the correct time of a pendulum's vibration, it will not be sufficient merely to allow for the loss of weight caused by the fluid medium, but that it will likewise be requisite to conceive the density of the body augmented by a quantity proportional to the density of this fluid. The value of the quantity last named when the body of the pendulum is an oblate spheroid vibrating in its equatorial plane, has been completely determined, and, when the spheroid becomes a sphere, is precisely equal to half the density of the surrounding fluid. Hence in this last case we shall have the true time of the pendulum's vibration, if we suppose it to move *in vacuo*, and then simply conceive its mass augmented by half that of an equal volume of the fluid, whilst the moving force with which it is actuated is diminished by the whole weight of the same volume of fluid.

We will now proceed to consider a particular case of the motion of a non-elastic fluid over a fixed obstacle of ellipsoidal figure, and thence endeavour to find the correction necessary to reduce the observed length of a pendulum vibrating through exceedingly small arcs in any indefinitely extended medium to its true length *in vacuo*, when the body of the pendulum is a solid ellipsoid. For this purpose we may remark, that the equations of the motion of a homogeneous non-elastic fluid are

$$V - \frac{p}{\rho} = \frac{d\phi}{dt} + \frac{1}{2}\left\{\left(\frac{d\phi}{dx}\right)^2 + \left(\frac{d\phi}{dy}\right)^2 + \left(\frac{d\phi}{dz}\right)^2\right\} \text{ .... (1)},$$

$$0 = \frac{d^2\phi}{dx^2} + \frac{d^2\phi}{dy^2} + \frac{d^2\phi}{dz^2} \text{ .... (2)}.$$

Vide *Méc. Cél.* Liv. III. Ch. 8, No. 33, where ϕ is such a function of the co-ordinates x, y, z of any particle of the fluid mass, and of the time t that the velocities of this particle in the directions of and tending to increase the co-ordinates x, y, and z shall always be represented by $\frac{d\phi}{dx}$, $\frac{d\phi}{dy}$, and $\frac{d\phi}{dz}$ respectively. Moreover, ρ represents the fluid's density, p its pressure, and V a function dependent upon the various forces which act upon the fluid mass.

When the fluid is supposed to move over a fixed solid ellipsoid, the principal difficulty will be so to satisfy the equation (2), that the particles at the surface of this solid may move along this surface, which may always be effected by making

$$\phi = \left(\lambda + \mu \int_{\infty} \frac{df}{a^3bc}\right) x^* \text{ ....} (3),$$

supposing that the origin of the co-ordinates is at the centre of the ellipsoid; λ and μ being two arbitrary quantities constant with regard to the variables x, y, z: and a, b, c being functions of these same variables, determined by the equations

$$a^2 = a'^2 + f,\ b^2 = b'^2 + f,\ c^2 = c'^2 + f, \text{ and } \frac{x^2}{a^2} + \frac{y^2}{b^2} + \frac{z^2}{c^2} = 1 \text{ ....} (4),$$

in which a', b', c' are the axes of the given ellipsoid.

To prove that the expression (3) satisfies the equation (2), it may be remarked, that we readily get, by differentiating (3),

$$\frac{d^2\phi}{dx^2} + \frac{d^2\phi}{dy^2} + \frac{d^2\phi}{dz^2} = \frac{2\mu}{a^3bc}\frac{df}{dx} + \frac{\mu x}{a^3bc}\left(\frac{d^2f}{dx^2} + \frac{d^2f}{dy^2} + \frac{d^2f}{dz^2}\right)$$

$$- \frac{\mu x}{a^3bc}\left(\frac{3}{2a^2} + \frac{1}{2b^2} + \frac{1}{2c^2}\right)\left\{\left(\frac{df}{dx}\right)^2 + \left(\frac{df}{dy}\right)^2 + \left(\frac{df}{dz}\right)^2\right\}.$$

* In my memoir on the Determination of the exterior and interior Attractions of Ellipsoids of Variable Densities [1], recently communicated to the Cambridge Philosophical Society by Sir EDWARD FFRENCH BROMHEAD, Baronet, I have given a method by which the general integral of the partial differential equation

$$0 = \frac{d^2V}{dx_1^2} + \frac{d^2V}{dx_2^2} + \dots + \frac{d^2V}{dx_s^2} + \frac{d^2V}{du^2} + \frac{n-s}{u}\frac{dV}{du}$$

may be expanded in a series of peculiar form, and have thus rendered the determination of these attractions a matter of comparative facility. The same method applied to the equation (2) of the present paper has the advantage of giving an expansion of its general integral, every term of which, besides satisfying this equation, may likewise be made to satisfy the condition (6). The formula (3) is only an individual term of the expansion in question. But in order to render the present communication independent of every other, it was thought advisable to introduce into the test a demonstration of this particular case.

[1] [Vid. *supra*, p. 185.]

Moreover, by the same means, the last of the equation (4) gives

$$\frac{df}{dx} = \frac{\frac{2x}{a^2}}{\frac{x^2}{a^4} + \frac{y^2}{b^4} + \frac{z^2}{c^4}}, \quad \left(\frac{df}{dx}\right)^2 + \left(\frac{df}{dy}\right)^2 + \left(\frac{df}{dz}\right)^2 = \frac{4}{\frac{x^2}{a^4} + \frac{y^2}{b^4} + \frac{z^2}{c^4}},$$

and
$$\frac{d^2f}{dx^2} + \frac{d^2f}{dy^2} + \frac{d^2f}{dz^2} = \frac{\frac{2}{a^2} + \frac{2}{b^2} + \frac{2}{c^2}}{\frac{x^2}{a^4} + \frac{y^2}{b^4} + \frac{z^2}{c^4}},$$

which values being substituted in the second member of the preceding equation, evidently cause it to vanish, and we thus perceive that the value (3) satisfies the partial differential equation (2).

We will now endeavour so to determine the constant quantities λ and μ that the fluid particles may move along the surface of the ellipsoidal body of which the equation is

$$1 = \frac{x^2}{a'^2} + \frac{y^2}{b'^2} + \frac{z^2}{c'^2} \quad \text{......................(5).}$$

But by differentiation, there results

$$0 = \frac{xdx}{a'^2} + \frac{ydy}{b'^2} + \frac{zdz}{c'^2},$$

and as the particles must move along the surface, it is clear that the last equation ought to subsist, when we change the elements dx, dy, and dz into their corresponding velocities $\frac{d\phi}{dx}$, $\frac{d\phi}{dy}$, and $\frac{d\phi}{dz}$. Hence, at this surface

$$0 = \frac{x}{a'^2}\frac{d\phi}{dx} + \frac{y}{b'^2}\frac{d\phi}{dy} + \frac{z}{c'^2}\frac{d\phi}{dz} \quad \text{................(6).}$$

But the expression (3) gives generally

$$\frac{d\phi}{dx} = \lambda + \mu\int_\infty \frac{df}{a^3bc} + \frac{\mu x}{a^3bc}\frac{df}{dx}, \quad \frac{d\phi}{dy} = \frac{\mu y}{a^3bc}\frac{df}{dy}, \quad \frac{d\phi}{dz} = \frac{\mu z}{a^3bc}\frac{df}{dz} \quad \text{.... (7),}$$

and consequently at the surface in question, where $f=0$,

$$\frac{d\phi}{dx}=\lambda+\mu\int_{\infty}^{0}\frac{df}{a^3bc}+\frac{\mu x}{a'^3b'c'}\frac{df}{dx},\quad \frac{d\phi}{dy}=\frac{\mu y}{a'^3b'c'}\frac{df}{dy},\quad \frac{d\phi}{dz}=\frac{\mu z}{a'^3b'c'}\frac{df}{dz}.$$

These values substituted in (6) give, when we replace $\frac{df}{dx}$, $\frac{df}{dy}$ and $\frac{df}{dz}$ with their values at the ellipsoidal surface,

$$0=\lambda+\mu\int_{\infty}^{0}\frac{df}{a^3bc}+\frac{2\mu}{a'b'c'}\ \cdots\cdots\cdots\cdots\cdots(8),$$

which may always be satisfied by a proper determination of one of the constants λ and μ, the other remaining entirely arbitrary.

From what precedes, it is clear that the equation (2) and condition to which the fluid is subject may equally well be satisfied by making

$$\phi=\left(\lambda''+\mu\int_{\infty}\frac{df}{ab^3c}\right)y \text{ and } \phi=\left(\lambda''+\mu\int\frac{df}{abc^3}\right)z:$$

provided we determine the constant quantities therein contained by means of the equations

$$0=\lambda'+\mu'\int_{\infty}^{0}\frac{df}{ab^3c}+\frac{2\mu'}{a'b'c'} \text{ and } 0=\lambda''+\mu''\int_{\infty}^{0}\frac{df}{abc^3}+\frac{2\mu''}{a'b'c'}$$

respectively. The same may likewise be said of the sum of the three values of ϕ before given. However, in what follows, we shall consider the value (3) only, since, from the results thus obtained similar ones relative to the cases just enumerated may be found without the least difficulty.

Instead now of supposing the solid at rest, let every part of the whole system be animated with an additional common velocity $-\lambda$ in the direction of the co-ordinate x. Then it is clear that the equation (2) and condition to which the fluid is subject, will still remain satisfied. Moreover, if x', y', z' are now referred to three axes fixed in space, we shall have

$$x' = x - \int \lambda dt, \quad y = y', \quad z = z',$$

and if X' represents the co-ordinate of the centre of the ellipsoid referred to the fixed origin, we shall have

$$X' = -\int \lambda dt \quad \text{......................} \quad (9).$$

Adding now to ϕ the term $-\lambda x$ due to the additional velocity, the expression (3) will then become

$$\phi = \mu x \int_{\infty} \frac{df}{a^3bc},$$

and the velocities of any point of the fluid will be given, by means of the differentials of this last function. But ϕ and its differentials evidently vanish at an infinite distance from the solid, where $f = \infty$; and consequently, the case now under consideration is that of an indefinitely extended fluid, of which the exterior limits are at rest, whilst the parts in the vicinity of the moving body are agitated by its motions.

It will now be requisite to determine the pressure p at any point of the fluid mass. But, by supposing this mass free from all extraneous action, $V = 0$, and if the excursions of the solid are always exceedingly small, compared with its dimensions, the last term of the second member of the equation (1) may evidently be neglected, and thus we shall have, without sensible error,

$$-\frac{p}{\rho} = \frac{d\phi}{dt}, \quad \text{i.e. } p = -\rho \frac{d\phi}{dt},$$

or, by substitution from the last value of ϕ,

$$p = -\frac{d\mu}{dt} \rho x \int_{\infty} \frac{df}{a^3bc}.$$

Having thus ascertained all the circumstances of the fluid's motion, let us now calculate its total action upon the moving solid. Then the pressure upon any point on its surface will be had by making $f = 0$ in the last expression, and is

$$p_0 = -\frac{d\mu}{dt}\rho x \int^0 \frac{df}{a^3bc}.$$

Hence we readily get for the total pressure on the body tending to increase x

$$P = \int ds\,(p' - p_0'') = \frac{d\mu}{dt}\rho \int \frac{df}{a^3bc} \times \int 2x ds = \frac{d\mu}{dt}\rho v \int_\infty^0 \frac{df}{a^3bc};$$

v representing the volume of the body, p_0'' the pressure on that side where x is positive, p_0' the pressure on the opposite side, and ds an element of the principal section of the ellipsoid perpendicular to the axis of x.

If now we substitute for μ its value given from (8), the last expression will become

$$P = \frac{a'b'c'\rho v \int_0^\infty \frac{df}{a^3bc}}{2 = a'b'c' \int_0^\infty \frac{df}{a^3bc}} \frac{d\lambda}{dt} \quad \text{.................. (10).}$$

Having thus the total pressure exerted upon the moving body by the surrounding medium, it will be easy thence to determine the law of its vibrations when acted upon by an exterior force proportional to the distance of its centre from the point of repose. In fact, let $\rho_{,}$ be the density of the body, and, consequently, $\rho_{,}v$ its mass, gX' the exterior force tending to decrease X'. Then by the principles of dynamics,

$$0 = \rho_1 v \frac{d^2X'}{d^2t} + gX' - P.$$

If, now, in the formula (10) we substitute for λ its value drawn from (9), the last equation will become

$$0 = \left(\rho_1 + \frac{a'b'c' \int_0^\infty \frac{df}{a^3bc}}{2 - a'b'c \int_0^\infty \frac{df}{a^3bc}}\rho\right) v \frac{d^2X'}{dt^2} + qX',$$

which is evidently the same as would be obtained by supposing the vibrations to take place *in vacuo*, under the influence of the given exterior force, provided the density of the vibrating body were increased from

$$\rho_1 \text{ to } \rho_1 + \frac{a'b'c'\int_0^\infty \frac{df}{a^3bc}}{2 - a'b'c'\int_0^\infty \frac{df}{a^3bc}}\,\rho \quad \ldots\ldots\ldots\ldots\ldots(11).$$

We thus perceive, that besides the retardation caused by the loss of weight which the vibrating body sustains in a fluid, there is a farther retardation due to the action of the fluid itself; and this last is precisely the same as would be produced by augmenting the density of the body in the proportion just assigned, the moving force remaining unaltered.

When the body is spherical, we have $a' = b' = c'$, and the proportion immediately preceding becomes very simple, for it will then only be requisite to increase ρ, the density of the body, by $\frac{\rho}{2}$, or half the density of the fluid, in order to have the correction in question.

The next case in point of simplicity is where $a' = c'$; for then

$$\int_0^\infty \frac{df}{a^3bc} = \int_0^\infty \frac{df}{a^4b} = 2\int_{b'}^\infty \frac{db}{a^4} \quad \ldots\ldots\ldots\ldots\ldots\ldots(12).$$

If $a' > b'$, or the body is an oblate spheroid vibrating in its equatorial plane, the last quantity properly depends on the circular arcs, and has for value

$$(a'^2 - b'^2)^{-\frac{3}{2}}\left\{\frac{\pi}{2} - \text{arc}\left(\tan = \frac{b'}{\sqrt{(a'^2 - b'^2)}}\right)\right\} - \frac{b'}{a'^2(a'^2 - b'^2)}.$$

If, on the contrary, $a' < b'$, or the spheroid is oblong, the value of the same integral is

$$\frac{1}{2}(b'^2 - a'^2)^{-\frac{3}{2}}\log\frac{b' + \sqrt{(b'^2 - a'^2)}}{b' - \sqrt{(b'^2 - a'^2)}} + \frac{b'}{a'^2(b'^2 - a'^2)}.$$

Another very simple case is where $c' = b'$, for then the first of the quantities (12) becomes, if $a' > b'$,

$$(a'^2 - b'^2)^{-\frac{3}{2}} \log \frac{a' + \sqrt{(a'^2 - b'^2)}}{a' - \sqrt{(a'^2 - b'^2)}} - \frac{2}{a' (a'^2 - b'^2)},$$

and if $a' < b'$, the same quantity becomes

$$2\,(b'^2 - a'^2)^{-\frac{3}{2}} \left\{ \text{arc} \left(\tan = \frac{a'}{\sqrt{(b'^2 - a'^2)}} \right) - \frac{\pi}{2} \right\} + \frac{2}{a' (b'^2 - a'^2)}.$$

By employing the first of the four expressions immediately preceding, we readily perceive that, when an oblate spheroid vibrates in its equatorial plane, the correction now under consideration will be effected by conceiving the density of the body augmented from

$$\rho_{,} \text{ to } \rho_{,} + \frac{\frac{\pi}{2} a'^2 b' - a'^2 b' \text{arc} \left\{ \tan = \frac{b'}{\sqrt{(a'^2 - b'^2)}} \right\} - b'^2 \sqrt{(a'^2 - b'^2)}}{2(a'^2 - b'^2)^{\frac{3}{2}} - \frac{\pi}{2} a'^2 b' + a'^2 b' \text{arc} \left\{ \tan = \frac{b'}{\sqrt{(a'^2 - b'^2)}} \right\} + b'^2 \sqrt{(a'^2 - b'^2}}.$$

When b' is very small compared with a', or the spheroid is very flat, we must augment the density

$$\text{from } \rho_{,} \text{ to } \rho_{,} + \frac{\pi}{4} \frac{b'}{a'} \rho \text{ nearly;}$$

and we thus see that the correction in question becomes less in proportion as the spheroid is more oblate.

In what precedes, the excursions of the body of the pendulum are supposed very small compared with its dimensions. For if this were not the case, the terms of the second degree in the equation (1) would no longer be negligible, and therefore the foregoing results might thus cease to be correct. Indeed, were we to attend to the term just mentioned, no advantage would even then be obtained; for the actual motion of the fluid where the vibrations are large will differ greatly from what would be assigned by the preceding method, although this method consists in satisfying all the equations of the fluid's motion, and likewise the particular conditions to which it is subject.

It would be encroaching too much upon the Society's time to enter on the present occasion into an explanation of the cause of this apparent anomaly: it will be sufficient here to have made the remark, and, at the same time to observe, that when the extent of the vibrations is very small, as we have all along supposed, the preceding theory will give the proper correction to be applied to bodies vibrating in air, or other elastic fluid, since the error to which this theory leads cannot bear a much greater proportion to the correction before assigned, than the pendulum's greatest velocity does to that of sound.

APPENDIX.

APPENDIX.

Note to Art. 6, p. 36.

THE important theorem of reciprocity, established in Art. 6, may be put in a clearer light by the following demonstration, which is due to Professor Maxwell.

Let A, B be any two points on a closed conducting surface, and let a unit of positive electricity be placed at a point Q, within the surface, then a unit of negative electricity will be so distributed over the surface that there will be no electrical force outside the surface, and the potential outside it will be everywhere zero. The potential at any point P within the surface, due to the electricity on the surface, is a function of the positions of P, Q, and of the form of the surface.

Denoting this by $G_p^{(q)}$ it is required to shew that $G_p^{(q)} = G_q^{(p)}$, or that the potential at P, due to the distribution on the surface caused by a unit of positive electricity at Q, is equal to the potential at Q, due to the distribution on the surface caused by a unit of positive electricity at P.

Let X be any point outside the surface. The potential there is zero, hence

$$\Sigma\, dS_A \rho_A \frac{1}{AX} + \frac{1}{QX} = 0 \quad\text{.....................}\ (1),$$

where ρ_A is the density and dS_A the element of surface, at any point A of the surface, and the integration is extended over the whole surface.

Also, by definition,

$$\Sigma\, dS_A \rho_A \frac{1}{AP} = G_p^{(q)} \quad\text{..... ....}\ (2).$$

Now if we consider a unit of positive electricity placed at P, and if ρ_B be the density on an element dS_B at B, we shall have, similarly,

$$\Sigma\, dS_B \rho_B \frac{1}{BX} + \frac{1}{PX} = 0,$$

for all points outside the surface, or on it, since the potential is zero on the surface.

Let X be on the surface, say at A, this equation becomes

$$\Sigma\, dS_B \rho_B \frac{1}{BA} + \frac{1}{PA} = 0.$$

Hence, substituting in equation (2) we get

$$\Sigma\Sigma\, dS_A dS_B \rho_A \rho_B \frac{1}{AB} = G_p^{(q)},$$

and as this is the same as we shall obtain for $G_q^{(p)}$, the property is proved.

Note to Art. 10, pp. 50, 51.

The equation $\phi(r) = \frac{a}{r}\psi\left(\frac{a^2}{r}\right)$ proved on p. 51, may be expressed in words as follows. Let O be the centre of a sphere of radius a, and A, B, two points each of which is the electrical image of the other with respect to the sphere (i.e. let O, A, B be in the same straight line, and $OA \,.\, OB = a^2$), then, if electricity be distributed in any manner over the surface of the sphere, the potential at A is to the potential at B as a is to OA or as OB is to a.

For, if a point P move in such a manner that the ratio BP to AP is constant ($= \lambda$ suppose) it will describe a sphere, and if C, C' be the points in which this sphere cuts AB,

$$AC = \frac{AB}{\lambda + 1}, \quad AC' = \frac{AB}{\lambda - 1};$$

$$\therefore \; OA = \frac{1}{2}(AC' - AC) = \frac{1}{\lambda^2 - 1} AB,$$

and $$OC = AC + OA = \frac{AB}{\lambda + 1} + \frac{AB}{\lambda^2 - 1}$$

$$= \frac{\lambda}{\lambda^2 - 1} AB.$$

Hence $$OC : OA :: OB : OC :: \lambda : 1.$$

And potential at A : potential at B :: $\frac{1}{PA} : \frac{1}{PB} : \lambda : 1$.

Hence the theorem is proved.

The laws of the distribution of electricity on spherical conductors have been geometrically investigated by Sir William Thomson in a series of papers published in the *Cambridge and Dublin Mathematical Journal.* See also Thomson and Tait's *Natural Philosophy*, Arts. 474, 510.

Note to Art. 12, p. 68.

In the case of a straight line uniformly covered with electricity, the form of the equipotential surface, and the law of distribution of the electricity over the surface may be investigated as follows.

Denoting the extremities of the straight line by S, H, we know that the attraction of the line on p' may be replaced by that of a circular arc of which p' is the centre, and which touches SH, and has Sp', Hp' as its bounding radii. Hence the direction of the resultant attraction bisects the angle $Sp'H$, and the equipotential surface is a prolate spheroid of which S, H are the foci.

Again, $\frac{d\overline{V}}{dw'}$ is the resultant force exerted by the straight line, or by the circular arc, and therefore

$$\frac{d\overline{V}}{dw'} = \frac{2 \sin \frac{1}{2} Sp'H}{y}.$$

Now $$y = \frac{Sp' . Hp' \sin Sp'H}{2a};$$

$$\therefore \frac{d\overline{V}}{dw'} = \frac{2a}{Sp' \, . \, Hp' \cos \frac{1}{2} \, Sp'H}.$$

And by the properties of the ellipse

$$\cos \frac{1}{2} \, Sp'H = \frac{V}{(Sp' \, . \, Hp')^{\frac{1}{2}}};$$

$$\therefore \frac{d\overline{V}}{dw'} = \frac{2a}{V(Sp' . Hp')^{\frac{1}{2}}},$$

which agrees with the result in the text.

Note to p. 246.

To prove that the equilibrium of the medium will be unstable, unless $\frac{A}{B} > \frac{4}{3}$. We have

$$\iiint \rho \, dx \, dy \, dz \left(\frac{d^2 u}{dt^2} \delta u + \frac{d^2 v}{dt^2} \delta v + \frac{d^2 w}{dt^2} \delta w \right)$$

$$= \iiint \rho \, dx \, dy \, dz \, \delta\phi.$$

And ϕ, as shewn by Green, $= \phi_0 + \phi_2$.

Now, in order that equilibrium may be stable, it is necessary that ϕ_0 be a maximum value of ϕ, or that 0 be a maximum value of ϕ_2. In other words, that ϕ_2 should never be positive. But

$$\phi_2 = -A \left(\frac{du}{dx} + \frac{dv}{dy} + \frac{dw}{dz} \right)^2 + B \left\{ 4 \frac{dv}{dy} \frac{dw}{dz} + 4 \frac{dw}{dz} \frac{du}{dx} + 4 \frac{du}{dx} \frac{dv}{dy} \right.$$
$$\left. - \left(\frac{dv}{dz} + \frac{dw}{dy} \right)^2 - \left(\frac{dw}{dx} + \frac{du}{dz} \right)^2 - \left(\frac{du}{dy} + \frac{dv}{dx} \right)^2 \right\}.$$

Now

$$4 \frac{dv}{dy} \frac{dw}{dz} + 4 \frac{dw}{dz} \frac{du}{dx} + 4 \frac{du}{dx} \frac{dv}{dy}$$
$$= \frac{4}{3} \left(\frac{du}{dx} + \frac{dv}{dy} + \frac{dw}{dz} \right)^2$$
$$- \frac{2}{3} \left\{ \left(\frac{dv}{dy} - \frac{dw}{dz} \right)^2 + \left(\frac{dw}{dz} - \frac{du}{dx} \right)^2 + \left(\frac{du}{dx} - \frac{dv}{dy} \right)^2 \right\}.$$

Hence

$$\phi_2 = -\left(A - \frac{4}{3}B\right)\left(\frac{du}{dx} + \frac{dv}{dy} + \frac{dw}{dz}\right)^2$$

$$-\frac{2}{3}B\left\{\left(\frac{dv}{dy} - \frac{dw}{dz}\right)^2 + \left(\frac{dw}{dz} - \frac{du}{dx}\right)^2 + \left(\frac{du}{dx} - \frac{dv}{dy}\right)^2\right\}$$

$$-B\left\{\left(\frac{dv}{dz} + \frac{dw}{dy}\right)^2 + \left(\frac{dw}{dx} + \frac{du}{dz}\right)^2 + \left(\frac{du}{dy} + \frac{dv}{dx}\right)^2\right\}.$$

It thus appears that $A - \frac{4}{3}B$, $\frac{2}{3}B$, B are each of them the coefficient of an essentially negative expression. Hence, in order that ϕ_2 may always be negative, it is necessary and sufficient that B should be positive, and $A > \frac{4}{3}B$.

Note to p. 253.

Let P, Q be the positions of two particles of a medium in equilibrium distant from each other by a small interval. Let the medium receive a small displacement, in consequence of which these particles assume the positions P', Q', respectively.

Let the co-ordinates of P be $x, \quad y, \quad z,$

$Q \quad x + \delta x, \quad y + \delta y, \quad z + \delta z,$

$P' \quad x + u, \quad y + v, \quad z + w,$

then those of Q' will be

$$x + u + \left(1 + \frac{du}{dx}\right)\delta x + \frac{du}{dy}\delta y + \frac{du}{dz}\delta z,$$

$$y + v + \frac{dv}{dx}\delta x + \left(1 + \frac{dv}{dy}\right)\delta y + \frac{dv}{dz}\delta z,$$

$$z + w + \frac{dw}{dx}\delta x + \frac{dw}{dy}\delta y + \left(1 + \frac{dw}{dz}\right)\delta z.$$

Hence the co-ordinates of P' relatively to Q' are

$$\left.\begin{aligned} \left(1 + \frac{du}{dx}\right)\delta x + \frac{du}{dy}\delta y + \frac{du}{dz}\delta z &= \delta\xi \\ \frac{dv}{dx}\delta x + \left(1 + \frac{dv}{dy}\right)\delta y + \frac{dv}{dz}\delta z &= \delta\eta \\ \frac{dw}{dx}\delta x + \frac{dw}{dy}\delta y + \left(1 + \frac{dw}{dz}\right)\delta z &= \delta\zeta \end{aligned}\right\} \text{suppose.}$$

If then $\delta\xi^2 + \delta\eta^2 + \delta\zeta^2 = \rho^2$, a small given quantity, we have, neglecting powers and products of $\frac{du}{dx}, \frac{du}{dy} \cdots$

$$\rho^2 = \left(1 + 2\frac{du}{dx}\right)\delta x^2 + \left(1 + 2\frac{dv}{dy}\right)\delta y^2 + \left(1 + 2\frac{dw}{dz}\right)\delta z^2$$
$$+ 2\left(\frac{dv}{dz} + \frac{dw}{dy}\right)\delta y\,\delta z + 2\left(\frac{dw}{dx} + \frac{du}{dz}\right)\delta z\,\delta x + 2\left(\frac{du}{dy} + \frac{dv}{dx}\right)\delta x\,\delta y,$$

or $$\rho^2 = (1 + 2s_1)\,\delta x^2 + (1 + 2s_2)\,\delta y^2 + (1 + 2s_3)\,\delta z^2 + 2\alpha\,\delta y\,\delta z$$
$$+ 2\beta\,\delta z\,\delta x + 2\gamma\,\delta x\,\delta y.$$

It hence appears that all particles which, after displacement, lie at a given distance from P', must lie before displacement on the surface of a certain ellipsoid of which the centre is at P. Hence, in general, the force called into play by the displacement must be a function of the six coefficients involved in the equation of this surface referred to P as origin.

But, if the medium be homogeneous, the force thus called into play will be independent of the position of this ellipsoid, and will depend upon its form and magnitude only, that is, will be a function of the lengths of the axes of this ellipsoid. But the reciprocals of the squares on the semi-axes are the values of λ given by the equation

$$\begin{vmatrix} -\lambda\rho^2 + (1 + 2s_1), & \gamma, & \beta, \\ \gamma, & -\lambda\rho^2 + (1 + 2s_2), & \alpha, \\ \beta, & \alpha, & -\lambda\rho^2 + (1 + 2s_3) \end{vmatrix} = 0,$$

and are therefore expressible in terms of

$$s_1 + s_2 + s_3,$$
$$4(s_2 s_3 + s_3 s_1 + s_1 s_2) - (\alpha^2 + \beta^2 + \gamma^2),$$
$$4 s_1 s_2 s_3 - (s_1\alpha^2 + s_2\beta^2 + s_3\gamma^2) + \alpha\beta\gamma.$$

Hence, if the force function be called $\phi, \equiv \phi_1 + \phi_2 + \phi_3 + \ldots$ we have

$$\phi_1 = M(s_1 + s_2 + s_3),$$
$$\phi_2 = A(s_1 + s_2 + s_3)^2 + B\{4(s_2 s_3 + s_3 s_1 + s_1 s_2) - (\alpha^2 + \beta^2 + \gamma^2)\},$$
$$\phi_3 = C(s_1 + s_2 + s_3)^3 + D(s_1 + s_2 + s_3)\{4(s_2 s_3 + s_3 s_1 + s_1 s_2) - (\alpha^2 + \beta^2 + \gamma^2)\}$$
$$+ E\{4 s_1 s_2 s_3 - (s_1\alpha^2 + s_2\beta^2 + s_3\gamma^2) + \alpha\beta\gamma\},$$
$$\ldots = \ldots\ldots\ldots\ldots\ldots\ldots$$

M, A, B, C, D, E being certain arbitrary coefficients. By the reasoning of the text it appears that $\phi_1 = 0$ and that ϕ_3 may be neglected. The above value of ϕ_2 agrees with Green's result.

Note to p. 269.

The intensity of the reflected light attains a minimum value accurately when

$$\frac{a_1}{a} = \left\{\frac{4\mu^4 + (\mu^2 - 1)^2}{4 + (\mu^2 - 1)^2}\right\}^{\frac{1}{2}} = \left(\frac{5\mu^4 - 2\mu^2 + 1}{\mu^4 - 2\mu^2 + 5}\right)^{\frac{1}{2}},$$

which, by (27), gives

$$\frac{b}{a} = \left\{\frac{4\mu^2 - (\mu^2 - 1)^2}{4 + (\mu^2 - 1)^2}\right\}^{\frac{1}{2}} = \left(\frac{-\mu^4 + 6\mu^2 - 1}{\mu^4 - 2\mu^2 + 5}\right)^{\frac{1}{2}},$$

and, for the minimum value of $\frac{\beta^2}{\alpha^2}$,

$$\frac{\{4\mu^4 + (\mu^2 - 1)^2\}^{\frac{1}{2}}\{4 + (\mu^2 - 1)^2\}^{\frac{1}{2}} - (\mu^2 + 1)^2}{\{4\mu^4 + (\mu^2 - 1)^2\}^{\frac{1}{2}}\{4 + (\mu^2 - 1)^2\}^{\frac{1}{2}} + (\mu^2 + 1)^2},$$

or

$$\frac{(5\mu^4 - 2\mu^2 + 1)^{\frac{1}{2}}(\mu^4 - 2\mu^2 + 5)^{\frac{1}{2}} - (\mu^2 + 1)^2}{(5\mu^4 - 2\mu^2 + 1)^{\frac{1}{2}}(\mu^4 - 2\mu^2 + 5)^{\frac{1}{2}} + (\mu^2 + 1)^2}.$$

If $\mu = \frac{4}{3}$, as when the two media are air and water, we get $\frac{\beta^2}{\alpha^2} = \frac{1}{166}$ nearly.

If $\mu = \frac{3}{2}$, as when the two media are air and glass, we get $\frac{\beta^2}{\alpha^2} = \frac{1}{49}$ nearly.

And the minimum value of $\frac{\beta^2}{\alpha^2}$ can never be greater, even if $\mu = \infty$, than $\frac{\sqrt{5} - 1}{\sqrt{5} + 1}$ or $\frac{3 - \sqrt{5}}{2}$.

Again,

$$\tan(e_1 \mp e) = \frac{(\mu^2 - 1)^2}{\mu^2 + 1}\frac{b}{a}\,\frac{\dfrac{1}{\mu^2 - \dfrac{a_1}{a}} \pm \dfrac{1}{\mu^2 + \dfrac{a_1}{a}}}{1 \mp \dfrac{(\mu^2 - 1)^4}{(\mu^2 + 1)^2}\dfrac{b^2}{a^2}\dfrac{1}{\mu^4 - \dfrac{a_1^2}{a^2}}},$$

$$=\frac{(\mu^2-1)^2}{\mu^2+1}\,\frac{b}{a}\,\frac{\mu^2+\frac{a_1}{a}\pm\left(\mu^2-\frac{a_1}{a}\right)}{\mu^4-\frac{a_1^2}{a^2}\mp\frac{(\mu^2-1)^4}{(\mu^2+1)^2}\frac{b^2}{a^2}}.$$

Writing $\left(\frac{5\mu^4-2\mu^2+1}{\mu^4-2\mu^2+5}\right)^{\frac{1}{2}}$ for $\frac{a_1}{a}$, $\left(-\frac{\mu^4-6\mu^2+1}{\mu^4-2\mu^2+5}\right)^{\frac{1}{2}}$ for $\frac{b}{a}$, this gives

$$\tan(e_1-e)=\frac{(\mu^2-1)^2}{\mu^2+1}\left(-\frac{\mu^4-6\mu^2+1}{\mu^4-2\mu^2+5}\right)^{\frac{1}{2}}\frac{2\mu^2}{\mu^4-\frac{5\mu^4-2\mu^2+1}{\mu^4-2\mu^2+5}+\frac{(\mu^2-1)^4}{(\mu^2+1)^2}\frac{\mu^4-6\mu^2+1}{\mu^4-2\mu^2+5}}$$

$$=\frac{2\mu^2(\mu^2+1)(\mu^2-1)^2(-\mu^4+6\mu^2-1)^{\frac{1}{2}}(\mu^4-2\mu^2+5)^{\frac{1}{2}}}{\{\mu^4(\mu^4-2\mu^2+5)-(5\mu^4-2\mu^2+1)\}(\mu^2+1)^2+(\mu^2-1)^4(\mu^4-6\mu^2+1)}$$

$$=\frac{2\mu^2(\mu^2+1)(\mu^2-1)^2(-\mu^4+6\mu^2-1)^{\frac{1}{2}}(\mu^4-2\mu^2+5)^{\frac{1}{2}}}{(\mu^2-1)^3(\mu^2+1)^3+(\mu^2-1)^4(\mu^4-6\mu^2+1)}$$

$$=\frac{2\mu^2(\mu^2+1)(-\mu^4+6\mu^2-1)^{\frac{1}{2}}(\mu^4-2\mu^2+5)^{\frac{1}{2}}}{(\mu^2-1)\{(\mu^2+1)^3+(\mu^2-1)(\mu^4-6\mu^2+1)\}}$$

$$=2\,\frac{\mu^2+1}{\mu^2-1}\,\frac{(-\mu^4+6\mu^2-1)^{\frac{1}{2}}(\mu^4-2\mu^2+5)^{\frac{1}{2}}}{2(\mu^4-2\mu^2+5)}$$

$$=\frac{\mu^2+1}{\mu^2-1}\left(\frac{-\mu^4+6\mu^2-1}{\mu^4-2\mu^2+5}\right)^{\frac{1}{2}}.$$

And $\tan(e_1+e)=\frac{(\mu^2-1)^2}{\mu^2+1}\left(-\frac{\mu^4-6\mu^2+1}{\mu^4-2\mu^2+5}\right)^{\frac{1}{2}}$

$$\frac{2\left(\frac{5\mu^4-2\mu^2+1}{\mu^4-2\mu^2+5}\right)^{\frac{1}{2}}}{\mu^4-\frac{5\mu^4-2\mu^2+1}{\mu^4-2\mu^2+5}-\frac{(\mu^2-1)^4}{(\mu^2+1)^2}\frac{\mu^4-6\mu^2+1}{\mu^4-2\mu^2+5}}$$

$$=\frac{2(\mu^2+1)(\mu^2-1)^2(-\mu^4+6\mu^2-1)^{\frac{1}{2}}(5\mu^4-2\mu^2+1)^{\frac{1}{2}}}{(\mu^2-1)^3(\mu^2+1)^3-(\mu^2-1)^4(\mu^4-6\mu^2+1)}$$

$$=\frac{2(\mu^2+1)(-\mu^4+6\mu^2-1)^{\frac{1}{2}}(5\mu^4-2\mu^2+1)^{\frac{1}{2}}}{(\mu^2-1)\{(\mu^2+1)^3-(\mu^2-1)(\mu^4-6\mu^2+1)\}}$$

$$=2\,\frac{\mu^2+1}{\mu^2-1}\,\frac{(-\mu^4+6\mu^2-1)^{\frac{1}{2}}(5\mu^4-2\mu^2+1)^{\frac{1}{2}}}{2(5\mu^4-2\mu^2+1)}$$

$$=\frac{\mu^2+1}{\mu^2-1}\left(\frac{-\mu^4+6\mu^2-1}{5\mu^4-2\mu^2+1}\right)^{\frac{1}{2}}.$$

Note to pp. 301, 302.

These results may be otherwise obtained by the consideration that if one of the waves be propagated by normal vibrations, the corresponding values of α, β, γ must be proportional to a, b, c. We thus obtain, from equations (6),

$$\frac{A'a + F'b + E'c}{a} = \frac{F'a + B'b + D'c}{b} = \frac{E'a + D'b + C'c}{c} = \rho e^2.$$

Now replacing A', B', C', D', E', F' by their values, we see that $\frac{A'a + F'b + E'c}{a}$ is equal to

$$(G + A)\, a^2 + (N + B)\, b^2 + (M + C)\, c^2 + (N + R)\, b^2 + (M + Q)\, c^2,$$

or $$Ga^2 + (2N + R)\, b^2 + (2M + Q) c^2 + Aa^2 + Bb^2 + Cc^2.$$

The second and third members of the above equation being similarly transformed, we see that

$$\begin{aligned} &Ga^2 + (2N + R)\, b^2 + (2M + Q)\, c^2 \\ &= Hb^2 + (2L + P)\, c^2 + (2N + R)\, a^2 \\ &= Ic^2 + (2M + Q)\, a^2 + (2L + P)\, c^2, \end{aligned}$$

for all values of a, b, c; which leads at once to the equations given at the foot of p. 302; and also proves that the normal velocity of propagation e is equal to $\left(\frac{\mu - Aa^2 - Bb^2 - Cc^2}{\rho}\right)^{\frac{1}{2}}$, which, when the system is free from extraneous pressure, becomes $\left(\frac{\mu}{\rho}\right)^{\frac{1}{2}}$.

If the values of A', B', C', D', E', F' in terms of μ, L, M, N be substituted in equations (6), we obtain the following equation for the determination of ρe^2, the system being supposed free from extraneous pressure:

$$\begin{vmatrix} \mu a^2 + Nb^2 + Mc^2 - \rho e^2, & (\mu - N)\, ab, & (\mu - M)\, ca \\ (\mu - N)\, ab, & \mu b^2 + Lc^2 + Na^2 - \rho e^2, & (\mu - L)\, bc \\ (\mu - M)\, ca, & (\mu - L)\, bc, & \mu c^2 + Ma^2 + Lb^2 - \rho e^2 \end{vmatrix} = 0,$$

which, when evaluated, becomes

$$(a^2+b^2+c^2)^2\,(\mu-\rho e^2)\,\{a^2\,(M-\rho e^2)\,(N-\rho e^2)$$
$$+\,b^2\,(N-\rho e^2)\,(L-\rho e^2)+c^2\,(L-\rho e^2)\,(M-\rho e^2)\}=0.$$

It thus appears that the velocities of propagation of the two vibrations, whose directions are in front of the wave, are given by the equation

$$\frac{a^2}{L-\rho e^2}+\frac{b^2}{M-\rho e^2}+\frac{c^2}{N-\rho e^2}=0,$$

the same equation as we should obtain for the determination of the axes of the section of the ellipsoid

$$Lx^2+My^2+Nz^2=1,$$

by the plane

$$ax+by+cz=0;$$

agreeably to equation (10).

CAMBRIDGE: PRINTED BY C. J. CLAY, M.A., AT THE UNIVERSITY PRESS.

Printed in the United States
By Bookmasters